Continuities and Discontinuities in Development

TOPICS IN DEVELOPMENTAL PSYCHOBIOLOGY
Series Editor: Robert N. Emde
University of Colorado School of Medicine
Denver, Colorado

The Development of Attachment and Affiliative Systems
Edited by Robert N. Emde and Robert J. Harmon

Continuities and Discontinuities in Development
Edited by Robert N. Emde and Robert J. Harmon

Continuities and Discontinuities in Development

Edited by

Robert N. Emde

and

Robert J. Harmon

University of Colorado School of Medicine
Denver, Colorado

Plenum Press • New York and London

Library of Congress Cataloging in Publication Data

Main entry under title:

Continuities and discontinuities in development.

(Topics in developmental psychobiology)
Includes bibliographies and index.
1. Developmental psychobiology—Congresses. 2. Child psychopathology—
Congresses. 3. Pediatrics—Psychological aspects—Congresses. I. Emde, Robert N. II.
Harmon, Robert John, 1946–   . III. Series.
RJ131.C595   1984   155   84-8225
ISBN 0-306-41563-1

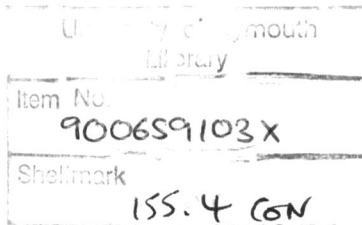

Contributors

Karen Caplovitz Barrett • Department of Psychology, University of Denver, University Park, Denver, Colorado

Elizabeth Bates • Department of Psychology, University of California-San Diego, LaJolla, California

Bennett I. Bertenthal • Department of Psychology, Gilmer Hall, University of Virginia, Charlottesville, Virginia

Inge Bretherton • Department of Human Development and Family Studies, Colorado State University, Fort Collins, Colorado

Daniel Bullock • Department of Psychology, University of Denver, University Park, Denver, Colorado

Joseph J. Campos • Department of Psychology, University of Denver, University Park, Denver, Colorado

J. P. Connell • Graduate School of Education and Human Development, University of Rochester, Rochester, New York

Linda S. Crnic • Departments of Pediatrics and Psychiatry, University of Colorado School of Medicine, Denver, Colorado

Robert N. Emde • Department of Psychiatry, University of Colorado School of Medicine, Denver, Colorado

Kurt W. Fischer • Department of Psychology, University of Denver, University Park, Denver, Colorado

Wyndol Furman • Department of Psychology, University of Denver, University Park, Denver, Colorado

Theodore Gaensbauer • Department of Psychiatry, University of Colorado School of Medicine, Denver, Colorado

H. H. Goldsmith • Department of Psychology, University of Texas, Austin, Texas

William T. Greenough • Department of Psychology, University of Illinois, Champaign, Illinois

Robert J. Harmon • Department of Psychiatry, University of Colorado School of Medicine, Denver, Colorado

John Horn • Department of Psychology, University of Denver, University Park, Denver, Colorado

Jerome Kagan • Department of Psychology and Social Relations, Harvard University, Cambridge, Massachusetts

William J. Kimberling • Boys Town Institute for Communication Disorders in Children, Omaha, Nebraska

Mary D. Klinnert • Department of Psychiatry, University of Colorado School of Medicine, Denver, Colorado

Michael E. Lamb • Department of Psychology, University of Utah, Salt Lake City, Utah

Herbert A. Lubs • Mailman Center for Child Development, University of Miami Medical Center, Miami, Florida

Jack McArdle • Department of Psychology, University of Denver, University Park, Denver, Colorado

Linda L. McCabe • Department of Psychology, University of Colorado, Denver, Colorado

George A. Morgan • Department of Human Development and Family Studies, Colorado State University, Fort Collins, Colorado

David A. Mrazek • National Jewish Hospital and Research Center, Denver, Colorado

Bruce F. Pennington • Department of Psychiatry, University of Colorado, Denver, Colorado

Sandra L. Pipp • Department of Psychiatry, University of Colorado School of Medicine, Denver, Colorado

Michael Rutter • Department of Child and Adolescent Psychiatry, Institute of Psychiatry, London, England

Harris D. Schwark • Department of Psychology, University of Illinois, Champaign, Illinois

Robert Short • Department of Psychology, University of Denver, University Park, Denver, Colorado

David W. Shucard • Brain Sciences Laboratories, Department of Pediatrics, National Jewish Hospital and Research Center, and Department of Psychiatry, University of Colorado School of Medicine, Denver, Colorado

Janet L. Shucard • Brain Sciences Laboratories, Department of Pediatrics, National Jewish Hospital and Research Center, and Department of Psychiatry, University of Colorado School of Medicine, Denver, Colorado

Shelley D. Smith • Boys Town Institute for Communication Disorders in Children, Omaha, Nebraska

James F. Sorce • Bell Laboratories, Holmdel, New Jersey

Craig Stenberg • Department of Psychology, University of Denver, Denver, Colorado

David G. Thomas • Brain Sciences Laboratories, Department of Pediatrics, National Jewish Hospital and Research Center, and Department of Psychiatry, University of Colorado School of Medicine, Denver, Colorado

Ross A. Thompson • Department of Psychology, University of Nebraska, Lincoln, Nebraska

Preface

"Continuities and Discontinuities in Development" was the theme for the Second Biennial DPRG Retreat, a three-day meeting held at Estes Park, Colorado, in June 1982. The meeting was sponsored by the Developmental Psychobiology Research Group (DPRG) of the Department of Psychiatry at the University of Colorado School of Medicine. The DPRG is a group of individuals conducting research in many areas of development who meet on a regular basis to present and discuss their work and receive feedback and encouragement. In 1974, this group was awarded an endowment fund by the Grant Foundation, the aims of which were to facilitate the research of young investigators, to encourage new research, and to provide seed money for collaborative ventures. Much of the work reported in this volume and in the earlier volume from the First DPRG Retreat is the result of that support.

In addition to the work of the members of the DPRG, a select group of guests was invited to participate in the meeting and contribute to this volume. The chapters by William Greenough, Jerome Kagan, and Michael Rutter result from the participation of these scholars at the retreat.

We would like to acknowledge the support of a number of individuals who have been instrumental in supporting the DPRG as a whole, as well as those who contributed directly to the Second Biennial Retreat and to the volume. Robert Harmon and Linda Crnic served as co-chairpersons of the retreat and they, along with Gordon Farley, Marshall Haith, George Morgan, Bruce Pennington, and Martin Reite, arranged the scientific meeting. Francine Butler, in her role of DPRG Executive Secretary, served ably in organizing and coordinating the meeting.

Last, but certainly not least, we wish to acknowledge the administrative and secretarial assistance of Maxine Conlon, for her role in both the retreat meeting arrangements and with aspects of this volume, and Debra Becker for her role in coordinating the final chapters of the volume.

Dr. Emde is supported by Research Scientist Award 5-KO5-MH 36808 and National Institute of Mental Health Project Grant 2-RO1-MH 22803.

Dr. Harmon is supported by Research Scientist Development Award 5-KO1-MH 00281 and Research Grant 5-RO1-MH 34005 from the National Institute of Mental Health.

Preparation of this volume was supported in part by the Developmental Psychobiology Research Group (DPRG) of the Department of Psychiatry, University of Colorado School of Medicine, through funding from the Grant Foundation.

ROBERT N. EMDE
ROBERT J. HARMON

Contents

Chapter 3

Continuities and Discontinuities in Socioemotional
Development: Empirical and Conceptual Perspectives 41

Michael Rutter

Chapter 4

Age-related Aspects of Experience Effects upon Brain Structure   69

William T. Greenough and Harris D. Schwark

PART II   STUDIES OF DISCONTINUITIES AND DEVELOPMENTAL
           TRANSITIONS

Chapter 5

**Detecting Developmental Discontinuities: Methods and
Measurement 95**

Kurt W. Fischer, Sandra L. Pipp, and Daniel Bullock

Chapter 8

**Self-produced Locomotion: An Organizer of Emotional,
Cognitive, and Social Development in Infancy................. 175**

Bennett I. Bertenthal, Joseph J. Campos, and Karen Caplovitz Barrett

Chapter 9

Effects of Hospitalization on Early Child Development......... 211

David A. Mrazek

PART III   STUDIES OF CONTINUITIES

Chapter 10

**The Development of Representation from 10 to 28 Months:
Differential Stability of Language and Symbolic Play** **229**

Inge Bretherton and Elizabeth Bates

Chapter 11

**Developmental Transformations in Mastery Motivation:
Measurement and Validation** **263**

George A. Morgan and Robert J. Harmon

Chapter 12

**The Development of Cerebral Specialization in Infants:
Electrophysiological and Behavioral Studies** **293**

David W. Shucard, Janet L. Shucard, and David G. Thomas

Chapter 13

**Continuity and Change in Socioemotional Development during
the Second Year** **315**

Ross A. Thompson and Michael E. Lamb

Chapter 14

**Continuities and Change in Early Emotional Life: Maternal
Perceptions of Surprise, Fear, and Anger** 339

*Mary D. Klinnert, James F. Sorce, Robert N. Emde, Craig Stenberg,
and Theodore Gaensbauer*

Chapter 15

Early Experience Effects: Evidence for Continuity? 355

Linda S. Crnic

PART IV   CROSS-DISCIPLINARY STRATEGIES IN SEARCHING FOR
           CONTINUITIES

Chapter 16

Mathematical-Statistical Model Building in Analysis of Developmental Data 371

Robert Short, John Horn, and Jack McArdle

Chapter 17

Continuity of Personality: A Genetic Perspective 403

H. H. Goldsmith

CHAPTER 1

Entering a New Era in the Search for Developmental Continuities

Robert N. Emde and Robert J. Harmon

After reading this volume, we think the reader will agree that we are on the threshold of a new era in searching for continuities in behavior from early development to later. This statement requires an appreciation that our earlier expectations about continuities received a jolt from two decades of incontrovertible research findings. During this time developmental investigators engaged in longitudinal study were disappointed in finding little predictability from infancy to later ages. This was true for behavior related to cognition (see McCall, 1979; Kagan, Chapter 2, this volume) as well as for behavior presumably related to temperament (Plomin, 1983). During this time clinicians, who had assumed indelible effects from early experience, were repeatedly surprised by well-documented instances of resiliency following major infantile deficit and trauma (for review see Clarke & Clarke, 1977; Kagan, Kearsley, & Zelazo, 1978; Emde, 1981). Thus, in spite of our expectations for connectivity and continuity, research evidence supporting such expectations has been meager.

This situation has left us uneasy. Some have noted that behavioral continuities begin to become prominent after the second year, when self-awareness and an organized executive self become consolidated (see Kagan, Chapter 2; Rutter, Chapter 3). From age two or three continuities might be expected; perhaps there are no continuities from earlier. Still, one senses something different. The recent two decades of developmental research have not simply been negative. We have

Robert N. Emde and Robert J. Harmon • Department of Psychiatry, University of Colorado School of Medicine, Denver, Colorado 80262.

learned a good deal about early developmental processes and, as the chapters in this volume illustrate, there are new views and emergent strategies. These combine to give a sense of excitement about the possibility of major discoveries in the next decade.

NEW VIEWS OF DISCONTINUITIES AND CONTINUITIES

Let us discuss discontinuities first. It is important to realize that in referring to *discontinuity* we are not simply addressing the *null*—that is, negative findings about identified continuities from early childhood to later. We understand *discontinuities* to refer to specific times of developmental transformation or change. Thus, this concept encompasses major spurts in development and times of transition; however, it also includes times of qualitative shift with new patternings in behavior. Investigators have increasingly come to appreciate that there are normal developmental acquisitions (i.e., discontinuities) that are only partially connected with the past and are worthy of study in their own right. While, from an intuitive point of view, biological processes of individuality and self-regulation would seem to guarantee some behavioral continuities from infancy, there are also organized behavioral systems which have relevance to particular periods of early life and which subsequently disappear. These have been labeled "transient ontogenetic adaptations" by Oppenheim and are taken up by Kagan in Chapter 2. Further, as Kagan also discusses with such vigor in his chapter, our earlier views about the cumulative value of early experience and about its linearity are now thought distorted. The clearer view which seems to be emerging has early development as an aspect of the life course. Development is viewed as multidimensional; further, early behavioral change processes occur which may or may not be connected with later life behavioral change processes. Accordingly, novel behavioral systems can emerge not only in early childhood but in later childhood, adulthood, and old age (Baltes, 1983).

Thus, discontinuities are not trivial. We referred in a previous work to two of these times in infancy as times of *biobehavioral shift*, emphasizing their biological bases (Emde, Gaensbauer, & Harmon, 1976). Genes "switch on" and "switch off" in development and, as Goldsmith points out in his chapter, much of the genetic variance at age seven is not common to that of age four; thus, not only can genetic mechanisms form the basis for stabilities in development, they can also account for discontinuities.

It might be useful to think of some broad types of developmental discontinuities or transitions. As Connell and Furman point out in

Chapter 7, these can be endogenously determined or exogenously determined. The former can be seen in the onset of self-produced locomotion and fearfulness on the visual cliff; the latter can be seen in school entrance. But there are likely to be significant individual–environment interaction effects in each. Bertanthal, Campos, and Barrett, in Chapter 8, demonstrate this elegantly in showing us the effects of locomotor experience and its variations on visual cliff behavior; similarly, we know that there are major organismic shifts which correspond to school entrance (White, 1965). Further, as the life span psychologists have taught us, it is useful to speak of normative transitions or discontinuities (again the example of school entrance) and nonnormative transitions or discontinuities (for example, hospitalization or divorce).

Discontinuities do not necessarily imply a lack of connectedness between early development and later. Indeed, as Rutter reminds us in Chapter 3, even when transformations are extraordinarily dramatic, as in the case of metamorphosis in the insect, it has been shown that the experience of the larva can affect the subsequent behavior of the moth. Further, when one looks at human behavior over sufficiently long time spans, one sees an unevenness in development with transformations, regressions, dips, and spurts (Fischer, Pipp, & Bullock, Chapter 5, this volume; Bever, 1982). Clearly those investigators now addressing themselves to searching for continuities must employ strategies which take account of such phenomena.

Broadly speaking, *continuities* refer to connectedness in development, to the linkage of early behavior to later behavior. As Bretherton and Bates point out, we can think of two different types of continuities in longitudinal research: (1) those concerning style (temperament, the dyadic relationship) and (2) those concerning competence or skill (cognition, etc.). A number of our chapters describe continuities in style, including those referring to inhibition in temperament (Kagan, Chapter 2), emotional signaling in the mother–infant dyad (Klinnert *et al.*, Chapter 14), attachment in the mother–infant dyad (Thompson & Lamb, Chapter 13) and some form of childhood psychopathology (Rutter, Chapter 3). As for competence or skill, chapters discuss early developmental continuities in language acquisition (Bretherton & Bates, Chapter 10) and mastery motivation (Morgan & Harmon, Chapter 11). There is also discussion of a weakly demonstrated continuity in the area of symbolic play (Bretherton & Bates).

Some forms of childhood psychopathology show more evidence of continuity than others. Rutter points out that conduct disorders and aggression showed more continuity from age 10 to 15 in the Isle of Wight Study than did emotional disorders. A related point is that the environment is crucial in consideration of continuity, with a continuity of environment making a continuity of development more likely. Paradoxically, one can look on the effects of environment stressors either from the point of view of discontinuity (the stress event itself causing unexpected changes) or from the point of view of continuity (the stress event leading to chronic alterations in behavior). This in turn evokes questions concerning the age-dependent nature of emotional stressors. As Mrazek discusses in his chapter, the effects of hospitalization experience before ages 4–5 are quite different with respect to subsequent adverse behavior than hospitalization afterwards. Concerning a more general role of environment in setting a continuity, Rutter also discusses how acute events (e.g., the birth of a sibling) or particular opportunities (e.g., entering a good school) can result in enduring change. Usually such effects are mediated by an enduring change in the experienced environment (e.g., new patterns of family interaction or school interactions).

This volume also contains two other views of continuities, views which are likely to be somewhat controversial but which may be useful in broadening our thinking. One concerns *lateral continuities* and is used in a genetic investigation of developmental dyslexia (Pennington *et al.*, Chapter 6). A lateral continuity is based on genetic inference in a single generation, whereas a *cross-generational continuity* refers to the continuity of a phenotype across generations in an expected manner. This view is quite useful to these authors by virtue of the remarkable new technique of genetic linkage analysis. The other view is of Shucard, Shucard, and Thomas, who see continuities in terms of relative dependence or concordance of electrophysiological and other functions in development. In this, these authors appear to adopt a method implied by Fischer, Pipp, and Bullock in Chapter 5.

Changing Biological Models

Our earlier model for behavioral development appeared to be based on physical growth. This included the idea of a steadily decreasing velocity from early infancy to midchildhood and then a growth spurt in adolescence which more or less completed this developmental function. Such a model had little place for discontinuities, gave a predominant emphasis to early development, and had no place for significant development after adolescence. Aside from the fact that a closer look at the growth process may reveal evidence of more spurts (Lampl & Emde, 1983), physical growth is only a surface aspect of developmental biology. As chapters in this volume illustrate, appropriate biological models for today are hardly simple, unidimensional, or linear with respect to age. Instead, models based on current knowledge are complex and extraordinarily dynamic.

We have already commented on emerging models from modern genetics in which regulator genes apparently switch on and off in the course of development and genetic influence continues through life. There is, thus, a genetic basis for developmental discontinuities (or new behaviors) as well as continuities. Further, because of gene–environment interactions, genetic influences in later development are more dynamic than had been appreciated. As Goldsmith points out, current research in developmental genetics ponders such questions as: What mechanisms determine which of the 1% to 15% of the structural genes in the human are operative at any given time in development?

The chapters of Crnic and Greenough and Schwark, in considering the "critical period" hypothesis for effects of early experience, both point out the dynamic nature of age-related processes. No longer can we assume that the adult brain does not undergo structural alterations as a result of experience, and, as Crnic discusses, relevant adult controls for adult responses must be instituted in experimental studies of animals altering early nutrition or stimulation.

The biological model suggested by the work of Greenough and Schwark is worthy of special note, since it emerges from a new direction in neurosciences. In their chapter these authors discuss how brain plasticity, as evidenced from experience manipulation studies, is present in both young and mature animals. The adult brain in rats retains the capacity to form new synaptic connections as a consequence of exposure to an intense learning situation. Thus, evidence detected on a behavioral level for earlier critical or sensitive periods in development may be a reflection of the *rate* at which connections are being established and the *primacy* of early patterns, rather than a reflection of unique mechanisms in the immature brain. The authors also discuss a possible mediating mechanism for unevenness in development. According to their view, selective preservation of synapses operates as a continuing process in brain functioning. Evidence suggests that there is an initial overproduction of synapses, followed by a subsequent competition for survival based on use—in other words, a sort of "sunset law" for synapses. On the basis of their research and the findings of others, the authors speculate that this process continues throughout life in humans as well as in animals. Unevenness over the course of development would stem from two sources. First, there is a presumed massive overproduction of synapses in early development (for example, in human frontal cortex the number of synapses at 1 to 2 years of age is estimated to be 50% greater than adult levels), and this may be associated with greater plasticity. Second, there is regionalization such that synapse overproduction peaks occur at different ages for different brain areas (human frontal cortex at 1 to 2 years, human visual cortex at 8 to 12 months, for example).

Greenough and Schwark emphasize that their developmental model of brain plasticity, although data-based, is highly speculative. What makes this especially exciting to those in behavioral research, however, is the consistency of this model with the picture emerging from the recent corpus of human research findings. As with work in contemporary genetics, we have the promise of arriving at an understanding of the biological basis for both discontinuities and continuities in development, and not only for such processes in early development, but in later development as well.

New Strategies for Longitudinal Study

As the contributions to this volume document, there are now a host of new strategies for longitudinal studies, strategies that can search for continuities by taking into account discontinuities or times of developmental transition.

One such strategy is to focus on the time of transition itself. As Connell and Furman point out, this strategy requires at least three points in time and can profit by a comparison of those individuals who show more change and those individuals who show less change. Such newer statistical techniques as structural equation modeling may be useful in uncovering latent or process variables through the time of change; this may indicate some sort of regulatory continuity in the midst of changes more surface variables. This strategy of focusing on times of discontinuity or transition may represent the promise of finding particular kinds of continuities. It may be that during a time of transition the individual is more open to the effects of new experience and such experience may have more long-lasting effects than experience during times of relative stability. Further, special times of transition might be highlighted along this line. The chapters of Kagan and Rutter each suggest possibility in terms of the transition time when the representational self emerges and internalized sets of intentions and rule structures begin to appear. It may be no accident that, even with our relatively crude techniques of searching for continuities, evidence of behavioral connectedness begins after this time.

Another strategy, related to the one just described, emphasizes that we are interested not merely in stability of behavior over time but also in antecedent-conseqent relations. As Kagan (1971) pointed out some time ago with his discussion of "heterotypic continuity," any given behavior of interest may appear different at a later age, particularly if the child passes through a time of developmental transformation. If connectedness and a relationship are established, one then looks for a common underlying process, or a "latent variable" in the terminology of the new structural equation modeling (see the discussion in Chapter 16 by Short, Horn, and McArdle).

Another strategy has to do with attention to sampling. One must find out what relevant sample of behavior at a given developmental age is appropriate. We may have sampled many behaviors during all too brief observation periods and we may have inappropriately combined some behavioral variables so as to obscure antecedent-consequent relations. Bretherton and Bates, in their chapter, show rather striking continuities in language from 10 to 28 months, continuities which are of a much higher order than those previously found for cognitive development. This is especially dramatic as this continuity spans sensorimotor and representational intelligence! In discussing this, the authors reflect on sampling; perhaps developmental tests such as the Bayley Scales combine too many sorts of behaviors and reflect inappropriate sampling. One is also reminded of the discussion of Fischer, Pipp, and Bullock, in their chapter, that asserts that testing for optimal performance will give best estimates of times of discontinuity; it may be that spontaneous behavior (with longer time observations) is needed to demonstrate stability.

Another strategy which is increasingly appreciated in longitudinal study has to with subgrouping. In any sample under longitudinal review, some groups are more likely to show stability than others. It is often possible to specify "moderating variables" which are the basis for subgrouping. Thus, in the Fels' longitudinal (Kagan & Moss, 1962), continuities were found through childhood according to gender subgroups (with aggressiveness for males; with dependency for females), and in the Berkeley growth study (Block, 1971), important differences were found for adolescents who showed stability in personality variables over time as compared with those who showed instability. But perhaps most dramatic are the recent findings of Kagan, reported in his chapter, of a subgroup of inhibited or shy toddlers who remain shy in early childhood, showing a remarkable continuity in their behavior in social situations. This subgrouping is based on those inhibited infants who have a physiological pattern of high, unvarying heartrate when processing information during a cognitive task. Thus, the moderating variable for this subgroup appears to be a psychophysiological-temperamental one. Other chapters illustrate the subgrouping strategy. Rutter discusses this in terms of the importance of subgrouping for conduct disorders and for depressions in childhood. Thompson and Lamb make use of a subgrouping of attachment classifications according to stability or change from 12½ to 19½ months; sociability and stranger reactions show continuity when there is stability of attachment but not when there is not.

Another strategy has to do with the usefulness of an organizational viewpoint. This looks at the relationships among a number of variables over time, rather than single variables or groups of variables. The newer structural modeling techniques discussed by Short, Horn, and McArdle and by Connell and Furman, and to some extent by Goldsmith, reflect this strategy. Klinnert *et al.* see an organizational continuity in emotional signaling at the infant–mother dyadic level. Even though channels of emotional expression may vary across individual infants, and to some extent across ages, there is a signaling system showing continuity so far as mothers are concerned and this is used for caregiving. Thompson and Lamb use this strategy in studying continuity and change in attachment and socioemotional development during the second year; patterns of related dimensions of behavior show continuity, whereas isolated variables do not.

Another strategy has to do with searching for developmental continuities within an individual. Up to now developmental studies have used a correlative approach in which an individual's relative position in a group at one time is assessed in relation to that individual's relative position in the same group later. This is a rank-order form of continuity search, one which describes an individual's position over time within an arbitrary group distribution. Although this has been the predominant way of searching for longitudinal continuities, this is often not relevant to our interests. Clinicians, especially, are interested in meaningful continuities within an individual over time (ipsative continuities). Newer methods seek ipsative continuities over time; these characterize meaningful constant relationships among intraindividual variables of behavioral organization. Block (1971) pioneered such an approach (using the Q-sort) and found major ipsative continuities in adolescence; more recently, Martin found ipsative continuities in preschool children (as cited in Moss, 1983). This approach has yet to be applied directly to infancy, although a similar way of thinking has been used by Sroufe and his colleagues (Arend, Gove, & Sroufe, 1979) in finding continuities in the organization of attachment from the beginning of the second year into preschool times.

Still another new strategy for longitudinal study concerns the appreciation of context, environmental transactions, and matches in development. It is clear that continuity appears under certain environmental circumstances and not others. Many have discussed the differences in the behavior of the young child at home and at school. Context-related role expectations at the time may have led to aggressiveness showing continuity in male children and dependency in female children in the Fels' longitudinal study. A point made earlier about developmental transitions is also relevant here: there may be times when environmental input is especially important in generating long lasting effects. Thus, differential family socialization experiences during the time of self-awareness emergence might be particularly worthy of study.

Indeed, the last-mentioned strategy represents a major cross-cutting theme in our volume. All investigators appear to agree that an important consideration in longitudinal research today is to specify the conditions of environment under which developmental continuities will appear. Particular mention can be made of Rutter, Klinnert *et al.*, and Thompson and Lamb, who refer to aspects of the family and caregiving environment; Fischer, Pipp, and Bullock refer to the measurement environment in terms of optimal versus spontaneous performance in cognitive testing.

Plan of the Volume

The three chapters grouped in Part I present overarching research perpsectives related to continuities and discontinuities. As intended, each author reviews research from a particular viewpoint: Kagan from his viewpoint in developmental psychology, Rutter from that of child psychiatry, and Greenough and Schwark from a stance in the neurosciences. These chapters are presented first to orient the reader to changing views and current challenges.

Following this, Parts II and III group chapters dealing with research focused either on discontinuities or continuities. The reader should be aware, however, that this division is somewhat arbitrary. In Part II investigators are studying discontinuities in the midst of continuities and in Part III investigators are studying continuities in the midst of change and reorganization. This fact is illustrated by the inclusion of the chapter by Morgan and Harmon in the continuities section. As their title indicates, they are dealing with developmental transformations in mastery motivation, yet they discuss continuities across these times of transformation. This is a good example of the problem of heterotypic continuity. This situation is also illustrated by the inclusion of Crnic's chapter in this same section. Her discussion of early experience research in rats examines the critical period hypothesis. This hypothesis represents a discontinuity in the way the environment is experienced; still, in spite of this, evidence for the differential impact of early experience to later represents a form of continuity or connectedness if it is demonstrable.

Part IV contains two chapters which offer special integrative statements related to the search for continuities. Goldsmith's chapter is conceptual, discussing two major themes in the developmental literature which reflect different strategies for research and, quite often, different subdisciplines. One theme concerns searching for *linear stability* and the other for *organizational continuity*. These two themes represent alternative views on the nature of longitudinal continuity of behavior and, as Goldsmith points out, there has been an "unrecognized correspondence" between these two views and the views of two subfields in genetics, namely, quantitative genetics and developmental genetics.

We expect that the chapter by Short, Horn, and McArdle may appear strange at first to many interdisciplinary readers of this volume, but we strongly recommend it. Referring to study design and data analysis in other chapters, these authors describe mathematical statistical model-building procedures to be applied both for exploration and confirmation of relationships. In particular, structural equation modeling and related techniques appear to be quite appropriate for longitudinal data sets which search for continuities in the midst of discontinuities. This is so because multiple causes and multiple outcomes are searched for along with multiple possible links between them; one does this as one generates a mathematical model. An additional feature of this approach, which could well be the wave of the future for this kind of research, is graphic representation of a sort which allows the researcher to conceptualize multiple processes and their interrelationships. Indeed, many relationships may be seen for the first time because of the mathematical modeling itself.

In sum, the chapters that follow present a variety of new views of continuity and discontinuity, as well as new strategies for research. The fact that these chapters build on previous developmental research, use new methods, and provide findings consistent with emerging biological models lends a sense of excitement to current directions.

REFERENCES

Arend, R., Gove, F. L., & Sroufe, L. A. Continuity of individual adaptation from infancy to kindergarten: A predictive study of ego-resiliency and curiosity in preschoolers. *Child Development*, 1979, *50*, 950–959.

Baltes, P. B. Life-span developmental Psychology: Observations on history and theory revisited. In R. M. Lerner (Ed.), *Developmental Psychology: Historical and Philosophical Perspectives*. Hillsdale, New Jersey: Lawrence Erlbaum, 1983.

Bever, T. G. Regression in the service of development. In T. G. Bever (Ed.), *Regressions in mental development: Basic phenomena and theories*. Hillsdale, New Jersey: Lawrence Erlbaum, 1982.

Block, J. *Lives through time*. Berkeley, California: Bancroft, 1971.

Clarke, A. M., & Clarke, A. D. B. *Early experience: Myth and evidence*. New York: The Free Press, 1977.

Emde, R. N. Changing models of infancy and the nature of early development: Remodeling the foundation. *Journal of the American Psychoanalytic Association*, 1981, 29(1), 179–219.

Emde, R. N., Gaensbauer, T. J., & Harmon, R. J. Emotional expression in infancy: A biobehavioral study. *Psychological Issues, Monograph Series*, 1976, 10(37).

Kagan, J. *Change and continuity in infancy*. New York: Wiley, 1971.

Kagan, J., & Moss, H. A. *Birth to Maturity*. New York: Wiley, 1962.

Kagan, J., Kearsley, R., & Zelazo, P. *Infancy: Its place in human development*. Cambridge, Massachusetts: Harvard University Press, 1978.

Lampl, M., & Emde, R. N. Episodic growth in infancy: A preliminary report of length, head circumference and behavior. In K. Fischer (Ed.), *New directions for child development: Levels and transitions in children's development*. San Francisco: Jossey-Bass, 1983.

McCall, R. B. The development of intellectual functioning in infancy and the prediction of later IQ. In J. D. Osofsky (Ed.), *Handbook of infant development*. New York: Wiley, 1979.

Moss, H. *Review of longitudinal research*. Commissioned state-of-knowledge paper for NIMH Division of Extramural Research Programs review of Behavioral Sciences Research Branch programs, 1983.

Plomin, R. Childhood temperament. In B. Lahey & A. Kazdin (Eds.), *Advances in clinical child psychology*, 1983, 6, 1–78.

White, S. Evidence for a hierarchical arrangement of learning processes. In L. Lipsitt & C. Spiker (Eds.), *Advances in child development and behavior*, 1965, 2, 187–220.

PART I

CURRENT RESEARCH PERSPECTIVES RELATED TO THE CONTINUITY OF EARLY EXPERIENCE

CHAPTER 2

Continuity and Change in the Opening Years of Life

Jerome Kagan

Each historical period is characterized by a small number of intellectual concerns that become the titles of celebrated essays. Understanding, emotion, reason, morality, and freedom have many entries because each has been regarded as a quintessential property of human nature. Each remains a puzzle. Although these impenetrable themes continue to dominate the brooding of modern scholars, the puzzle of human psychological development is a relatively recent addition to the list. The reasons for its emergence as an enigma attracting serious study include historically novel apprehensions, demographic changes in the society, and the discovery of information about the child and its plan of growth.

When European citizens were less socially mobile than they are today, and infant mortality approached four of ten children, understanding of childhood did not seem to be of much value. If an infant survived the first two years, his or her future was determined in large measure by parental social position. But by the end of the seventeenth century both ascent and descent in status had become a real possibility for a significant proportion of youth whose families had acquired some measure of freedom and an expectation of altering their lives, as well as those of their children. Historical events had made the child's eventual position less knowable and, therefore, a source of parental worry. The community now needed an explanation of the obvious variation in life histories and a set of prophylactic practices that could be implemented

Jerome Kagan • Department of Psychology and Social Relations, Harvard University, Cambridge, Massachusetts 02138. Preparation of this paper was supported in part by research funds from the Foundation for Child Development and the John D. and Catherine T. MacArthur Foundation.

during childhood in order to reduce parental uncertainty over the child's future.

Of all the causes thoughtful commentators might have invented to explain the dramatic differences in adult talent, happiness, success, and morality, almost all Western theorists, at least since the Enlightenment, made early caretaking experiences (either quality or physical care or specific encounters with objects and people) the most relevant. The moods, values, skills, and habits created during the first half dozen years were presumed to persist indefinitely and to give form to the adult's character, competence, and capacity for joy. Many contemporary American parents are certain that adult happiness, which has become the most popular criterion for a successful life, is influenced in a non-trivial way by the events of infancy and early childhood. This connected interpretation of development, so central to the theoretical writings of Freud and Piaget, comprises the central reason for the average citizen's deep curiosity about development.

DEVELOPMENT AS A CONNECTED SEQUENCE OF STAGES

The course of a life, like the cycle of the seasons, consists of a small number of inevitable phenomena which, depending on time and place, are embedded in a script the details of which can take on remarkable surface variety. The Eskimo infant living near Hudson Bay remains physically close to its mother for most of the first year; the Israeli infant born on a kibbutz is cared for by a metapelet; the Mayan infant in northwest Guatemala spends much of the first year in a hammock wrapped in tattered pieces of woven cloth. But despite the differences in affectivity, alertness, and maturity of motor coordination produced by these rearing conditions, by three years of age all three classes of children share the capacity to recognize the past, to become frightened by the unusual, and to understand the significance of an adult prohibition—a trio of talents that emerges in that order.

The belief that the transformation of a babbling, poorly coordinated newborn into a loquacious ballerina involves passage through a series of hidden stages unites the two separate characteristics of stage. The least controversial quality is that each stage is characterized by a set of correlated competences. The more controversial premise is that the characteristics of one stage are dependent on or derivative of those of the preceding one. Most Western scholars presume a structural connection between all major phases of development with no breaks in the story line. William Stern (1930), a nineteenth-century observer of children, saw a connection between the babbling of the five-month-old and the speech of the two-year-old. Sigmund Bernfeld, an early Freudian disciple, supposed that all of adult cognition had its origin in early infancy: "Historically all phenomena of adult mental life must be traceable to birth" (Bernfeld, 1929, p. 213). Some theorists held that the adult's proprietary motive was derived from the grasp reflex of the infant; others suggested that a person's aesthetic sense was sculpted by early experiences with attractive toys (Rand, Sweeny, & Vincent, 1930). Modern authors have argued for a connection between the harshness of toilet training imposed during the second year and conformity during adulthood, or between multiple caretakers during infancy and a fragile emotional security in adolescence.

Piaget's interpretation of the rich description of his own children's behavior is a celebration of connectivity, for he insists that in order to study "the beginnings of intelligence, we were forced to go as far back as the reflex" (Piaget, 1951, p. 6). Jonas Langer (1980), who has taken Piaget's assumptions seriously, asserts that the origins of adult logic can be found in the manipulations of presymbolic infants. Langer observes a six-month-old raising an object to his face with his left hand while his right hand slides a second object to the right. His description follows: "The most advanced binary mappings produced at this stage consist of two-step sequences of simultaneous but different transformations" (Langer, 1980, p. 29).

These declarations of faith in connectivity have typically been based on superficial similarities between one aspect of the behavior of an infant and a single aspect of an action noted in older children or adults. But the human mind is remarkably adept at inventing similarities between fundamentally different phenomena by simply noting one feature that is shared, or seems to be shared, by the two events. The invention of a single thread of similarity between two temporally separated events differing in pattern and context is not sufficient to prove or even to imply their true functional relation. Nineteenth-century observers saw a similarity between the infant's grasp reflex and the adult's tight clasping of coins. A century ago the protesting cry of the one-year-old following maternal departure was classified as similar to the willful disobedience of the adult. Today the same act is grouped with the anxiety and sadness that follow loss of a sweetheart, spouse, or parent.

As this century began, Havelock Ellis postulated a similarity between a nursing infant and adult sexual intercourse to which Freud referred in "The Three Essays on Sexuality."

> The erectile nipple corresponds to the erectile penis: the eager watery mouth of the infant to the moist and throbbing vagina, the vitally albuminous milk to the vitally albuminous semen. The complete mutual satisfaction, physical and psychic, of mother and child in the transfer from one to

the other of a precious organized fluid, is the one true physiological analogy
to the relationship of a man and woman at the climax of the sexual act. (Ellis,
1900, p. 250)

Why did this analogy seem reasonable in 1900 but far less compelling today? Why has the event of nursing become, in this generation,
analogous to adult trust rather than sexual passion: I suggest that the
reasonableness of an analogy between two disparate domains requires,
first, belief in a fundamental similarity between the two phenomena and
second, uncertainty or concern over the larger category of experience to
which the analogy refers. Many scholars at the turn of the century
believed that sex and hunger were the two major human drives around
which all of society moved, and, additionally, that sexuality was a node
of conflict. Hence, the mind of the citizen was prepared to regard nursing as sexual.

Contemporary scholars regard the desire for a trusting, loving relationship with another and control of anxiety as two of the potent forces
guiding human behavior, and trust between adults has become a major
source of uncertainty. Hence, Erikson's (1963) suggestion that the relation between the nursing infant and its mother contains elements that
are similar to the bond of trust between adults is more attractive to
contemporary minds than Ellis's equally poetic idea.

The Challenge to Connectivity

Recent empirical results and theoretical essays provoke a more critical attitude toward connectivity. For example, even though the initiation of play behavior in juvenile squirrel monkeys is believed to facilitate
adaptive social behavior later, the absence of play behavior in some
troops of juvenile squirrel monkeys did not prevent the appearance of
normal social behavior, suggesting that the juvenile play did not make a
serious contribution to the adult profile (Baldwin & Baldwin, 1973).
Similar examples have led some biologists (Oppenheim, 1981) to suggest
that each life stage requires special temporary structures and functions
to facilitate maximal adaptation for that period. But when the next stage
is reached many of these adaptive functions are no longer needed and
are either inhibited, replaced, or lost.

This conclusion contrasts with the more traditional one which assumed a future purpose for all the major milestones and preserved
connectivity with the future. In modern psychological writing, the crying of one-year-olds to separation from the caregiver is regarded as an
index of the infant's quality of attachment, an emotional state that is
presumed to persist. The infant's attachment is regarded as neither as a
temporary process that serves to keep the mother close during the help-

lessness of the first year, nor as an epiphenomenon to the maturation of new cognitive talents which have separation distress as an unexpected and unnecessary consequence. For either of these scripts the phenomenon called separation distress has no important future role. It is similar to the actions of birds who are about to hatch, for special behaviors accompanying hatching are used once and once only and have no future function.

Each life phase makes certain demands on the animal and so each is accompanied by particular biological potencies. Succeeding phases have a slightly different biological substrate and a different set of demands. Hence, some of the past is inhibited or discarded and a new pattern actualized, not unlike the sequences of gene activation and deactivation that occur with growth.

Ontogenetic Adaptation

Application of these ideas to human psychological development implies that the fears to strangers and to separation seen in the first year, the single-word speech of the 18-month-old, and the absolute definition of right and wrong held by the three-year-old may be temporary adaptations. They may disappear with development and not be connected with future qualities that appear similar. Perhaps they might even be omitted without harming the child's development. Consider the following thought experiment for two universal developmental sequences, the growth of speech and the capacity for guilt. The traditional, connected story assumed that in order to speak multiword utterances the child must first pass through a stage of speaking only in single words. Similarly, popular theory holds that in order to experience guilt at age five, the child must pass through an earlier stage, around age two, of fearing punishment from parents. But is it unreasonable to suggest that the earlier stage could be eliminated and the child still acquire the later capacity? Multiword utterances and the capacity for guilt require a level of cognitive maturation not yet attained by two-year-olds, so they are not possible events. And the ability to communicate to parents in single words and the receptivity to anxiety over adult disapproval are adpative during the second year of life. But those conclusions do not demand that the earlier competence is necessary, or even that it makes a substantial contribution to the more mature one.

I am not claiming that there are no structural links between developmental phenomena—there must be some; I am only suggesting that it is unlikely that every actor in the first scene of the play has a role in the second act. Some milestones are likely to be either temporary adaptations to that stage of development or epiphenomenal to the maturation of cognitive competences adaptive for that stage.

THE ATTRACTIVENESS OF CONNECTIVITY

There are, at least, a half dozen good reasons for being receptive to the idea of connectivity in development. First, this doctrine renders original forms useful. If the origins of important adult properties occur during late childhood or adolescence, the first years of life would appear to have no future purpose, much like the embryonic notochord which vanishes after its mission is completed. The possibility that the products of a developmental era might be temporary or transitional is bothersome to many who want to believe that all psychological products are permanent and that "everything we learn is permanently stored in the mind" (Loftus & Loftus, 1980, p. 410).

Second, connectivity seems amenable to a mechanistic explanation of growth. When each new function is preceded by another which makes a substantial contribution to it, it is easier to state cause–effect sequences than if a function emerges relatively rapidly as a result of an endogenous change. In the second instance, the mind is left with an explanatory gap.

In Carmichael's (1927) classic experiment, salamander larvae were prevented from practicing any swimming movements by being anesthetized. When the anesthetic was removed and the animals transferred to fresh water, they swam as well as those that had never been anesthetized. The swimming movements had to wait upon maturation of the central nervous system and were not a derivative of the opportunity to practice the immature swimming movements of the normally growing larvae.

A third possible source of the persuasiveness of connectivity is more speculative. During each historical period there is a dominant philosophical view which most scholars avoid confronting. From the Renaissance to the nineteenth century, philosophers and scientists were reluctant to deduce or infer propositions that would refute or contradict biblical statements on man and nature. Although few contemporary scientists worry about the implications of their work for Christian teaching, many social scientists are concerned, often unconsciously, with the implications of their data and ideas for the doctrine of egalitarianism. The criticisms of Arthur Jensen (1973) and E. O. Wilson (1975) have less to do with the data they discuss than with the implications of their statements for the idea of equality. The assumption of connectedness is in accord with egalitarian principles because it implies that if all young children had psychologically benevolent experiences the extraordinary variation in competence and motivation—especially the differences between children from poor and privileged social classes—would be narrowed considerably. Legitimizing discontinuities in development, due either to biological maturation, genetic variation, or new social arrangements, implies that the benevolent products of early experience might be abrogated by peer groups, quality of schooling, neighborhood, or changes in physiology. Awarding power to the latter forces is regarded by some (incorrectly, I might add) as inconsistent with egalitarian suppositions, for it seems to make it more difficult to arrange similarly benevolent experiences for all young children.

Fourth, the English language reflects a bias toward continuity in an individual's qualities. The adjectives used to describe young children rarely refer to the age of the actor or the context of action. Like the names of colors, they imply a permanence over time and place. Adjectives like passive, irritable, intelligent, or labile are applied to infants, children, and adults as if the meaning of these terms were not altered by growth. This is not true in all languages. In Japanese, for example, different words are used to describe the quality of intelligence in a child and adult. On the island of Ifaluk in the Western Carolines, the emotion called *fago*, which only approximates our word compassion, is never applied to children under six years of age (Lutz, 1982). There is a strong temptation to assume that entities with the same name are of the same essence; hence, simply using the same word to describe a characteristic in children and adults makes us receptive to believing in a hidden disposition, part of which survives, unchanged, over the years. Thus, the belief in the permanence of characteristics is helped by our language habits.

A fifth reason for maintaining a belief in the maintenance of the consequences of early experience is the adult ranking of children on valued traits, both in and out of school. This practice, which sensitizes every parent of a preschool child to the fact that the child's talent would be ranked at the end of the first grade, influences the quality of education the child will receive from that time forward and hence builds in a preservation of a child's rank, at least for academic competences. Few societies practice such a severe grading of children with such zeal. Our commitment to a meritocratic system forces us to select candidates from the best trained, a decision that is made early, perhaps by age 10 or 11. Most parents know this sequence or sense it, and, from their perspective, the goal is to guarantee that their child is ahead early in the race. Most people believe that a 7-year-old who is relatively more talented, in comparison to other children his age, in reading, art, or mathematics, is more likely to remain more skilled, rather than plummet. That belief is true to some extent; the school-age child who gets off to a good start is likely, other things equal, to remain ahead because of the establishment of expectations that affect motivation and the opportunity for enriching school experiences. Parents assume that the half dozen years prior to school determine the teacher's initial evaluation and interpret the profile at age 7 as a complex derivative of all that has gone before.

Finally, nineteenth-century psychological science also made a contribution to the doctrine of infant determinism. Most scientists believe that psychological experiences can be translated into sentences that have purely physiological content. Experience can affect the weight of the brain, and early stimulation can add dendritic spines to brain cells or alter the sensitivity of the visual cortex to vertical or horizontal lines (see Greenough, this volume). These findings lead many to view the central nervous system as similar to Locke's tablet, a soft surface which accepts material marks that are difficult to erase. The belief that experience produces a permanent change in the brain, wedded to the premise that the brain directs thought and behavior, leads to the conclusion that since the structures first established are apt to direct the later ones, early experience must be important.

THE ARGUMENT FOR DISCONTINUITY

Although a majority of developmental psychologists favor a continuist interpretation of change, the most revolutionary scientific ideas during the last 75 years have been discontinuist in their essence. The photon has replaced the wave; the discrete gene has replaced the continuous effects of climate, diet, and environmental challenge; and the possible disintegration of protons may replace steady state theories of the universe and imply an end, far, far in the future, of what we would like to regard as a continuous cosmos. But if the mind creates prototypes—representations of events that were not experienced—from the mental average of encountered events, it follows that the structures of mind are continually subject to change. Like the proverbial ship's planks replaced one by one until no piece of wood in the original was present in the restoration, it is possible for a child at age 12 to possess no structure that existed at age 2, even though there is always overlap between the structures created a moment ago and those established sometime earlier. This metaphor implies connectivity of structures over short rather than long periods. I rather like this idea because it is in accord with evolutionary theory. Although biologists recognize theoretically a connection between all of today's living forms and ancient forms of life, many of the latter are now extinct.

The present faith in the preservation of qualities resembles the pre-Darwinian belief in the fixity of species. Darwin's great insight was to see the animal world as dynamic with some species vanishing and new ones emerging. The psychological development of an individual may share important metaphorical qualities with the branching tree of evolution. However, even in evolution, some basic structures and functions are preserved—the nuclear DNA in each cell and the biological processes of irritability, reproduction, and energy metabolism are obvious examples. If we take the analogy between evolution and ontogeny seriously, we should expect a structural process to be preserved only if it is adaptive. The demands of the first two years are so different from those of later childhood and adolescence, one might expect many of the reactions of the early period to disappear. The 2-year-old is not expected to behave autonomously, to be moral, to show sympathy, guilt, or justified anger. All of these demands are imposed later. The human infant is born less mature than many other animals; hence, we might expect many early characteristics to vanish, as do those of the tadpole and the newborn kangaroo, because the properties of the infant are designed to help it adapt to this special period of growth. When the special demands of infancy have passed, these behaviors should either be inhibited or eliminated, like the stepping reflex of the fetus, the presumed function of which is to ensure normal position of the head for delivery.

This discussion of connectivity among psychological stages is analogous to debates among philosophers of history on the nature of societal growth (Almond, Chodorow, & Pearce, 1982). Those debates have centered on five themes—the forces that facilitate change, and the direction, rate, content, and bearer of change. These questions are also appropriate to human development and are being answered in new ways. The view of development promoted by theorists of this century assumed that development had a progressive direction—the child was always getting better. Freud's scheme emphasized the growing control of basic desires and the gradual attainment of what Emerson called self-reliance. Piaget made rationality the perfectible canvas. But both were faithful to the nineteenth-century assumption that the products we call growth were actualized slowly and cumulatively, dependent on both endogenous changes and interaction with people and objects, and additionally, stored in a structure that lay outside of motive, emotion, belief, and skill.

The emergent view fine-tunes four of these premises by awarding greater potency to endogenous changes in the central nervous system during the early years, insisting that, on occasion, the rate of change can be rapid, and making cognitive functions both the major product and reservoir of development. But the issue of direction is altered more radically. The new facts suggest that an Escher staircase may be a better metaphor for development than a highway, for some desirable qualities of children are lost completely and others replaced by less desirable ones, rendering the idea of maturity as a terminus more obscure than it ever has been.

THE PRESERVATION OF INDIVIDUAL DIFFERENCES: THE EVIDENCE

Because scientific study of the stability of differences among children and adults has such a short history, the evidence is meager, and the conclusions must be viewed with caution. But the existing data, however crude and vulnerable to criticism, do not support the belief that the variation in psychological characteristics displayed during the first two years of life provides a sensitive preview of the future. Although the results of each investigation are too weak to bear the burden of proof alone, the findings of many related studies can be woven into a fabric with more persuasive power (Brim & Kagan, 1980).

After the Second World War an international social service agency arranged for middle-class American families to adopt homeless children who had led uncertain lives in Europe during the war. When the children arrived in the United States, their ages ranged from about 5 to 10 years. A group of 38 of these children were followed to their new homes. About 20% of the children displayed initially severe signs of anxiety, such as overeating, sleep disturbance, and nightmares. But over the years all of these symptoms vanished, the vast majority made good school progress, and there was no case of academic difficulty among them. The authors of this study wrote:

> The thing that is most impressive is that with only a few exceptions they do not seem to be suffering either from frozen affect or the indiscriminate friendliness that Bowlby describes. As far as can be determined their relationships to their adopted families are genuinely affectionate. . . . The present results indicate that for the child suffering extreme loss the chances for recovery are far better than had previously been expected. (Rathbun, DiVirgilio, & Waldfogel, 1958, pp. 413–414)

In a later but similar study 229 Korean girls who had been adopted by middle-class American families when they were between 2 and 3 years of age were followed for at least six years. The children were divided into three groups based on their degree of malnutrition at the time of admission. The children were studied again about six years after admission to their foster homes, when all were in elementary school. The average IQ of the severely malnourished group (IQ of 102) was 40 points higher than the scores reported for similar Korean samples who had been returned to their original home environments. This degree of recovery of intellectual skills, as well as physical health, does not occur among malnourished children who are returned to their originally poor environments after a period of rehabilitation (Winick, Meyer, & Harris, 1975).

Psychologists have been following the development of a girl named Genie who at 13½ years of age was removed from a home where she had been immobilized, isolated from contact with others, and physically beaten for most of the first 13 years of her life. When discovered, she was malnourished, unable to stand erect, and without language, an unsocialized victim of extreme deprivation. After only four years in a normal environment she developed some language, learned some social skills, was able to take a bus to school, and began to express some of the basic human affects. On some of the scales of a standardized intelligence test she obtained scores that were close to those of the average child. Although she is still markedly different from an average California 18-year-old, Genie has grown remarkably in a short period of time (Curtiss, Fromkin, Rigler, Rigler, and Krashen, 1975).

Studies of normal children reveal that behaviors, motives, and beliefs normally undergo alteration with development. The results of several longitudinal investigations of working and middle-class American children, most of whom were growing up in relatively stable and supportive homes, indicated that variation in psychological qualities during the first three years of life was not very predictive of variation in culturally significant and age-appropriate characteristics 5, 10, or 20 years later, while variation during the early school years, especially 5 to 10 years of age, was predictive of adolescent and adult profiles.

For example, children from working and middle-class parents living in the Boston area had been evaluated extensively four times during the first three years of life. In the first part of the study the infants' attentiveness, vocal excitability, activity level, tempo of play, irritability, and disposition to smile were evaluated at 4, 8, 13, and 27 months of age. With only one exception, variation in these behaviors prior to one year was not highly predictive of a small set of theoretically related qualities at 27 months of age (Kagan, 1971). These children were evaluated again when they were 10 years old in order to determine whether variation in the infant qualities would predict intelligence test scores, reading ability, or a tendency toward reflection–impulsivity at age 10. The educational level of the family, and therefore the family's social class, was the only robust predictor of the child's IQ and reading skill. When children from the same social class who were markedly different on both reading ability and IQ were compared, there were no major differences between them during the period of infancy (Kagan, Lapidus, & Moore, 1978).

A major long-term study of 71 children who had grown up in southwestern Ohio during the period 1929 to 1957 (the Fels Longitudinal Study) also revealed little relation between variation in activity, dependency, tantrums, or aggression during the first three years of life and variation in a variety of culturally relevant qualities during adolescence and early adulthood. Infants who were extremely irritable did not differ in a major way from those who were less irritable when they were of school age; active infants did not turn out differently than less active ones (Kagan & Moss, 1962).

The Preservation of Inhibition

The only hint of preservation of a particular quality appearing in infancy held for a characteristic we have come to call "inhibition to unfamiliar or unexpected situations." Some infants, especially during the period 9 to 36 months, initially withdrew from the unfamiliar, from threat, and from obstacles, whereas others were more disposed to approach, attack, or retaliate to threat or domination. When the inhibited children were of school age, they continued to display behaviors that suggested inhibition in situations that might provoke uncertainty. For example, they avoided dangerous activities, were less aggressive with peers, and less spontaneous with other people. As adults, the boys who had been most inhibited as children chose vocations that minimized competitiveness and were less traditionally masculine. By contrast, those that had been least inhibited chose more competitive and traditionally masculine vocations (athletic coach, salesman, and engineer) (Kagan & Moss, 1962).

There is also more recent evidence favoring preservation of a quality of inhibition. In a recent study in our laboratory the behavior of over 100 21-month-old children to varied incentives was observed. The incentives included encounter with an unfamiliar woman, exposure to a robot, separation from the mother, and the modeling behavior of an adult. The coding of discrete behavioral signs of inhibition from videotape records permitted us to select the 28 most- and 30 least-inhibited children. These 58 children returned to the laboratory one month later and were retested in the same five situations. There was good stability for the frequency of inhibited behaviors across the two months—the correlation was +.63. Ten months later 40 of these 58 children who were still living in the area were visited at home by two unfamiliar women. The children who had been inhibited at 21 months were far less likely to interact with the visitors and stayed much closer to their mothers. The correlation between degree of inhibition displayed at 21 months and at 31 months was +.39. Additionally, 26 of these children were observed with an unfamiliar peer who had the opposite behavioral style. Once again, the children who had been classified as inhibited at 21 months remained extremely inhibited with the more outgoing peer (Garcia-Coll, Kagan, & Reznick, 1984).

One of the most intriguing characteristics of extremely inhibited children is that they show higher and more stable heart rates while looking at pictures or listening to a sound or speech that is unfamiliar and difficult to understand. When a child or adult is watching interesting pictures or listening to speech under relaxed conditions, the heart rate displays a cycling that is in phase with the person's breathing (respiratory sinus arrhythmia). As the child inspires, the heart rate rises; with expiration, the rate drops. The decrease in heart rate is mediated by the vagus nerve which is under parasympathetic control. However, when its complement, the sympathetic nervous system, discharges as an accompaniment to psychological arousal, it can inhibit the parasympathetic control of the heart and produce a slight rise in rate and a more stable rate—that is, a less variable heart rate. This fact suggests that children with higher and more stable heart rates to information that they find difficult to understand may be more sympathetically aroused by that information. Because the information is not startling or psychologically threatening in content, it is likely that the arousal is due to a mental set. One possibility is that these children make a greater effort to understand the unfamiliar. We know from studies of adults that when a person invests a great deal of mental effort in a problem, heart rate rises and stabilizes. Sitting in a laboratory room watching slides of strange pictures is not an ordinary experience, but an odd and inexplicable one. I believe that some infants show loss of sinus arrhythmia because they are trying harder to understand what is going on and cannot. They are also likely to become inhibited in a host of situations the central characteristic of which is the fact that they, too, are unexpected, unfamiliar, and not easily understood. Thus, the temperamental quality is a cognitive one that we might call vigilance.

The small group of adult males in the Fels Longitudinal Study who were extremely inhibited during the first three years of life and throughout childhood and adolescence also had more stable heart rates during a 10-minute relaxation period prior to the administration of test materials. Because this is the only evidence available that implies continuity of this temperamental disposition from infancy through adulthood, it should be viewed with caution. But it invites the guess that this temperamental style, which is seen clearly during the second year, might influence later behavioral choices in subtle ways.

In sum, although most long-term longitudinal studies fail to find preservation of individual differences in popular molar behaviors like irritability, activity level, dependence on parents, it is possible that some dispositions are preserved. A tendency toward initial behavioral inhibition to the unfamiliar may be one of these special qualities.

THE MATURATION OF COGNITIVE COMPETENCES

The evidence just reviewed on the stability of individual differences might lead one to question a strong version of connectivity between phases of development. But the evidence available on uniform growth sequences during the opening years provides an even more persuasive basis for a reexamination of this idea. It now appears likely that major cognitive competences emerge as a consequence of maturation of the central nervous system, rather than gradually acquired through a long series of interactions. I now present two examples of such competences. One is the enhancement of retrieval memory during the first year; the second is an appreciation of standards and the emergence of self-awareness during the last half of the second year.

The Growth of Memory

The increased ability to relate an experience in the present to relevant schemata is one of the central maturing functions of the first two years of life. In common parlance, we say the child is able to remember the past. But that simple phrasing hides at least three quite different functions. During the first six months the infant is able to recognize that an event in its perceptual field shares properties with its schemata but is incapable—or finds it difficult—of retrieving a schema without any incentive in the field. It needs prodding. And a new schema created from a 15-second encounter fades if it is not renewed or the infant has similar experiences which interfere with its ability to recognize the event. Even when the infant only has to recognize a change in scene, if the delay between the original and the transformed scenes is as short as seven seconds, 6-month-olds seem unable to remember some original events. We say "seem," because it is not possible to be certain about a lack of competence (Kagan & Hamburg, 1981).

Infants often announce their memory for an event by looking back and forth between a new and old object. One group of infants tested monthly from 6 to 11 months of age was first shown a card containing three identical toys arranged in a triangular pattern for 15 seconds. For some cards, the three identical objects were ducks, for others, dolls. The card was then replaced by a second card after either a 1- or a 7-second delay. Sometimes the new card replaced only one object, sometimes two objects, and sometimes all three. On some occasions the second card was unchanged. Although the 6-month-olds did not alternate their gaze very much between the new and old elements, the 8-month-olds moved their eyes and head from the new to the old object about six times per trial. However, memory for the prior event decayed quickly in these infants. The 8-month-olds showed more shifting when the second card appeared after a 1-second delay than after a 7-second delay, implying that the 8-month-olds forgot the form of the elements on the first card. But 11-month-olds looked back and forth among the three objects as often following a 7-second as after a 1-second delay, suggesting they still remembered the original card (Kagan & Hamburg, 1981).

Later, the same infants were given a toy to examine. After the toy was removed, the child was shown simultaneously the old toy and a new one after delays of either 1, 3, or 15 seconds. Again, the younger infants—7 to 9 months old—looked back and forth between the two toys only when the delay was short, but by 11 months the shifting of gaze between the two toys was equivalent at all three delays.

A second component of memory begins to grow soon after the middle of the first year. It is the ability to retrieve a schema created from past experience with minimal clues—or hints—in the immediate field. Four-month-olds can recognize that a face is or is not similar to one they have seen before, but they are far less able to retrieve a schema for an event without some relevant incentive present. This phenomenon is reminiscent of the distinction between recognition and recall memory that we apply to older children. The latter capacity is enhanced in a major way after 8 months of age. Consider some examples that demonstrate this fact.

A convenient and intuitively reasonable procedure to evaluate retrieval ability requires the infant to remember where an attractive object was hidden, especially when he or she is forced to wait a while after the hiding before being allowed to reach. A group of infants was tested on such a problem once a month from 8 to 12 months of age. The children had to find the toy that was hidden under one of two identical cloths in front of them. On separate trials there was a delay of 1, 3, or 7 seconds before the child was permitted to reach for the toy, and in addition, there was either no screen, a transparent screen, or an opaque screen separating the child from the toy during the delay intervals. The infants improved steadily in their ability to remember the location of the toy across the four months of observation. No 8-month-old was able to remember the toy's location with a 1-second delay when the opaque screen was lowered during that interval. But by 1 year all infants could find the toy when the opaque screen was lowered for 3 seconds, and a majority could solve the problem when the screen was lowered for as long as 7 seconds (Kagan, Kearsley, & Zelazo, 1978).

Additionally, at 9 to 10 months of age, the child makes what has been called the "A not B" error. Once the child can reliably reach for a toy she watched being covered under one cloth—a victory shown around 8 months of age—she still remains vulnerable for a month or two to another class of error. The child is now shown two cloths, let us call them A and B. The examiner hides a toy under cloth A (on the left or right) permitting the baby to retrieve it two or three times. The examiner then hides the toy under the other cloth—cloth B. The puzzle is that most 9-month-old babies go to cloth A. The conditions that reduce the probability of this error provide a clue to its origin. If the delay between the hiding at cloth B and the opportunity to reach is short (less than a second or two), the error is much less likely to occur. And if the 9-month-old child does not make the error with a two-second delay, if the delay is increased to 6 seconds, the error will occur. Indeed, throughout the second half of the first year one can increase the likelihood of the error simply by increasing the delay between the hiding at location B and the time when the child is permitted to reach. At 8 months a delay of 5 seconds is sufficient; by 12 months a delay of 10 or 12 seconds is necessary to produce the error (Diamond, 1983).

There seems to be a steady improvement during the last part of the first year in the ability to hold or remember experience. Any procedure that helps the child remember the hiding at location B reduces the probability of error. Thus, if the object hidden is familiar or emotionally arousing—like a favorite toy or food—the error is less likely to occur at the same delay that produced the error with less familiar objects. And if there are five covers instead of two (three covers lie in a row between A and B located at each end), the infant is more likely to go to a cover closer to B than to A. But with each succeeding month the infant is better able to remember the location with increasing delays. By 18 months he rarely makes an error at any reasonable delay, even if an opaque screen is lowered for 10 seconds, and the child has no visual access to the locus of the hiding until the screen is raised. Novel experiences remain retrievable for a much longer time after the first birthday because of a major improvement in the ability to hold representations of prior experiences in memory and to recall them.

A related ability also emerges at this time. When older children or adults listen to a conversation, they are able to integrate the incoming information with their knowledge over a period of time that can last as long as a half-minute. The hypothetical process that permits this integration over time, called active memory, appears to be enhanced around 8 months of age, permitting the infant to relate incoming information to knowledge over an extended interval.

Many phenomena that normally appear during the last few months of the first year become understandable if we assume that active memory is being improved. The universal fears of infancy represent one example of the clarity that is gained by relating them to the growth of active memory. The two most popular fears of the first year have been called fear of strangers and fear following separation from the caretaker. Even blind one-year-olds cry when they hear their mother leave the room. The enhancement of active memory seems necessary for the appearance of these behaviors. As the stranger approaches, the 8-month-old studies the new face, retrieves her schemata for the familiar faces she knows, and compares the two ideas in active memory, trying to relate them and resolve their inconsistency. If she cannot relate the new person to her knowledge of the familiar ones, despite an attempt to do so, she becomes uncertain. If there is no behavior she can issue to deal with the state of uncertainty, she may cry. At the least, she will turn away from the stranger and stop playing or vocalizing.

A similar analysis may hold for the distress that often follows separation from the mother. The argument is that following the mother's departure the one-year-old generates from memory the schema of the parent's former presence in the room and compares that knowledge with the present situation and attempts to assimilate the inconsistency inherent in the two schemata. If the child cannot resolve the inconsistency—that is, assimilate the fact of mother absence with the retrieved schema of her former presence—the child becomes uncertain. If the uncertainty persists, the child may cry. The child's ability to remember that the mother had been present a minute earlier—a competence the six-month-old does not have—provokes her to try to understand the differences between past and present and therefore makes her vulnerable to anxiety if she fails. Distress often occurs when an infant is presented with a discrepant event that cannot be assimilated (Kagan, 1971).

I believe the state of uncertainty produced by a comparison of retrieved events and events in the perceptual field depends on the maturation of the brain. Support for that claim is found in Mason's (1978) demonstration that rhesus monkeys raised only with an inanimate object for the opening months of life showed maximal signs of anxious arousal to a novel context when they were 4 months old. The signs of anxious arousal were distress calls and acceleration of heart rate. Because the rhesus grows at a rate approximately three times that of the human, the comparable value for the human infant would be 12 months. Although infant monkeys raised with dogs also showed the same two behavioral signs of arousal, but with larger magnitudes, the important fact is that both groups showed their maximal values for these two responses at the same age—about 130 days of age.

The cognitive interpretation of separation distress differs from the two traditional explanations. An old one assumed that the child cried when the mother left because the infant anticipated pain or danger as a result of the conditioning of fear when the mother was absent. A more recent and very popular explanation holds that separation anxiety reflects the child's emotional attachment to the mother. The child's cry is its way of bringing the beloved mother back to the threatened infant. It was assumed that the more affectionate contact an infant had with its mother, the more attached it would be and therefore the more likely to show separation distress. Although these explanations seem reasonable, they will not work. One-year-old infants all over the world show separation anxiety beginning around 8 to 12 months—blind infants, !Kung Bushman infants who are on their mother's body for most of the day, infants raised in nuclear families, on kibbutzim in Israel, barrios in Guatemala, Indian villages in Central American highlands, or daycare centers. All infants begin to show separation distress around 8 months of age, maximal distress at about 15 months, and by the third birthday fail to show this response any more (Kagan, Kearsley, & Zelazo, 1978). This regularity has all the characteristics of a maturationally inevitable event.

Another class of fear appears during the early part of the second year. This time the apprehension is to an unfamiliar child (Kagan, Kearsley, & Zelazo, 1978). Regardless of the form of early rearing, 8-month-olds rarely show any signs of timidity or inhibition when put in a room with an unfamiliar infant of the same age. But by the middle of the second year they show obvious signs of apprehension: inhibition of play, clinging to the mother, and reluctance to interact with the other (Kagan, Kearsley, & Zelazo, 1978). These behaviors decrease by the third birthday. This apprehension occurs in children raised in nuclear families without any siblings, as well as in those raised on kibbutzim who have been with other children since the opening weeks of life (Zaslow, 1977). We suspect that the basis for the fear is the fact that 18-month-olds are mature enough to ask and wonder about the actions of the other child, but are still unable to answer their own inquiries.

This cognitive interpretation of the fears of infancy illustrates nicely the new conceptual frame used to describe early development. The fears appear because growth of the central nervous system has permitted new mental capacities to emerge. We cannot understand these phenomena by looking only to the child's past experience.

THE GROWTH OF SELF-AWARENESS

The period between 17 and 24 months is also a time of change. Children around the world first appreciate that certain actions are evaluated as right and wrong, and they become aware of some of their properties, feeling states, and ability to act. In popular language we say that the child is now self-conscious. Observations of children, in our own and other cultures, during the last six months of the second year reveal the appearance of a set of behaviors that invites the label self-awareness as a useful description (Kagan, 1981). I shall now list five important responses that appear during this interval.

A Sense of Right and Wrong

One of the milestones of this era is the first appreciation of right and wrong. Children now point to broken objects, torn clothing, and missing buttons and reveal in their voice and face a mood of concern. They point to a tiny crack in a plastic toy telephone and declare, "Oh-oh," or "yukky." In one experiment 14- and 19-month-old children were allowed to play with a set of 22 toys. Ten of the toys were unflawed without irregularities, while another set of 10 toys was purposely flawed in some way. Examples included a boat with holes in the bottom, a broken telephone, a white handkerchief with an irregular green streak, and a broken pencil. Additionally, two of the toys were odd-shaped meaningless wooden pieces that were unflawed. The purpose of including these two toys was to see whether special concern with the flawed toys was due to the fact that the integrity of the toys was violated, rather than their discrepant quality. Not one of the 14-month-olds behaved in any special way toward the flawed toys, while 57% of the older children showed unambiguous signs of concern with one or more of the flawed toys, but no special behavior toward the meaningless forms (Kagan, 1981).

This apparently new appreciation of the proper and improper appearance of objects, which seems to occur in children growing up in many different cultural settings, is part of a more general recognition that there are standards for behavior and that adults will react disapprovingly if these standards are violated. Of course, the child must learn that breaking, tearing, and dirtying objects are disapproved by adults. One function of the infant's dependence on adult care, which most suppose to have an adaptive purpose, is to make the growing child attentive to the caretaker's behavior and to prepare him to react with uncertainty when the caretaker raises her voice, frowns, hits the child, or otherwise reacts in ways that are unexpected when the child violates a norm. But the child does not have to be punished for every violation. Once they have acquired some information about the undesirability of a few behaviors, they go beyond that information and generate ideas of appropriate states and their contrasting violations. As with the fears of the first year, the regular appearance of this sensitivity to standards of right and wrong in children living in settings as divergent as Boston and atolls in the Fiji chain suggests that a new cognitive competence has emerged (Kagan, 1981). Parents have disapproved of soiling and destruction long before the infant is 17 months old, but the child's appreciation of those standards is delayed until this time.

The Role of Inference

I believe a critical new function that matures at this time is the ability to infer the cause of an event. The child now expects events to have antecedents and automatically generates causal hypotheses for unexpected events. Hence, when the child sees a broken toy on the floor, he infers that its flawed state was caused by someone's action. Because those actions are associated with adult disapproval, the child responds emotionally. It is of interest that children learning sign language first begin to combine two signs in the middle of the second year. This fact suggests that this is the time when the child begins to relate two ideas, which is what he does when he infers the cause of a broken toy.

Empathy

A second competence that matures at this time helps the child appreciate that physical aggression is wrong. Two-year-olds are now able to infer the psychological state of another person. In plainer language, they now show empathy. Longitudinal observations provided by mothers specially trained to record their children's reactions to the distress of others reveals that during the latter half of the second year children behave as if they are inferring the victim's state. They hug the victim or give him a toy or food (Zahn-Waxler, Radke-Yarrow, & King, 1979). As a result of this new empathic competence, children realize that if they strike another or seize a toy the other will feel the discomfort they feel when they are victims of the same action. That fresh realization acts as an important inhibiting force on aggressive behavior.

The Adaptiveness of Standards

The suggestion that by the end of the second year children become prepared to evaluate actions and events as right and wrong, good or bad, because of the maturation of the ability to infer cause and empathize with the state of another has an analogue in the evolutionary hypothesis that species differ in their biological readiness. Birds are prepared to learn the song of their species with just a little exposure to that song at the proper time in development. Children will learn a language with a little exposure to human speech. I am suggesting that children inherently move into an evaluative set toward the end of the second year—a suggestion confirmed by interviews with Fijian mothers who commented, without the help of formal knowledge of child development, that children naturally become more responsible after their second birthday, when they have acquired what the Fijians call *vakayalo,* best translated as "sense." The Utku Eskimo of Hudson Bay call this same quality *ihuma* or "reason" (Briggs, 1970). Many American parents will say about their child after the second birthday, "He has become a person now," or "He is now mature." The regular appearance of a sensitivity to right and wrong is one reason why nineteenth-century European observers believed that young children were innately moral. James Sully (1896), a popular nineteenth-century writer, suggested that the child has an "inbred respect for what is customary and wears the appearance of a rule of life, . . . there is in the child from the first a rudiment of true lawabidingness" (Sully, 1896).

Sully believed that all children must, because they are human, recognize that causing harm to another is immoral. Such knowledge could never be lost, regardless of any subsequent cruelty the child might experience. There are obvious advantages to an early appreciation of the fact that some acts are disapproved. Young children have the capacity for resentment toward persons and a desire for personal property, together with the ability both to plan and to plot to attain these goals. Thus, it may have been necessary in human evolution to make sure that early in development inhibitory functions would emerge to curb these disruptive qualities. In most parts of the world, for most of our history, mothers gave birth to the next offspring about three years after the last child was born. A 3-year-old both has the strength to inflict injury on a younger sibling and, more important, is able to retain hostile thoughts long after anger has subsided. The aggressive behavior that can be the product of jealousy a 3-year-old feels toward the younger sibling must be held in check. It would be adaptive, therefore, if 3-year-olds appreciated that aggression was wrong. The importance of a strong inhibition on aggression to younger children is seen in a rare event noted in a San Francisco newspaper last December. A 30-month-old killed his 22-month-old regular playmate by pounding the victim's head on the floor and striking his skull with a heavy glass vase. The fact that this is a freak phenomenon indicates that most children learn that such actions are improper by their third birthday.

Anxiety to Mastery Failure

A third phenomenon that occurs during the second year is the appearance of anxiety to possible task failure—what parents would call "fear of failure" in an older child. The child is playing happily on the floor when an adult comes to the child, asks if she can play, and then acts out some brief behaviors that are on the threshold of the child's sphere of mastery. She may pick up some dolls and a plate and say she is making dinner, or pick up two or three animals and take them for a walk. After completing these acts the woman says, "Now it's your turn to play" and returns to the couch to talk with the mother. Beginning around 18 months and peaking around the second birthday, children show extreme degrees of upset and distress after the woman returns to the couch. They fret, cry, run to their mother, stop playing, protest, or may insist they want to go home. This reaction occurs in Cambridge children, children growing up in huts on islands in the Fiji chain, as well as those who have recently arrived in northern California from their homes in Vietnam (Kagan, 1981). It is unlikely that all these children were punished for failing to imitate their parent or another adult. Hence, "learned fear of punishment" could not be the cause of this distress. It is likely that the child experienced an obligation to duplicate the adult's actions and recognized that he was unable to do so. Why should the child feel an obligation to imitate the woman when she did not ask the child to do so? I suggest that the child privately asks, "What does the adult want?" and concludes he is supposed to duplicate the adult behavior. But the more important idea is that the distress would not occur unless the child had some dim awareness of his lack of ability to imitate the woman. The child is now capable of anxiety if he believes he cannot meet a self-imposed standard. This phenomenon is the first sign of what will become guilt a year or two later. Several months later, when speech has become more mature, the children verify our suggestion that they have an awareness of their inability to perform the model's action, for when faced with a difficult problem they will say, "I can't do that." Children now set themselves problems they can master and draw adult attention to their products, because they want others to know they have met the standards they generated. By age 2, anxiety to failure and pride over success are part of every child's repertoire.

Motivation on Problem Tasks

As a result of this new set of characteristics, we might expect a dramatic improvement in performances on problems set by adults. This prediction is affirmed by the noteworthy increase in performance on a memory for locations task. The child watched an adult hide an object under 2, 4, 6, or 8 containers and varied the delay between the hiding and the child's reaching for intervals lasting from 1 to 10 seconds. Children growing up in divergent settings showed their largest gain in performance in the months before the second birthday (Kagan, 1981).

Self-descriptive Utterances

When the child begins to speak two-morpheme sentence with predicates, usually after the second birthday, he begins to describe his own actions as he is performing them. The child says, "Up," as he climbs up on a chair; "Go out," as he runs outside; "I fix," as he tries to rebuild a fallen tower of blocks. Because 2-year-olds are more likely to describe their own activities than the behavior of others, I suggest children are preoccupied with their own behaviors. The 2-year-old has a fresh insight. He has become aware of his ability to act, to influence others, and to meet self-imposed standards. These new ideas excite the child, and as a consequence he describes his behaviors as he performs them.

Recognition of Self

Finally, children seem to recognize themselves during the months before the second birthday. Children between 1 and 3 years of age were brought to a room which contained a large mirror. The mother surreptitiously rubbed a little rouge on the child's nose and then brought the child to the mirror. Infants under one year did not touch their faces, whereas most 2-year-olds put their hands directly to their noses, suggesting they recognized that the reflection in the mirror belonged to them (Lewis & Brooks-Gunn, 1979).

It is likely that maturing brain structures and functions permit the child to display these new behaviors. One scientist who has reviewed existing information has concluded that the period between 15 and 24 months is a special one because almost all the layers of the cortex reach, for the first time, a similar state of maturation (Rabinowicz, 1979). It may not be a coincidence that the period from 15 to 24 months corresponds to the interval when the child displays the behavior we have called signs of self-awareness.

Observers from different historical eras, as well as parents in different cultures, have used different names for the changes that are salient during this interval. Dietrich Tiedemann (1897) remarked that the child develops *Eigenliebe* (love of self) during this period; the nineteenth-century German psychologist Preyer (1888) suggested that *Ichheit* (selfhood) appeared during the second year. But recall that the Utku of Hudson Bay believe the child develops *ihuma* (reason) and the Fijians claim *vakayalo* (sense) appears at this time (Kagan, 1981).

Important premises are hidden in each of these phrases. Although the Europeans assumed that the changes reflected a new appreciation of individuality, the Utku and Fijians, like most non-Western societies, emphasized the child's ability to appreciate the difference between right and wrong. It is understandable that urban Europeans would wonder about the origins of the autonomous and actualizing ego, while small, isolated hunting or agricultural communities would be more concerned with the time when the child began to adhere to social norms. The word chosen by a scientist to name a natural phenomenon has nontrivial connotations for future scientific work.

SUMMARY

The corpora of empirical data summarized in this paper are united by their common concern with the degree of connectedness in early psychological development. The empirical evidence from longitudinal studies does not favor a strong version of connectedness, and the data on the emergence of active memory and self-awareness imply a more rapid, rather than a gradual and cumulative, appearance of new qualities during the opening years of growth. Now that both sides of the issue have some degree of reasonableness, future inquiry can be open to both suppositions.

REFERENCES

Almond, A. G., Chodorow, M., & Pearce, R. H. (Eds.), *Progress and its discontents*. Berkeley: University of California Press, 1982.

Baldwin, J. D., & Baldwin, J. I. The role of play in social organization. *Primates*, 1973, *14*, 369–381.

Bernfeld, S. *The psychology of the infant*. New York: Brentano, 1929.

Briggs, J. L. *Never in anger*. Cambridge: Harvard University Press, 1970.

Brim, O. G., & Kagan, J. (Eds.). *Constancy and change in human development*. Cambridge: Harvard University Press, 1980.

Carmichael, L. A further study of the development of behavior in vertebrates experimentally removed from the influence of external stimulation. *Psychological Review*, 1927, *34*, 34–47.

Curtiss, S., Fromkin, V., Rigler, D., Rigler, M., & Krashen, S. An update on the linguistic development of Genie. In D. P. Data, (Ed.), *Georgetown University Roundtable on Language and Linguistics*. Washington, D.C.: Georgetown University Press, 1975, pp. 145–157.

Diamond, A. *Cognitive development in the first year*. Unpublished doctoral dissertation, Harvard University, 1983.

Ellis, H. The analysis of the sexual impulse. *The Alienist and the Neurologist*, 1900, *21*, 247–262.

Erikson, E. H. *Childhood and society* (2nd. ed.). New York: W. W. Norton, 1963.

Garcia-Coll, C., Kagan, J., & Reznick, J. S. Behavioral inhibition in young children. *Child Development*, 1984, *55*.

Jensen, A. R. Level I and level II abilities in three ethnic groups. *American Journal of Education Research*, 1973, *10*, 263–276.

Kagan, J. *Change and continuity in infancy.* New York: Wiley, 1971.

Kagan, J. *The second year.* Cambridge: Harvard University Press, 1981.

Kagan, J., & Hamburg, M. The enhancement of memory in the first year. *Journal of Genetic Psychology*, 1981, *138*, 3–14.

Kagan, J., & Moss, H. A. *Birth to maturity.* New York: Wiley, 1962 (New Haven: Yale University Press, 1983).

Kagan, J., Lapidus, D., & Moore, M. Infant antecedents of cognitive functioning. *Child Development*, 1978, *49*, 1005–1023.

Kagan, J., Kearsley, R., & Zelazo, P. *Infancy: Its place in human development.* Cambridge: Harvard University Press, 1978.

Langer, J. *The origins of logic.* New York: Academic Press, 1980.

Lewis, M., & Brooks-Gunn, J. *Social cognition and the acquisition of self.* New York: Plenum Press, 1979.

Loftus, E. F., & Loftus, G. R. On the permanence of stored information in the human brain. *American Psychologist*, 1980, *35*, 409–420.

Lutz, C. The domain of emotion words on Ifaluk. *American Ethnologist*, 1982, *9*, 113–128.

Mason, W. A. Social experience in primate cognitive development. In G. M. Burghardt & M. Bekoff, (Eds.), *The development of behavior: Comparative and evolutionary aspects.* New York: Garland Press, 1978, pp. 233–251.

Oppenheim, R. W. Ontogenetic adaptations and retrogressive processes in the development of the nervous system and behavior: A neuroembryological perspective. In K. J. Connolly & H. F. R. Prechtel (Eds.), *Maturation and development: Biological and psychological perspectives.* Philadelphia: Lippincott, 1981, pp. 73–109.

Piaget, J. *Play, dreams, and imitation in childhood.* Trans. C. Gattegno & F. M. Hodgson. London: Routledge & Kegan Paul, 1951.

Preyer, W. *The mind of the child. Part 1: The senses and the will.* New York: D. Appleton, 1888.

Rabinowicz, T. The differentiate maturation of the human cerebral cortex. In F. Falkner & J. M. Tanner (Eds.), *Human growth* (vol. 3). New York: Plenum Press, 1979, pp. 97–123.

Rand, W., Sweeny, M. E., & Vincent, E. L. *Growth and development of the young child.* Philadelphia: W. B. Saunders, 1930.

Rathbun, C., DiVirgilio, L., & Waldfogel, S. A restitutive process in children following radical separation from family and culture. *American Journal of Orthopsychiatry*, 1958, *28*, 408–415.

Stern, W. *Psychology of early childhood* (6th ed.). Trans. A. Barwell. New York: Henry Holt, 1930.

Sully, J. *Studies of childhood.* New York: Appleton, 1896.

Tiedemann, D. *Beobachtungen über die Entwicklung der Seelenfähigkeiten* (1st ed. 1787). Altenburg: Oskar Bonde, 1897.

Wilson, E. O. *Sociobiology: The new synthesis.* Cambridge: Harvard University Press, 1975.

Winick, M., Meyer, K. K., & Harris, R. C. Malnutrition and environmental enrichment by early adoption. *Science*, 1975, *190*, 1173–1175.

Zahn-Waxler, C., Radke-Yarrow, M., & King, R. A. Child rearing and children's prosocial initiations toward victims of distress. *Child Development*, 1979, *50*, 319–330.

Zaslow, M. A. *A study of social behavior.* Unpublished doctoral dissertation, Harvard University, 1977.

CHAPTER 3

Continuities and Discontinuities in Socioemotional Development

EMPIRICAL AND CONCEPTUAL PERSPECTIVES

Michael Rutter

INTRODUCTION

The general question of the extent to which psychological development is continuous or discontinuous remains one that excites both interest and controversy. In view of the vast body of developmental research that has been undertaken over the last half century, one might suppose that it is high time that a satisfactory answer was obtained in order to settle the issue one way or the other. But no such answer is available and, still, investigators disagree vigorously on the conclusions to be drawn. For example, Lipsitt (1983, p. 182) claims, "The assumption that the cumulation of early life experiences, *beginning* in infancy, is critical for and determinative of later development and behavior has been almost universally embraced by human development scholars." He accepts that intervening conditions can disrupt anticipated events but points out that that does not negate continuity in development: "Apparent non-continuities may be instances of continuities not yet fully revealed" or of "continuities not yet sufficiently investigated."

In sharp contrast, Kagan (1981) observes that "many instances of developmental change can be characterized by replacement of an old structure or process by a new one, with little or no connectedness be-

---

Michael Rutter • Department of Child and Adolescent Psychiatry, Institute of Psychiatry, London SE5 8AF, England.

tween the two hypothetical structures. This suggestion implies that some structures and processes vanish" (p. 68). Further, "some historical sequences that seem to be contingently related are not part of a chain of derivatives but rather form a set of non-contingent replacements" (p. 45). Kagan, of course, accepts that there is some continuity in development, but he emphasizes that many capacities that seem to be derived from that which went before in fact represent new phenomena showing discontinuity with the past: "The early form did not determine the new one; the new entity was not contingent on the earlier structure or process" (p. 39).

As Hall and Lindzey (1978) point out in their review of theories of personality, "some . . . imply the key to adult behavior is to be found in the events that have taken place in the earliest years of development, while other theories state quite explicitly that behavior can be understood and accounted for solely in terms of contemporary or ongoing events" (p. 23). The former theories "tend to view the individual as a continuously developing organism. The structure that is observed at one point in time is related in a determinant manner to the experiences that occurred at an earlier point." In contrast, the latter theories "tend to consider the organism as going through stages of development that are relatively independent and functionally separated from the earlier stages of development" (p. 23). It follows that theories that emphasize continuity in development in terms of an orderly, consistent process generally posit a single set of principles for the mechanisms underlying that process. On the other hand, theories that emphasize discontinuity usually suggest that somewhat different principles may be needed to account for what takes place at different stages of development.

This sounds hopelessly contradictory. Can scientists really be so completely at odds with one another on the issue of what is involved in the process of development? Perhaps different aspects or features of development are being considered in these opposing views, or perhaps disparate concepts of continuity and discontinuity are being employed. The purpose of this chapter is the critical examination of these, and other, possibilities in terms of both conceptual and empirical considerations, with particular reference to some of the research undertaken by my colleagues and myself.

MEANINGS OF CONTINUITY

The first point that requires emphasis is that the general notion of continuity includes several quite different meanings of the term (Emmerich, 1964, 1968; Kagan, 1980; Magnusson & Endler, 1977; Wohlwill,

1980). First, there is the concept of absolute invariance. This may mean the unchanging persistence of a quality; for example, the suggestion that once a child has acquired object permanence that capacity cannot subsequently be lost (except as a consequence of gross disease). The same would apply to the ability to form selective bonds or the quality of self-awareness. Alternatively, it may mean that a skill has reached an asymptote such that there is little subsequent change; for example, the expectation that intelligence remains fairly stable once early adulthood is reached.

Second, there is the concept of regularity in the *pattern* of development or in the form of change. Thus, twin comparisons show that monozygotic pairs show a greater concordance than dizygotic pairs in both the timing and patterning of ups and downs in intellectual development (Wilson, 1977). Considered in isolation, intellectual development shows substantial variability or instability over time, but the twin data suggest that such variability may itself be genetically determined to an important extent. Another example is provided by schizophrenia, in which about half the cases of psychosis with an onset in adult life have been preceded by nonpsychotic abnormalities in childhood (Rutter & Garmezy, 1983). These childhood precursors are not sufficiently distinctive in form for diagnosis to be possible at the time in most cases, but there is no doubt that they represent continuities in the disease process.

Third, there is the matter if ipsative stability; that is the notion that there is a persistent pattern *within an individual* with respect to the pattern of relationships among different personality features (Emmerich, 1968; Thomas, Birch, Chess, Hertzig, & Korn, 1963). In other words, it may be meaningful to describe someone as predominantly cheerful in the sense that, although his mood may fluctuate considerably from moment to moment or situation to situation, nevertheless, there is a predictable and lasting expectation that he is more likely to be happy than sad at any given moment.

Fourth, there is the feature of normative stability, or the constancy of hierarchical or relative positions in the population with regard to some attribute. Thus, between 6 and 16 years intelligence is said to show a moderate degree of stability—not because intellectual abilities are unchanging (obviously they are not) but rather because rank correlations over this time span are moderately high, with more intelligent children tending to remain above average in their performance, and vice-versa.

Fifth, there is the question of continuity in structure or process or mechanism. For example, it is commonly supposed that close personal relationships serve a psychologically protective function throughout life—such that bereavements leads to grief and that the presence of a loved one enhances resilience in the face of stress or adversity (Bowlby,

1969, 1973, 1980). It is not suggested that the *form* of relationships at age 2, 20, and 82 years is the same; obviously there are differences determined both by developmental alterations and by changes in social context. Also, it is not necessarily argued that the quality of a relationship at one age is predictive of the quality of another at a later age. Furthermore, it is not postulated that the strength of relationships shows any particularly consistency. Rather, what is hypothesised is that the relationships at all these various ages represent the same psychological processes and hence that continuity is to be found in the underlying structure, not in the surface representations of that structure. Of course, such a hypothesis demands evidence that there *is* some underlying structure, process, or mechanism.

Sixth, continuity may be thought of in terms of a predictable pattern of associations between events or happenings or experiences at an early phase of development and some type of psychological outcome at a later age. Thus, the concept of maternal deprivation suggests that a lack of a stable family life during the preschool years predisposes to both immediate and later psychological disturbance (see Rutter, 1981b). That hypothesis does not necessarily imply any correlation between *behavior* in infancy and that at maturity; nor does it necessarily demand any correlation between the *degree* of family instability and the degree of later disturbances. But, it does suggest that family discord or disruption in infancy substantially increases the *risk* of psychological disorder during the years that follow. It is this last concept of continuity to which I propose to pay most attention in this chapter. However, its discussion demands some attention to the previous five concepts when considering possible mechanisms.

METHODOLOGICAL ISSUES

Before turning to some of the empirical evidence on continuities and discontinuities in social-emotional development, it is necessary to note a few of the crucial methodological issues. Thus, it is apparent that different statistical representations of the findings may convey quite different impressions of the strength of continuities, even when the data used in the analyses are identical (see Rutter, 1977b). For example, correlation coefficients probably constitute the most frequently employed statistic for the quantification of continuities. Broman and her colleagues (1975), in their analyses of the collaborative Perinatal Project, reported a correlation of .076 between Down's syndrome and IQ! At first sight, that would seem to mean that there is no predictable relationship between the two but, of course, it does not mean anything of the kind. We know that almost all Down's syndrome individuals are mentally retarded to some degree. The near-zero correlation simply reflects the fact that there were only 12 cases of Down's syndrome in a total population of 25,000! Although the condition accounts for a trivial proportion of the population variance in IQ, it constitutes a major and overwhelming influence on intelligence in those children who happen to have the syndrome. Correlation coefficients constitute a fair measure of population variance, but they are a completely inappropriate measure of association or continuity when (a) the independent variables involves a very small proportion of the population and/or (b) when many different factors contribute to the overall variance.

Also, continuities in extreme groups or subpopulations of various kinds may be quite different from those in the population as a whole. For example, overactivity as measured in any one setting carries a very low level of predictability for children's functioning a few years later, but *pervasive* overactivity (i.e., that similarly present in several different settings shows a much stronger association with persistence of behavioral difficulties (Schachar, Rutter, & Smith, 1981). Measures of conduct disturbance in middle childhood show only quite modest correlations (.2–.4) with similar measures a few years later (Rutter, 1977b). But this is because many children show transient behavioral problems. At any one time, these ephemeral difficulties account for about half of all conduct disorders in the population. But the other half consists of persistent disturbances. The overall impression of low consistency is misleading in that there are two rather subpopulations with different characteristics. The children with persisting delinquency or conduct disturbance show deviance in many aspects of their functioning; those with transient difficulties do not (Rutter & Giller, 1983).

A further crucial point is that continuities "looking forward" are not necessarily the same as those "looking backward" (Rutter, Quinton, & Liddle, 1983). That is to say, the continuity statement that all individuals with a particular behavior at age X had already shown that behavior at age Y some ten years earlier (i.e. looking backward) is not at all the same as the continuity statement that all individuals with that behavior at age Y will still show the same behavior ten years later at age X (i.e., looking forward). Thus, Robins (1978) and others (see Rutter & Giller, 1983) have shown that the great majority of adults with an antisocial personality disorder or sociopathy were delinquent or antisocial as children. But, on the other hand, only a minority of delinquent juveniles become adult sociopaths. That is because there is a massive drop in delinquent activities as people reach early adult life. Similarly, we found that most women showing a serious breakdown in parenting function had experienced multiple adversities during childhood; but of children experiencing such adversities only a minority later failed as parents (Rutter, Quinton, & Liddle, 1983). The relative base rates of problems at different ages will influence measures of continuity.

A different issue is raised by the observation that there may be continuities in the *form* of behavioral functioning in spite of discontinuities in the *rates*. For example, it is clear that most adults with depressive conditions were not depressed as children, and indeed most have shown no emotional or behavioral disturbance of any type when young (Zeitlin, 1983). Also, although the empirical data are weak, it is likely that many depressed children do not go on to exhibit depression in adult life. But it appears that within the subgroup of individuals who suffer psychiatric disorder as *both* children and adults there is quite strong continuity in that those with depressive symptomatology in childhood tend to show the same symptom patterns in adult life (Zeitlin, 1983).

It is apparent that there are major problems in measurement when one tries to assess the same attributes at very different ages. This is especially the case when predicting from the infancy period (Rutter, 1980). Crying in a 6-month-old and crying in a 16-year-old, for example, are both real reflections of emotional expression. But do they have the same functioning meaning, and is it sensible to regard correlations between the two as if they were reflections of continuity or discontinuity in development (Rutter, 1982b)? Conversely, because behaviors appear different in form, does it follow that in functional (or genotypic) terms they are dissimilar? Obviously not (see Kagan, 1980; Moss & Susman, 1980)—developmental changes may modify or alter the particular manner in which a characteristic is manifest. But, then, how does one determine what constitutes valid functional equivalence for temperamental or personality features?

Lastly, there is the most crucial issue of all—namely, the testing of cause and effect relationships (see Rutter, 1981a; 1982a). Obviously, the observation that A preceded B does not mean that A caused B or even that both were linked with some common process or mechanism (see Kagan, 1981). The history of psychology and of medicine contains numerous examples of false causal inferences. Of course, there are well-established means available for the testing of causal hypotheses, but in practice such testing is a difficult matter when there are complex interacting variables and multifactorial determination of outcomes.

STRENGTH OF CONTINUITIES IN BEHAVIOR

Since there have been many reviews of the the extensive literature on different forms of continuity in behavior over time (see, e.g., Brim &

Kagan, 1980; Kohlberg, LaCrosse, & Ricks, 1972; Olweus, 1979; Richman, Stevenson, & Graham, 1982; Robins, 1979; Rutter, 1972, 1982a; Rutter & Giller, 1983), I will not consider the evidence in any detail. Rather, the main conclusions can be summarized quite briefly. First, with almost all variables the correlations between measures obtained during the first two years and measures obtained in later childhood or adult life are near-zero. However, this does not necessarily mean that there are no continuities between infancy and later life. To a considerable extent, the low level of prediction from infancy reflects the extreme difficulty of *assessing* adult-like qualities at that age. Second, from age 3 years onwards there are moderately strong continuities over periods of several years in some forms of behavior, provided that the measures of behavior reflect the child's performance on several occasions or in several settings. This is evident for characteristics such as aggressivity and it is also apparent for overt disturbances in behavior—as shown, for example, by the Richman *et al.* (1982) epidemiologically based longitudinal study from the age of 3 to 8 years. Of the children with behavioral disturbances at 3 years, 62% still had a handicapping disorder 5 years later and only 14% were entirely free from behavioral problems. Third, continuities are stronger for some variables than for others. For example, aggression and conduct disturbance show stronger consistencies over time than do emotional difficulties or temperamental features other than aggressivity. Between 10 and 15 years in the Isle of Wight study nearly three-quarters of the conduct disorders persisted, whereas less than half of the emotional disorders did so (Graham & Rutter, 1973). Fourth, with the exception of the infancy period, the strength of continuities within the years of childhood and adolescence tends to be more a function of the duration of the time interval than the particular phase of development. However, there is some tendency for correlations (over the same time period) to be stronger in later childhood than in middle or early childhood. It makes little sense to give any overall figure for consistencies in emotional and behavioral functioning, but it would give a reasonable impression of the level to state that corrected correlations over a time span of 5 years or more during middle childhood tend to be in the range .3 to .7 for composite measures. Whether that is taken to mean high or low consistency depends on what standard is employed. Obviously, it means that there is *some* meaningful consistency in individual differences in behavior but, equally clearly, it means that there must be considerable flux and change.

REASONS FOR EXPECTING DISCONTINUITY

To what extent should we expect continuity rather than discontinuity? As Kagan (1980, 1981) has clearly pointed out, contemporary psychologists show a strong commitment to the notions of continuity, connectivity, and gradual change in their concepts of the developmental process. He argues that to a large extent this commitment is mistaken because many behaviors represent acute reactions to transient external provocations and many reflect the differential maturation of various cognitive capacities and functions. Both these sets of phenomena lead to an expectation of *dis*continuity rather than continuity.

There is an abundance of evidence to support the view that people remain vulnerable to environmental influences not only throughout the whole of childhood but during adult life as well (Rutter, 1981a, 1984b; Rutter & Giller, 1983). This is shown, for example, by the evidence on the importance of school influences, of the effects of bereavement, and of the psychological sequelae of divorce—three rather different types of major life changes. It is also apparent in the behavioral changes following the birth of a sib—an example of a very common, normal life event. But is it correct to view these behavioral changes as purely acute reactions to transient external provocations, as Kagan suggests? Or may the effects be more lasting? If the sequelae persist, then there should be continuities of some degree. We may accept the evidence that personality is not fixed by early life experiences, or by early maturational effects for that matter, but that does not necessarily mean that all effects of life experiences are evanescent—a point to which I shall return.

Also, there are data to support the argument that some crucial changes in behavior reflect new maturational forces rather than the accumulation of prior life experiences or any derivation from previously acquired skills or capacities. For example, puberty is associated with a marked increase in depressive feelings and a reversal of the sex ratio for depressive disorders (Rutter, 1984a). The same age period is accompanied by an upsurge in sexual feelings and by a rise in aggressivity. There are reasons for supposing that these striking changes are directly or indirectly a function of the physical and hormonal changes of puberty, although the precise mechanisms may not be understood. In a real sense, the changes represent a type of discontinuity with the past although, of course, their timing and extent are influenced by both physical status and psychosocial circumstances (and, of course, puberty represents a meaningful continuation of physical maturation). But we should distinguish here between *whether* there is a change in behavior (which is strongly dependent on maturational forces) and *how* the maturational stage is negotiated. Most young people show emotional changes during puberty, but there is considerable individual variation in the extent and quality of the changes and in the responses to those changes. It is likely that developmentally influenced individual factors may be more important in these matters (although it has to be added that little is known on their operation).

Other reasons, too, for expecting discontinuity may be put forward. Thus, the very fact that much of maturation is still to come reduces the likelihood of finding strong continuity from the preschool years. Of course, it is not that the course of maturation lacks pattern. As I have noted, genetic factors are known to influence the timing and course of the maturational process. But still that process brings about change, which is both universal and individual, and moreover change which is not well predicted by the child's earlier behavioral style or level of cognitive skills. Thus, a child's IQ score at age 6 years is a fair predictor of his IQ score at 12 years in large part because his cognitive capacity at 6 *forms part* of his capacity at 12; but his score at 6 is not a good predictor of *changes* between 6 and 12.

Also, there is a strong and general biological tendency toward healing and restoration of normal function. The physically stunted, malnourished child shows a "catch-up" in growth if he is once again given access to an adequate diet. The broken bone mends and fibrous tissue rejoins the lacerated skin surface. Of course, whether or not the healing is complete depends on the severity and duration of the trauma and the quality of the circumstances for growth after the injury. But nevertheless it is usual for the restoration of function to be very considerable. If this applied to physical damage, it is plausible that the same may apply to psychosocial traumata. It seems that in child development there are important self-correcting mechanisms that lie both in the children themselves and in their relationships with others (Hinde, 1982). It should not, however, be assumed that this resilience is independent of environmental circumstances. Kagan and Klein's (1973) Guatemalan research was taken by some people to imply this assumption (and hence to suggest that efforts at intervention with disadvantaged children were unnecessary because inherent resilience would enable them to outgrow their difficulties without the need for help; see Greenbaum, 1979). Their more recent research (Kagan, Klein, Finley, Rogoff, & Nolan, 1979) indicates that this inference was unwarranted.

SITUATIONISM

Perhaps the most extreme rejection of the continuity view of personality development comes from behaviorists who see performance as a consequence of antecedent, current, and consequent reinforcing conditions without the need to invoke either the developmental process or personality as mediating variables (Krasner & Ullman, 1973) and from those espousing situationism, in which situational effects stemming from environmental forces are seen as the main source of behavioral variation (Mischel, 1968). Both sets of views involve a range of postulates, some of which are well supported but others less so.

Situational Effects

The first, of course, is the suggestion that there *are* powerful situational determinants of behavior (Mischel, 1968, 1979). That there are has been well demonstrated through numerous different types of studies. For example, it has been found that infants' patterns of attachment (i.e., whether secure or insecure) tend to be relatively specific to the person with whom they are interacting—father attachment often differs from mother attachment (Lamb, 1978; Main & Weston, 1981). There is evidence that how mothers interact with their children is influenced by whether or not the father is also present (Clarke-Stewart, 1978; Parke, 1978) and by the content of the activities that constitute the context of the interaction (Dunn, Wooding, & Hermann, 1977). It is also apparent in the changes in mother–child interaction that take place with the first-born when a second child arrives in the family (Dunn & Kendrick, 1982). The findings on the differences in delinquent activities according to opportunities in the community and according to institutional characteristics also point to the importance of the influence of circumstances (Rutter & Giller, 1983). The repeated observation that many children behave differently at home and at school (Mitchell & Shepherd, 1966; Rutter, Tizard, & Whitmore, 1970) provides further evidence on situational variation. We may accept the fact that there *are* strong situational effects.

Individual Differences

But that does not necessarily mean that there will not be strong continuities over time. As both Mischel (1979) and the Eysencks (1980) have pointed out, albeit from quite different theoretical standpoints, the questions of consistency over time and stability over situation are rather separate. The empirical findings indicate that both occur but deal with somewhat different phenomena. Trait theories recognize situational variability, but investigators interested in the study of traits take that into account by using measures that average responses across situations in order to study consistency in individuality *over time* (Epstein, 1979). Such research, as I have noted, has shown substantial temporal consistency. Situationism, in contrast, is less concerned with constancy and change over time than with behavioral variations *over place and situation*. As Mischel (1968, p. 282) put it, "although behavior patterns often may be stable, they usually are not highly generalised across situations." The interest here, then, is not in the consistency of individual styles over time but rather with the demonstrated fact that some behaviors are much more easily elicited by some situations than others. During the last decade, it has become apparent that *both* situational and trait influences are important. The importance of personality traits is shown by the great individual variation in people's responses to any one situation; but equally the need to invoke environmental determinants is evident in the extent to which any person's mode of functioning alters from situation to situation.

Person–Situation Interactions

But in addition the evidence has shown that in many instances these trait and situation effects do not simply summate; rather, there is an *interaction* between persons and situations (Rutter, 1983b). Thus, the genotype may shape an individual's responsiveness to particular environmental influences by means of either ordinal or disordinal gene–environment interactions (Shields, 1980). Temperamental variables (as derived from both genetic and nongenetic influences) not only may determine how an individual responds to a particular environment but also may change the *effective* environment itself (Dunn, 1980; Porter & Collins, 1982; Rutter, 1977a). In some circumstances sex differences may be crucial too. Thus, many studies have shown that boys are more likely than girls to develop psychological disturbance as a result of family discord (Emery, 1982; Rutter, 1982a). Also, Martin, Maccoby, and Jacklin (1981) have shown that boys and girls respond in opposite ways to maternal responsiveness. Although person–situation interactions have been demonstrated in only a few special circumstances, it is obvious that the *possibilities* of interactional effects must be incorporated into any adequate theory of personality development (Magnusson & Endler, 1977). Nevertheless, it must be added that although interactionist perspectives have had a major impact on the study of behavior, so far they have had a rather limited influence on concepts or theories of development *per se*.

Persistence of Environmental Effects

The demonstrated importance of situational effects raises the further question of whether the persistence of sequelae of adverse experiences is no more than a reflection of the persistence of the adversities themselves. It has been observed that children subjected to deprivation or disadvantage in the early years often go on to show chronic psychological or psychiatric disorders. Usually this is taken to mean that lasting damage has taken place. However, this hypothesis can be tested only in the unfortunately rare circumstance of serious early adversities being followed by a good psychosocial situation. When these circumstances are examined it is clear that very substantial recovery, or at least very marked improvement, is the rule if the adversities cease (Rutter, 1981b). It is apparent that, to an important extent, the persistence of sequelae is indeed a function of the persistence of the bad experiences. Similarly, as shown by attachment data, behavioral continuities vary according to whether or not life circumstances remain the same (Vaughn, Egeland, Sroufe, and Waters, 1979). But is that the whole story? Several pieces of evidence indicate that it is not. Hinde's experimental studies with monkeys indicated that acute separation experiences lasting no more than a week or two had effects that were detectable as long as two years later in some cases (Spencer-Booth & Hinde, 1971). The Harlow and Suomi experiments involving total social isolation in infancy also demonstrated severe deficits which persisted right into adult life (Ruppenthal, Arling, Harlow, Sackett, & Suomi, 1976). In humans, Tizard and Hodges' (1978) study of late-adopted children showed behavioral sequelae at school at age 8, several years after the children left the institution in which they had spent their infancy. The change in environment was massive but still the behavioral consequences persisted long after the change. Our own long-term follow-up of institution-reared children showed major ill-effects (in terms of personality disorder, criminality, and parenting deficiencies) lasting through the mid-20s some years after the young people left the institution to go into an entirely different environment (Rutter, Quinton, & Liddle, 1983). Also, both Douglas's (1975) study and ours (Quinton & Rutter, 1976) found lasting effects following multiple hospital admissions in early childhood. We can firmly reject the hypothesis that the persistence of environmental effects is merely a reflection of the continuing presence of the same adverse environment. That is an important part of the explanation, but only part.

Developmental Considerations

The findings, then, show some persistence of environmental effects but perhaps this is just a question of habits of behavior having been established which take some time to change. Is there any need to invoke developmental considerations? Three rather different pieces of evidence suggest that there is. First, some of the effects are strongly influenced by the child's developmental level. For example, the protest–despair–detachment sequence following admission to hospital or some other institution is rarely seen before 6 months of age and it becomes progressively less frequent after age 4 years (Rutter, 1981b). Grief reactions following bereavement appear milder and of shorter duration in young children than in adolescents or adults (Rutter, 1984a). The total social isolation which has such devastating effects in some primate species when experienced in infancy does not give rise to the same sequelae if the isolation is experienced at an older age (Davenport, Manzel, & Rogers, 1966). Altogether, the findings support the notion of age-dependent differential sensitivities to environmental stressors, even though they run counter to the notion of fixed critical periods (Bateson, 1983).

Second, it appears that the late effects of early adversities may be different in form (and, in some respects, more severe in character) than the immediate effects. Thus, in our follow-up into adult life of institution-reared girls we found that just over half showed emotional and behavioral disturbance during childhood and adolescence (as assessed by contemporaneous measures). Not surprisingly many of these disturbed girls continued to show problems in adult life. But even among those *without* overt disturbance when young the rate of adult problems (including severe difficulties in parenting) was substantially raised (Rutter, Quinton, & Liddle, 1983). Of course, it is possible that these girls had had earlier difficulties that were not reflected in our measures but nevertheless it is apparent that sometimes there were late effects that could not be accounted for in terms of the residue of gradually diminishing situational effects.

Third, although the evidence on this point is less secure, it appears that in some cases there may be either sensitizing or steeling effects from early stressors that do not result in immediate disorder but which, by altering vulnerabilities or by modifying styles of coping, protect from or predispose toward disorder in later life only in the presence of later stress events (Rutter, 1981c). We may conclude that some kind of developmental consideration must be introduced into any explanation of the persistence of environmental effects in childhood.

LUCK, OR THE ROULETTE WHEEL VIEW OF LIFE'S CHANCES

One other concept of discontinuity in development must be noted—namely, that of luck, the roulette wheel view of life's chances. For example, Jencks, Smith, Acland, Bane, Cohen, Gentis, Heyns, and Michelson (1972) found that none of the variables they measured accounted for much of the variation in men's income in adult life. They concluded that chance constituted one of the major influences on people's life course:

Chance acquaintances who steer you into one line of work rather than another, the range of jobs that happen to be available in a particular community when you are job hunting, the amount of overtime work in a particular plant, whether bad weather destroys your strawberry crop, whether the new superhighway has an exit near your restaurant, and a hundred other unpredictable accidents. (p. 27)

Of course, chance is a real phenomenon and necessarily its operation introduces an essential degree of unpredictability into the process of development. But, as Bandura (1982) has noted in a recent essay on the topic, although a science of psychology cannot shed much light on the occurrence of fortuitous encounters, it can provide the basis for predicting the impact they will have on human lives. An essential part of that basis will concern the processes by which the products of early development foster continuities in behavioral patterns either through the *selection* of environments or through their *production*. Moreover, the skills, values, emotional ties, and self-concepts that people bring to chance encounters will play a major role in determining how they respond to the encounters, as well as the extent to which they can exert control over their future, grasping and utilizing the opportunities that chance provides and avoiding or overcoming the hazards that also come with luck or ill-luck. In the remainder of this chapter I consider some of the possible ways in which continuities in development stemming from environmental effects may come about.

POSSIBLE MECHANISMS UNDERLYING CONTINUITIES

Continuities in the Environment

The most obvious mechanism, of course, resides in the constancy of environmental forces. As I have mentioned, some of the continuities in behavior reflect the persistence of environmental conditions. Insofar as that is the case, it does not necessarily represent any intrinsic continuity in development so far as the child is concerned. To some extent it will mean no more than that the immediate situational effects of environmental conditions are much the same throughout childhood. But also it may imply developmental continuity if there is a consistent environmental shaping of the developmental process throughout the phase of growth.

Selection of Environments

However, there may be rather different types of continuities in the environment—continuities stemming from links between different en-

vironments rather than from a lack of change. For example, in our follow-up study of girls reared in institutions (Rutter, Quinton, & Liddle,
1983), we found that the strongest effect on the quality of parenting was
provided by the characteristics of the women's spouses. At first sight,
that seems to provide a discontinuity with the effects of early childhood
experiences. But the impression is misleading because the conditions of
rearing exerted a powerful effect on the *choice* of spouse. Just over half
(51%) of the institution-reared women married men with psychosocial
problems (criminality, psychiatric disorder, and the like)—a rate nearly
four times that (13%) in the general population comparison group. The
immediate effect on parenting came from the spouse, but this represented a long-term indirect effect of the childhood experiences in that
they strongly influenced the type of men the women married. The same
applied to the women's social circumstances in adult life. Social conditions had an effect on the quality of parenting, but the conditions of
rearing played a part in determining whether the women experienced
poor social conditions. Thus, institutional rearing was associated with
an increased rate of behavioral disturbance which in turn was associated
with a greater likelihood of drop-out from education and dismissal from
jobs, which then led to low social status work or unemployment. Similarly, the institution-reared girls who left the institution to return to
discordant families were more likely than other girls to have babies in
their teens; teenage pregnancy, in turn, was then associated with an
increased risk of a poor social outcome. The environments change, but
the experience of one unfavorable environment makes it more likely that
the individual will go on to experience other unfavorable environments.

Opportunities

A further mechanism leading to continuities concerns the effects
stemming from the opening up or closing down of opportunities. For
example, in our study of inner London secondary schools (Rutter,
Maughan, Mortimore, Ouston, & Smith, 1979) we found quite powerful
effects of the school environment on pupil behavior and attainments (as
reflected in such variables as poor attendance, dropping out of school
before taking school-leaving examinations, the level of performance in
those examination, and the likelihood of the child's voluntary staying on
in the sixth form beyond the period of compulsory schooling in order to
obtain higher educational attainments). We found no direct effects of
schooling on the young people's employment, as assessed one year after
leaving school (Gray, Smith, & Rutter, 1980). But there were more
important *indirect* effects as a result of the school effects on attendance
and attainment. Young people with good scholastic attainments were

likely to be in better jobs than others of similar IQ and social background but with poor attainments. The school influence on exam qualifications opened or closed down employment opportunities and in this way produced more lasting chain effects.

In the same way, the already mentioned associations between institutional rearing and teenage pregnancy may reflect a limitation of opportunities that will affect social functioning in adult life. Early pregnancies tended to tie the women to the same disadvantaged environment and also made it less likely that they would go on with further education or occupational training. The Pederson, Faucher, and Eaton (1978) study of the long-term effects on children of being taught by one particularly effective teacher tells much the same story. A critical paths analysis showed no direct effects on the children's scholastic achievement beyond the second year. But indirect effects lasted right into adult life. It appeared to be the case that the children's early experiences with the good teacher led to educational attitudes, work habits, and a sense of self-esteem that both increased their capacity to *profit* from later education and also made them more rewarding students for the teachers, thus increasing the likelihood that they would *receive* better teaching.

Effects on the Environment

A further type of linkage over time is provided by effects *on* the environment. For example, Dunn and Kendrick's (1982) study showed quite marked behavioral reactions in the firstborn following the birth of a sibling. Most showed signs of disturbance and unhappiness for a while after the birth, although both the pattern of disturbance and its severity showed marked individual variation. This new event introduced an important element of discontinuity but there were links with the past and with the future. Both the child's temperamental characteristics and the mother's style of parenting had an effect on how the child responded to this new event. But, perhaps more important, the combination of these variables led to a different pattern of parent–child interaction with the firstborn. Confrontations between mother and child increased, there was less joint play, and mothers were less likely to initiate interactions. This change was brought about by the birth of a sibling, but the form of the change was shaped by the characteristics of mother and child before the birth. Moreover, the fact that the arrival of a sibling altered the overall pattern of family interaction meant that this supposedly acute event had chronic consequences.

Perhaps this explains why the quality of the firstborn's relationship with his sibling showed the high level of consistency over several years that was found by Dunn and Kendrick (1982). Perhaps, too, such a changed pattern of family interaction may constitute part of the explanation for the surprisingly lasting effects of recurrent hospital admission (Douglas, 1975; Quinton & Rutter, 1976). Children tend to be more clinging and difficult on return home after hospitalization and, as with the birth of a sib, this change in the children's behavior toward the parents is likely to have an affect on the parents' behavior toward the child and hence on the patterns of interaction between them. Hinde's experimental studies with rhesus monkeys have shown that, to a considerable extent, the infant's disturbance following separation is a consequence of the fact that separation serves to disturb and increase tensions in the mother–infant relationship (Hinde & McGinnis, 1977).

Another example is provided by our study of children suffering head injuries (Rutter, Chadwick, & Shaffer, 1983), in which we found an effect on subsequent behavior due to the brain damage and to the associated family responses to the accident. But, also, there are links with the past in that the children's preinjury behavior was associated with the likelihood that they would *experience* an accident. Impulsive, daring, overactive, disobedient children were more likely to put themselves in situations where there was a high risk of an accident resulting in head injury. The child's behavior changed the environment which then changed the behavior. The resulting picture provides meaningful continuities over time even though it does not provide constancy of behavior or normative stability.

Vulnerability, Resilience, and Coping Skills

So far, the possible mechanisms underlying continuities that have been considered have involved one or other aspect of the environment. Continuities may also reflect features in the child. One postulated mode of effect by which relatively short-lived events in childhood may have an enduring influence concerns styles of response to stress. The point here is that the experience of adversities or life changes provides both physiological and psychological learning opportunities. For example, in adults it has been shown that novices making their first parachute jump show a marked rise in blood cortisol (Ursin, Baade, & Levine, 1978). However, after a few jumps the physiological response alters in both pattern and timing—in that the main reaction comes to take place in the anticipation phase rather than in the period following, and in that the rise in cortisol is much reduced, although the heart rate and epinephrine response remain much the same. However, there is marked individual variation in physiological responses to potentially stressful events, and it may be that the sequelae will vary according to whether the person acquires an adaptive or maladaptive form of response (a possibility little studies so far; Rutter, 1981c). Similarly, Sackett (1972) found that surrogate-reared rhesus monkeys differed from mother-reared infants in their exploration of the environment—the surrogate-reared animals tended to explore simple stimuli whereas the mother-reared ones spent more time exploring the more complex stimuli. Levine (1982) and his colleagues have also shown that these two patterns of rearing resulted in a different pattern of physiological response to later separation experiences. Levine summarized the findings as showing that "early contingent relationships are required in order for the infant to learn certain cogent aspects of its environment that permit it to acquire adaptive coping responses" (p. 50).

Similarly, there is evidence suggesting that children who experience happy separations from their parents (as by staying with friends or relatives or through good baby-sitting arrangements) may be better able to cope later with the more stressful occurrence of hospital admission (Stacey, Dearden, Pill, & Robinson, 1970). Conversely, in our long-term follow-up of institution-reared women we found little difference in their quality of parenting compared with that in our general population comparison group—provided that, in adult life, they were living in good social conditions with a supportive spouse (Rutter, Quinton, & Liddle, 1983). On the other hand, if they lacked such supports in their twenties, the institution-reared group fared much worse than the women in the comparison group. It seemed that somehow their adverse experiences in childhood had left them without the resilience, resources, or coping mechanisms to deal successfully with later life hazards. Other studies, too, have produced findings pointing to the probability of some type of effect on vulnerability or coping—an effect that does not in itself lead directly to disorder but which increases or decreases some aspect of adaptability or coping (Rutter, 1981c; Werner & Smith, 1982).

It must be said that, so far, the evidence on this process remains fragmentary and inadequate for the drawing of any firm conclusions. However, there are strong grounds for assuming that the process takes place. Moreover, it is clear that this mechanism emphasizes individual differences in response to stress as well as both continuities and changes in behavior as a result of those differences. The findings emphasize that the issue of *continuities* in development is not the same as constancies or stabilities in behavior or in the developmental process.

Habits, Attitudes, and Self-Concepts

A second mechanism involving the child as a person concerns the possible effects of experiences on habits, attitudes, and self-esteem. Harter (1983) has recently provided a most thoughtful and informative review of issues and findings regarding children's development of self-concepts.

It is clear that several rather different dimensions are involved in the development of self-concepts; these include the notion of self-esteem (a feeling of one's own worth as an individual; Coopersmith, 1967), of self-efficacy or ability to control one's destiny (Bandura, 1977), and of ego control and resiliency (Block & Block, 1980). Empirical studies have shown associations between various patterns of rearing and differences in the development of self-concepts. Findings are very limited on the role of these features as mediating mechanisms or modulating influences in children's response to life experiences. However, some results suggest that they may play a significant role. For example, there is the evidence showing that the taking of a boy to court for theft, and hence his public labelling as a delinquent, serves to increase the likelihood that he will persist in delinquent activities (see Rutter & Giller, 1983). Other research, too, has shown that if people's social roles are altered their attitudes and behavior are likely to follow suit (Kuhn, 1964; Lieberman, 1956) and that, to some extent, teachers' expectations influence pupils' academic progress (see Pilling & Pringle, 1978). It is probable that some of the school effects on pupils' attainment that have been found stem from influences on habits, attitudes, and self-concepts, as well as from more direct effects on learning (Rutter et al., 1979; Rutter, 1983a).

Kagan (1980) has suggested that the operation of this mechanism may explain why some environmental effects seem not to persist from infancy, whereas they do from later childhood. Thus, he argued:

> A stable belief system about self and others, which is such an important determinant of later behavior, is not articulated by 2 years of age. The reason for the selective appearance of normative stability after 2 years of age may be that the executive functions we call *ego* and *self* do not emerge until later in the second year. Once these functions emerge, the child begins to interpret his experience, and the first expectations begin to be established. Prior to the second birthday, experience does not undergo this critical transduction and, as a result, may be of less consequence. The moderate stability of individual differences (found in later childhood and adolescence) is due, in part, to the consistency of belief systems that happen to find continuous affirmation in the environment. (p. 64)

Kagan's suggestions on the role of belief systems in producing behavioral stabilities after, but not before, age 2 years are plausible and important; but, as we have seen, some effects from infancy experiences may persist.

The Effects of Experiences on Neural and Neuroendocrine Systems

A further possible mechanism lies in the effect of experiences in bringing about alterations in somatic structures and functioning. Thus, animal studies have shown that a wide range of aversive experiences in infancy may lead to an enhanced resistance to later stress, a resistance that seems to be due to changes in the neuroendocrine system (Hennessy & Levine, 1979; Hunt, 1979; Thompson & Grusec, 1970). Also, it has been found that visual impoverishment in infancy is followed by changes in the visual cortex (Riesen, 1975) and very early environmental impoverishment by changes in brain chemistry and histology (Rosenzweig & Bennett, 1977). One practical consequence of these phenomena in humans is that if a strabismus is not corrected early, the child may be left thereafter without normal binocular vision, even if the squint is corrected later.

The findings are of considerable interest in understanding processes of development, but it is doubtful whether neural changes play a major role in environmental effects on socioemotional development. The limited available evidence suggests that the main neural effects follow a lack of crucial *physical* experiences rather than any kind of social stress or privation (Floeter & Greenough, 1979). Moreover, research findings suggest that at least some of the physical effects stemming from early sensory deprivation may be modifiable through later experiences (Bateson, 1982; Stein & Dawson, 1980). However, little is known of the neural consequences of psychosocial adversities.

Imprinting and Other Similar Phenomena

At one time, much reference was made to the concept of *imprinting* as a possible mechanism by which relatively brief experiences in infancy might have semipermanent effects. Strictly speaking, imprinting applies to the development of a following response in particular species of birds, but the term has been extended by some writers to other forms of supposedly specific, very rapid forms of learning that are restricted to a short phase of development and which apparently persist thereafter. For example, the notion has been used to apply to the process by which mothers develop emotional bonds with their infants (Hales, Losoff, Sosa, & Kennell, 1977; Klaus & Kennell, 1976). The general observation that there are sensitive periods in development during which there is an enhanced sensitivity to particular environmental stimuli remains valid, but the associated concepts of fixed and very narrow critical periods, of irreversibility, and of effects dependent on some highly specific stimulus (such as bodily contact in mother–infant attachment) have not stood the test of time (Bateson, 1979, 1982; Lamg & Hwang, 1982; Oyama, 1979; Rutter, 1981b; Svejda, Pannabecker, & Emde, 1982). Accordingly, there is no reason to suppose that imprinting mechanisms play any significant role in environmental effects on socioemotional development.

Effects on Personality Structure

Lastly, there is the possibility that adverse experiences in early childhood may cause some lasting interferences with the structure of personality development. Thus, psychoanalytically oriented clinicians describe children's emotional and behavioral disorders in terms of a *fixation* at some stage of early development as a result of some adverse experience at that phase—a process by which the experience is internalized (Stone & Koupernik, 1974; Wolff, 1973). Similarly, Anna Freud (1966) argues that the clinical assessment of children should be undertaken in terms of the structure of the personality, with the assumption that disorders are "initiated by libido regressions to fixation points at various early levels, [and thus] the location of these trouble spots in the history of the child is one of the vital concerns of the clinician" (p. 143). At least two rather separate issues are involved here. First, there is the question of whether personality development is most appropriately viewed in terms of the growth of a coherent structure with a series of stages, each of which builds on and is dependent on the preceding one. Second, there is the question of whether experiences act by "fixing" development at some stage or by somehow distorting the structure of development.

As Flavell (1982) noted in his recent review of Piagetian concepts of stages of cognitive development, the issues are complex, with the notions of *stage* and *sequence* difficult to test in a rigorous fashion. Nevertheless, the empirical findings run counter to the view that there is horizontal structure and high homogeneity within stages. Rather, development is often uneven, with progression advanced in some respects but retarded in others. The evidence that there are relatively predictable sequential progressions in each area of development is consistent with the notion of a vertical structure. However, the fact that phase 3 follows phase 2 does not necessarily mean that it *derives* from the earlier phase, nor that the progression through one phase has any relationship to the ways in which the next one is dealt with.

Although it cannot be said that the theories of personality development that posit a coherent internal structure with a predetermined set of stages linked one with another in a causal chain have been disproven, certainly, there is a lack of adequate supporting empirical evidence. In the ordinary course of events children learn to crawl before they walk, but there is no evidence that those skills are interdependent; some children never crawl at all. Similarly, parent–child attachments usually precede friendships with peers, but Freud and Dann's (1951) study of concentration camp children suggests that, in some circumstances, that sequence may be reversed.

The second issue of whether adverse experiences have their main impact in terms of developmental fixation or distortion of developmental lines is equally difficult to evaluate in the absence of accepted operational criteria for the hypothesized personality structures and stages. Nevertheless, the lack of evidence that adversities at one age have radically different effects to similar adversities at other ages, together with the evidence of the very substantial potential for later change without any apparent passage through missed stages, makes it implausible that this is how most experiences operate. As already noted, the empirical findings *do* suggest the importance of developmental considerations and do suggest that experiences may both alter styles of functioning and impair socioemotional (as well as cognitive) capacities. But the view of development that derives from the empirical data is much more fluid than the structural theories would allow.

OVERVIEW ON CONTINUITIES AND DISCONTINUITIES

The question posed in the introduction in this chapter concerned the extent to which socioemotional development is continuous or discontinuous. The main thesis put forward here has been that the notion of continuity is not the same as that of stability or constancy. The findings show significant correlations between measures of behavioral functioning at different phases of development separated by some years, but the correlations are of modest to moderate strength only, with those from the infancy period to maturity near-zero. Measures in early childhood provide weak predictors of development in adolescence, or of personality functioning in adult life. Moreover, concepts of continuity do not imply that early life experiences will have enduring consequences for personality development. Sometimes they do but often they do not. The long-term direct effects of even very serious adversities in early life tend to be quite minor provided that the later experiences are good ones.

Rather, the concept of continuity implies meaningful links over the course of development—not a lack of change. Even with this broader view of continuity, it is still necessary to ask, given a continuing interaction between a changing organism and a changing physical and social environment, how far development is to be seen as proceeding smoothly and gradually and how far by periods of marked changes in organization (Hinde, 1982). As Hinde (1982) put it from a biological perspective:

> Just because immature forms must be specialised for the tasks of growing and surviving to adulthood, we must expect some aspects of their structure or behaviour to be irrelevant to adult life: a caterpillar is adapted as a growing machine, and is also adapted to change dramatically to fulfill the functions of adulthood. In the same way, some of the behavioural characteristics of children can be seen as scaffolding, erected temporarily for the job of growing and subsequently demolished as the adult functions emerge in their own right. (p. 91)

The empirical evidence forces us to accept some important discontinuities in development. The dramatic changes around the time of puberty were given as an example but many others could have been cited (Hinde, 1982; Kagan, 1980, 1981).

Nevertheless, such discontinuities may still be accompanied by continuities. Hinde (1982) went on to note:

> But even where, as in infant metamorphosis, the tissues are almost completely broken down and the body is redeveloped in a new form, continuity is not totally absent: larval experience may effect the subsequent behaviour of the moth (Thorpe and Jones, 1937). Experience before metamorphosis is known to affect behaviour afterwards also in amphibians (Hershkowitz and Samuel, 1973). (p. 91)

If such continuities apply with changes as dramatic as the transformation of caterpillar to moth or the metamorphosis of the newt, they are likely to apply with greater force to human development.

In this chapter there has been an exploration of some of the mechanisms by which continuities in development may arise. The possible processes involve links within the environment as well as within the child. The evidence runs counter to the view that early experiences irrevocably change personality development (Rutter, 1981b) and also runs counter to the suggestion that any single process is involved (Sackett, 1982); nevertheless, in some circumstances the indirect effects may be quite long-lasting. Even so, such long-term effects are far from independent of intervening circumstances. Rather, the continuities stem from a multitude of links over time (Rutter, Quinton, & Liddle, 1983). Because each link is incomplete, subject to marked individual variation and open to modification, there are many opportunities to break the chain. Such opportunities continue right into adult life.

REFERENCES

Bandura, A. Self-efficacy: Toward a unifying theory of behavioral change. *Psychological Review*, 1977, *84*, 191–215.

Bandura, A. The psychology of chance encounters and life paths. *American Psychologist*, 1982, *37*, 747–755.

Bateson, P. How do sensitive periods arise and what are they for? *Animal Behaviour*, 1979, *27*, 470–486.

Bateson, P. The interpretation of sensitive periods. In A. Oliverio and M. Zapella (Eds.), *The behavior of human infants* New York: Plenum Press, 1983.

Block, J. H., & Block, J. The role of ego-control and ego-resiliency in the organization of behaviour. In W. A. Collins (Ed.), *Development of cognition, affect and social relations*. Minnesota Symposia on Child Psychology, Vol. 13. Hillsdale, N.J.: Lawrence Erlbaum, 1980.

Bowlby, J. *Attachment and loss: I. Attachment*. London: Hogarth Press, 1969.

Bowlby, J. *Attachment and loss: II. Separation, anxiety and anger*. London: Hogarth Press, 1973.

Bowlby, J. *Attachment and loss: III. Loss, sadness and depression*. New York: Basic Books, 1980.

Brim, O. G., Jr., & Kagan, J. (Eds.) *Constancy and change in human development*. Cambridge, Mass.: Harvard University Press, 1980.

Broman, S. H., Nichols, P. L., & Kennedy, W. A. *Preschool IQ: Prenatal and early developmental correlates*. Hillsdale, N.J.: Lawrence Erlbaum, 1975.

Clarke-Stewart, K. A. And daddy makes three: The father's impact on mother and young child. *Child Development*, 1978, *49*, 446–478.

Coopersmith, S. *The antecedents of self-esteem*. San Francisco: W. H. Freeman, 1967.

Davenport, R. K., Menzel, E. W., & Rogers, C. M. Effects of severe isolation on "normal" juvenile chimpanzees: Health, weight gain and stereotyped behaviors. *Archives of General Psychiatry*, 1966, *14*, 134–138.

Douglas, J. W. B. Early hospital admissions and later disturbances of behaviour and learning. *Developmental Medicine and Child Neurology*, 1975, *17*, 456–480.

Dunn, J. Individual differences in temperament. In M. Rutter (Ed.) *Scientific foundations of developmental psychiatry*. London: Heinemann Medical, 1980, pp 101–109.

Dunn, J., & Kendrick, C. *Siblings: Love, envy and understanding*. Cambridge, Mass: Harvard University Press, 1982.

Dunn, J., Wooding, C., & Hermann, J. Mothers' speech to young children: Variation in context. *Developmental Medicine and Child Neurology*, 1977, *19*, 629–638.

Emery, R. E. Interparental conflict and the children of discord and divorce. *Psychological Bulletin*, 1982, *92*, 310–330.

Emmerich, W. Continuity and stability in early social development. *Child Development*, 1964, *35*, 311–332.

Emmerich, W. Personality development and concepts of structure. *Child Development*, 1968, *39*, 671–690.

Epstein, S. The stability of behavior. I. On predicting most of the people much of the time. *Journal of Personality and Social Psychology*, 1979, *37*, 1097–1126.

Eysenck, M., & Eysenck, H. J. Mischel and the concept of personality. *British Journal of Psychology*, 1980, *71*, 191–204.

Flavell, J. H. Structures, stages, and sequences in cognitive development. In W. A. Collins (Ed.), *The concept of development*. Minnesota Symposium on Child Psychology, Vol. 15. Hillsdale, N.J.: Lawrence Erlbaum, 1982.

Floeter, M. K., & Greenough, W. T. Cerebellar plasticity: Modification of Purkinje cell structure by differential rearing in monkeys. *Science*, 1979, *206*, 227–229.

Freud, A. *Normality and pathology in childhood. Assessments of development*. London: Hogarth Press/Institute of Psychoanalysis, 1966.

Freud, A., & Dann, S. An experiment in group upbringing. *The Psychoanalytic Study of the Child*, 1951, *6*. 127–168.

Graham, P., & Rutter, M. Psychiatric disorder in the young adolescent: A follow-up study. *Proceedings of the Royal Society of Medicine*, 1973, *66*, 1226–1229.

Gray, G., Smith, A., & Rutter, M. School attendance and the first year of employment. In L. Hersov & I. Berg (Eds.), *Out of school: Modern perspectives in truancy and school refusal.* Chichester: Wiley, 1980, pp. 343–370.

Greenbaum, C. Commentary and discussion of Kagan, J. *et al.*, A cross-cultural study of cognitive development. *Monographs of the Society for Research in Child Development*, 1979, serial no. 180, *44*, No.5, 67–73.

Hales, D. J., Lozoff, B., Sosa, R., & Kennell, J. H. Defining the limits of the maternal sensitive period. *Developmental Medicine and Child Neurology*, 1977, *19*, 454–461.

Hall, C. S., & Lindzey, G. *Theories of personality* (3 ed.). New York: Wiley, 1978.

Harter, S. Developmental perspectives on the self-system. In E. M. Hetherington (Ed.), *Carmichael's manual of child psychology (vol. 4): Social and personality development.* New York: Wiley, 1983.

Hennessy, J. W., & Levine, S. Stress, arousal, and the pituitary-adrenal system: A psycho-endocrine hypothesis. In J. M. Sprague & A. N. Epstein (Eds.), *Progress in psychobiology and physiological psychology.* New York: Academic Press, 1979, pp. 133–178.

Hershkowitz, M., & Samuel, D. The retention of learning during metamorphosis of the crested newt (*Triturus eristatus*). *Animal Behaviour*, 1973, *21*, 83–85.

Hinde, R. A. *Ethology.* London: Fontana, 1982.

Hinde, R. A., & McGinnis, L. Some factors influencing the effect of temporary mother–infant separation: Some experiments with rhesus monkeys. *Psychological Medicine*, 1977, *7*, 197–212.

Hunt, J. McV. Psychological development: Early experience. *Annual Review of Psychology*, 1979, *30*, 103–143.

Jencks, C., Smith, M., Acland, H., Bane, M. J., Cohen, D., Gentis, H., Heyns, B., & Michelson, S. *Inequality: A reassessment of the effect of family and schooling in America.* New York: Basic Books, 1972.

Kagan, J. Perspectives on continuity. In O. G. Brim & J. Kagan (Eds.), *Constancy and change in human development.* Cambridge, Mass.: Harvard University Press, 1980, pp. 26–74.

Kagan, J. *The second year: The emergence of self-awareness.* Cambridge, Mass.: Harvard University Press, 1981.

Kagan, J., & Klein, R. E. Cross-cultural perspectives on early development. *American Psychologist*, 1973, *28*, 947–961.

Kagan, J., Klein, R. E., Finley, G. E., Rogoff, B., & Nolan, E. A cross-cultural study of cognitive development. *Monographs of the Society for Research in Child Development*, 1979, Serial no. 180, *44*, No. 5.

Klaus, M. H., & Kennell, J. H. *Maternal–infant bonding: The impact of early separation or loss on family development.* St. Louis: Mosby, 1976.

Kohlberg, L., LaCrosse, J., & Richs, D. The predictability of adult mental health from childhood behavior. In B. B. Wolman (Ed.), *Manual of child psychopathology.* New York: McGraw-Hill, 1972, pp. 1217–1284.

Krasner, L., & Ullman, L. P. *Behavior influence and personality: The social matrix of human action.* New York: Holt, 1973.

Kuhn, M. H. Major trends in symbolic interaction theory in the past twenty-five years. *Sociological Quarterly*, 1964, *5*, 61–84.

Lamb, M. E. Qualitative aspects of mother– and father–infant attachments. *Infant Behavior and Development*, 1978, *1*, 265–275.

Lamb, M. E., & Hwang, C.-P. Maternal attachment and mother–neonate bonding: A critical review. In M. E. Lamb & A. L. Brown (Eds.), *Advances in developmental psychology* (Vol. 2). Hillsdale, N.J.: Lawrence Erlbaum, 1982.

Levine, S. Comparative and psychobiological perspectives on development. In W. A.

Collins (Ed.), *The concept of development*. Minnesota Symposia on Child Psychology, Vol. 15. Hillsdale, N.J.: Lawrence Erlbaum, 1982.

Lieberman, S. The effects of changes in roles on the attitudes of role occupants. *Human Relations*, 1956, *9*, 385–402.

Lipsitt, L. P. Stress in infancy: Toward understanding the origins of coping behavior. In N. Garmezy and M. Rutter (Eds.), *Stress, coping and development in children*. New York: McGraw-Hill, 1983.

Magnusson, D., & Endler, N. S. (Eds.) *Personality at the crossroads: Current issues in interactional psychology*. Hillsdale, N.J.: Lawrence Erlbaum, 1977.

Main, M. B., & Weston, D. R. Security of attachment to mother and father: Related to conflict behavior and the readiness to establish new relationships. *Child Development*, 1981, *52*, 932–940.

Martin, J. A., Maccoby, E. E., & Jacklin, C. N. Mother's responsiveness to interactive bidding and nonbidding in boys and girls. *Child Development*, 1981, *52*, 1064–1067.

Mischel, W. *Personality and assessment*. New York: Wiley, 1968.

Mischel, W. On the interface of cognition and personality: Beyond the person–situation debate. *American Psychologist*, 1979, *34*, 740–754.

Mitchell, S., & Shepherd, M. A comparative study of children's behaviour at home and at school. *British Journal of Educational Psychology*, 1966, *36*, 248–254.

Moss, H. A., & Susman, E. J. Longitudinal study of personality development. In O. G. Brim, Jr., & J. Kagan (Eds.), *Constancy and change in human development*. Cambridge, Mass.: Harvard University Press, 1980, pp. 530–595.

Olweus, D. Stability of aggressive reaction patterns in males: A review. *Psychological Bulletin*, 1979, *86*, 852–875.

Oyama, S. The concept of the sensitive period in developmental studies. *Merrill-Palmer Quarterly*, 1979, *25*, 83–103.

Parke, R. D. Parent–infant interaction: Progress, paradigms and problems. In G. P. Sackett (Ed.), *Observing behavior (vol. 1): Theory and applications in mental retardation*. Baltimore: University Park Press, 1978.

Pedersen, E., Faucher, T. A., & Eaton, W. W. A new perspective on the effects of first-grade teachers on children's subsequent adult status. *Harvard Educational Review*, 1978, *48*, 1–31.

Pilling, D., & Pringle, M. K. *Controversial issues in child development*. London: Paul Elek, 1978.

Porter, R., & Collins, G. (Eds.), *Temperamental differences in infants and young children*. Ciba Foundation Symposium 89. London: Pitman Books, 1982.

Quinton, D., & Rutter, M. Early hospital admissions and later disturbances of behaviour: An attempted replication of Douglas' findings. *Developmental Medicine and Child Neurology*, 1976, *18*, 447–459.

Richman, N., Stevenson, J., & Graham, P. *Pre-school to school: A behavioural study*. London: Academic Press, 1982.

Riesen, A. H. *The developmental neuropsychology of sensory deprivation*. New York: Academic Press, 1975.

Robins, L. Sturdy childhood predictors of adult antisocial behaviour: Replications from longitudinal studies. *Psychological Medicine*, 1978, *8*, 611–622.

Robins, L. Follow-up studies. In H. C. Quay & J. S. Werry (Eds.), *Psychopathological disorders of childhood* (2nd ed.). New York: Wiley, 1979.

Rosenzweig, M. R., & Bennett, E. L. Effects of environmental enrichment or impoverishment on learning and on brain values in rodents. In A. Oliviero (Ed.), *Genetics, environment and intelligence*. Amsterdam: North-Holland, 1977, pp. 163–196.

Ruppenthal, G. C., Arling, G. L., Harlow, H. F., Sackett, G. P., & Suomi, S. J. A 10-year perspective of motherless-mother monkey behavior. *Journal of Abnormal Psychology,* 1976, *85,* 341–349.

Rutter, M. Psychological Development: Predictions from infancy. *Journal of Child Psychology and Psychiatry,* 1970, *11,* 49–62.

Rutter, M. Relationships between child and adult psychiatric disorders. *Acta Psychiatrica Scandinavica,* 1972, *68,* 3–21.

Rutter, M. Individual differences. In M. Rutter & L. Hersov (Eds.), *Child psychiatry: Modern approaches.* Oxford: Blackwell Scientific, 1977, pp. 3–21. (a)

Rutter, M. Prospective studies to investigate behavioral change. In J. S. Strauss, H. M. Babigian, & M. Roff (Eds.), *The origins and course of psychopathology.* New York: Plenum Press, 1977. (b)

Rutter, M. Epidemiological/longitudinal strategies and causal research in child psychiatry. *Journal of the American Academy of Child Psychiatry,* 1981, *20,* 513–544. (a)

Rutter, M. *Maternal deprivation reassessed* (2nd ed.). Harmondsworth, Middlesex: Penguin, 1981. (b.)

Rutter, M. Stress, coping and development: Some issues and some questions. *Journal of Child Psychology and Psychiatry,* 1981, *22,* 323–356. (c)

Rutter, M. Epidemiological-longitudinal approaches to the study of development. In W. A. Collins (Ed.), *The concept of development.* Minnesota Symposia on Child Psychology, Vol. 15. Hillsdale, N.J.: Lawrence Erlbaum, 1982. (a)

Rutter, M. Temperament: Concepts, issues and problems. In R. Porter and G. Collins (Eds.), *Temperamental differences in infants and young children.* Ciba Foundation Symposium 89. London: Pitman, 1982, pp. 1–16. (b)

Rutter, M. School effects on pupil progress: Research findings and policy implications. *Child Development,* 1983, *54,* 1–29. (a)

Rutter, M. Statistical and personal interactions: Facets and perspectives. In D. Magnusson & V. Allen (Eds.), *Human development: An interactional perspective.* London: Academic Press, 1983. (b)

Rutter, M. The developmental psychopathology of depression: Issues and perspectives. In M. Rutter, C. Izard, & P. Read (Eds.), *Depression in childhood: Developmental perspectives.* New York: Guilford Press, 1984. (a)

Rutter, M. Family and school influences: Meanings, mechanisms and implications. In A. R. Nicol (Ed.), *Longitudinal studies in child care and child psychiatry: Practical lessons from research.* Chichester: Wiley, 1984. (b)

Rutter, M., & Garmezy, N. Developmental psychopathology. In E. M. Hetherington (Ed.), *Handbook of Child Psychology: Vol. 4. Socialization, personality, and social development.* New York: Wiley, 1983.

Rutter, M., & Giller, H. *Juvenile delinquency: Trends and perspectives.* Harmondsworth, Middlesex: Penguin, 1983.

Rutter, M., Tizard, J., & Whitmore, K. (Eds.), *Education, health and behaviour.* London: Longmans, 1970 (Reprinted, Huntington, N.Y.: Krieger, 1981).

Rutter, M., Maughan, B., Mortimore, P., & Ouston, J., with Smith, A. *Fifteen thousand hours: Secondary schools and their effects on children.* London: Open Books; Cambridge, Mass.: Harvard University Press, 1979.

Rutter, M., Chadwick, O., & Shaffer, D. Head injury. In M. Rutter (Ed.), *Developmental neuropsychiatry.* New York: Guilford Press, 1983.

Rutter, M., Quinton, D., & Liddle, C. Parenting in two generations: Looking backwards and looking forwards. In N. Madge (Ed.), *Families at risk.* London: Heinemann Educational, 1983.

Sackett, G. P. Exploratory behaviour of rhesus monkeys as a function of rearing experiences and sex. *Developmental Psychology,* 1972, *6,* 260–270.

Sackett, G. P. Can single processes explain effects of postnatal influences on primate development? In R. N. Emde & R. J. Harmon (Eds.), *The development of attachment and affiliative systems*. New York: Plenum Press, 1982.

Schachar, R., Rutter, M., & Smith, A. The characteristics of situationally and pervasively hyperactive children: Implications for syndrome definition. *Journal of Child Psychology and Psychiatry*, 1981, *22*, 375–392.

Shields, J. Genetics and mental development. In M. Rutter (Ed.), *Scientific foundations of developmental psychiatry*. London: Heinemann Medical, 1980, pp. 8–24.

Spencer-Booth, Y., & Hinde, R. A. Effects of brief separations from mothers during infancy on behaviour of rhesus monkeys 6–24 months later. *Journal of Child Psychology and Psychiatry*, 1971, *12*, 157–172.

Stacey, M., Dearden, R., Pill, R., & Robinson, D. *Hospitals, children and their families: The report of a pilot study*. London: Routledge & Kegan Paul, 1970.

Stein, D. G., & Dawson, R. G. The dynamics of growth, organization, and adaptability in the central nervous system. In O. G. Brim, Jr., & J. Kagan (Eds.), *Constancy and change in human development*. Cambridge, Mass.: Harvard University Press, 1980.

Stone, F. H., & Koupernik, C. *Child psychiatry for students*. Edinburgh and London: Churchill Livingstone, 1974.

Svejda, M. J., Pannabecker, B. J., & Emde, R. N. Parent-to-infant attachment: A critique of the early 'bonding' model. In R. N. Emde & R. J. Harmon (Eds.), *The development of attachment and affiliative systems*. New York: Plenum Press, 1982.

Thomas, A., Birch, H. G., Chess, S., Hertzig, M. E., & Korn, S. *Behavioral individuality in early childhood*. New York: New York University Press, 1963.

Thompson, W. R., & Grusec, J. E. Studies of early experience. In P. H. Mussen (Ed.), *Carmichael's manual of child psychology*. New York: Wiley, 1970, pp. 565–654.

Thorpe, W. H., & Jones, F. G. W. Olfactory conditioning and its relation to the problem of host selection. *Proceedings of the Royal Society, Series B*, 1937, *124*, 56–81.

Tizard, B., & Hodges, J. The effect of early institutional rearing on the development of eight-year-old children. *Journal of Child Psychology and Psychiatry*, 1978, *19*, 99–118.

Ursin, H., Baade, E., & Levine, S. *Psychobiology of stress: A study of coping men*. New York: Academic Press, 1978.

Vaughn, B., Egeland, B., Sroufe, L. A., & Waters, E. Individual differences in the infant–mother attachment at 12 and 18 months. *Child Development*, 1979, *50*, 971–975.

Werner, E. E., & Smith, R. S. *Vulnerable but invincible: A longitudinal study of resilient children and youth*. New York: McGraw-Hill, 1982.

Wilson, R. S. Mental development in twins. In A. Oliverio (Ed.), *Genetics, environment and intelligence*. Amsterdam: North-Holland, 1977, pp. 305–334.

Wohlwill, J. F. Cognitive development in childhood. In O. G. Brim and J. Kagan (Eds.), *Constancy and change in human development*. Cambridge, Mass.: Harvard University Press, 1980, pp. 359–444.

Wolff, S. *Children under stress*. Harmondsworth: Penguin Books, 1973.

Zeitlin, H. *The natural history of psychiatric disorder in children*. M.D. thesis, University of London, 1983.

CHAPTER 4

Age-related Aspects of Experience Effects upon Brain Structure

William T. Greenough and Harris D. Schwark

Because this chapter appears in a book oriented primarily toward researchers in human development, we should point out that many of its major points would be considered speculative by neuroscientists. Interpretations placed on the data described below are not invariably those of the authors of the original papers. In this chapter we make three major points:

1. Brain plasticity, as reflected in estimates of synaptic connectivity numbers and patterns, is evident in experience manipulation studies in both developing and mature organisms. This suggests that similar, if not identical, mechanisms may be involved in both developmental brain organization processes and in adult information storage processes, including learning and memory.
2. Critical or sensitive periods in development may reflect the rate at which connectivity patterns are being established and the primacy of those patterns rather than mechanisms which are unique to the developing brain. In particular, sensitive periods in developing sensory systems and elsewhere may be underlaid by an overproduction of synaptic connections followed by a comparatively rapid loss to a stable or slowly ascending baseline.
3. A less rapid, but otherwise similar, process may continue in the adult brain. Ongoing generation of transient synapses, a subset

William T. Greenough • Department of Psychology, University of Illinois, Champaign, Illinois 61820.   Harris D. Schwark • Department of Psychology, University of Illinois, Champaign, Illinois 61820.

of which is selectively preserved as a result of brain activity, may be a mechanism of adult memory.

We first review data indicating brain plasticity in development and adulthood, then describe studies which have given rise to the selective preservation hypothesis of developmental plasticity, and finally show how this view may apply to the adult brain. Data from studies of brain structure suggesting efficacy changes at existing synapses are not discussed here (for a review see Greenough & Chang, 1984).

PLASTICITY IN BRAIN DEVELOPMENT

Much of the evidence for altered synaptic connectivity as a mediator of the effects of experience upon later behavior in developing organisms has come from two paradigms: sensory deprivation studies (particularly visual deprivation) and complex or enriched environment studies. (The latter paradigm should also be considered a relative deprivation technique, since even the most complex laboratory condition almost certainly falls short of the complexity which would be encountered in the natural environment. For this reason we prefer the term *environmental complexity* to *enriched condition*.) Visual deprivation procedures include dark-rearing, usually of young with the mother (which has possible hormonal confounds; see Mos, 1976), and monocular deprivation by lid suture. Monocular deprivation in species with largely binocularly overlapping visual projections has quite different effects from the same procedure in species like the rat, with largely lateralized projections of each eye. Details of these monocular deprivation studies are discussed in a later section. The environmental complexity paradigm typically involves a group of animals reared from weaning in individual cages (IC) and a group reared together in large cages equipped with toys which are changed each day, and often supplemented with experience in a separate toy-equipped arena (EC). A third group reared in pairs or triplets in social cages (SC) is also often included. In both paradigms, littermates are typically assigned to each condition to minimize preexperimental variance.

Light microscopic analyses of neural changes resulting from experience manipulations frequently are based on Golgi-stained tissue. In this procedure an apparently random subpopulation of neurons is fully impregnated with heavy metals, revealing the morphology of the cell body and entire dendritic field (where axons from other neurons terminate). Figure 1 is a photograph of a neuron stained by the Golgi procedure. Two light microscopic measures have been used to estimate the location

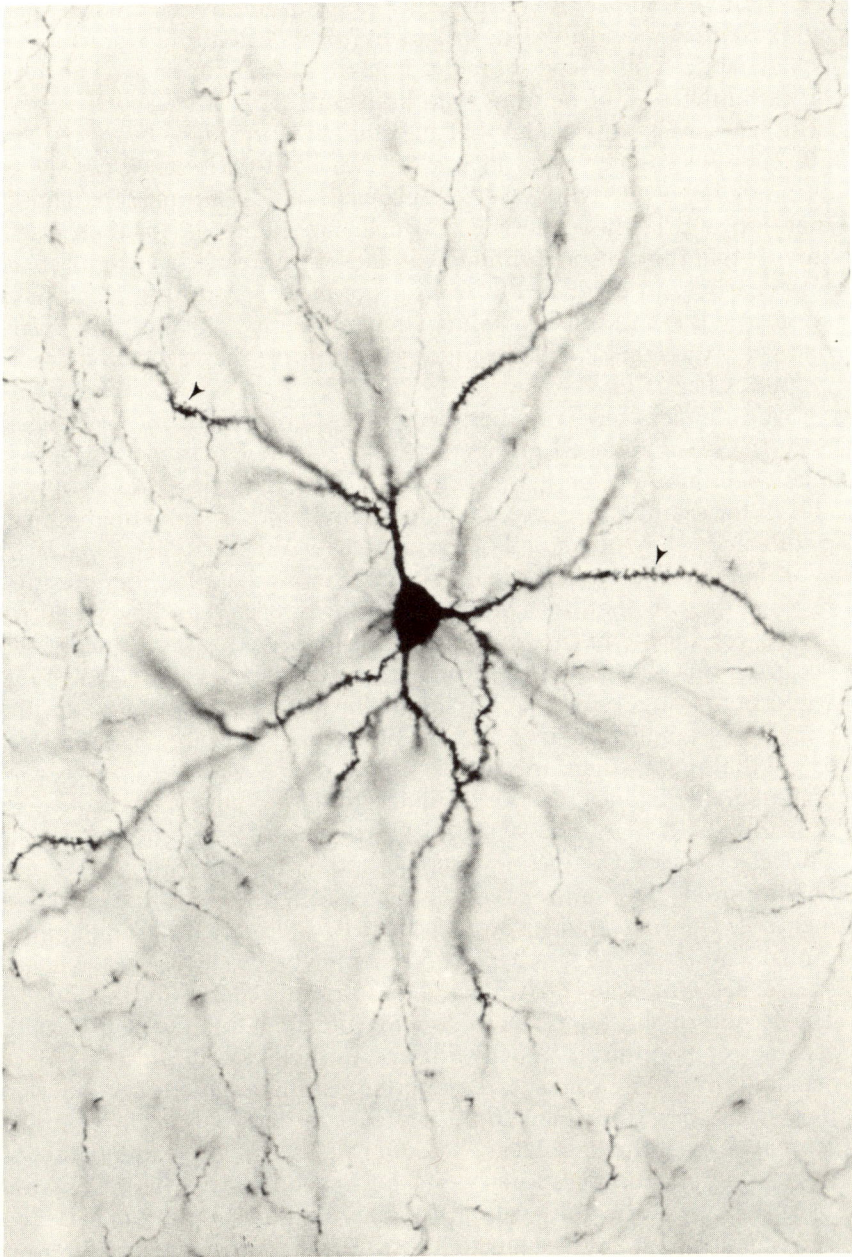

Figure 1. Photograph of a Golgi-stained neuron in the visual cortex of a rat. The dendrites of this neuron contain numerous spines (arrowheads).

and relative number of synapses per neuron: the frequency of spines along dendrites and the length and number of dendrites. Spines, the postsynaptic component of the predominant axodendritic synapse type on certain classes of neurons, can be counted per unit length of dendrite. This measure assumes that all spines are innervated by axons. The presence or absence of synapses on dendrites and spines cannot be assured, however, without electron microscopic confirmation. Thus a more appropriately conservative conclusion from both of these measures is that the neurons with more spines or dendrites have more space for synapses. In general, the electron microscopic studies which have addressed these issues have tended to confirm qualitatively, if not quantitatively, the results of light microscopic work (see review by Greenough & Chang, 1984).

A number of recent studies using these light microscopic techniques have indicated reduced numbers of synapses per neuron in deprived relative to more experienced groups. The classic work of Valverde (1971), for example, showed that in visually deprived mice the frequency of spines along the apical dendrite of layer V pyramidal cells in layer IV of the visual cortex (the main site of afferent thalamic fiber termination) was lower than in light-reared mice. Similar data have been reported for the hemisphere opposite the deprived eye in unilaterally occluded rats (Fifkova, 1970; Rothblat & Schwartz, 1979). Two recent studies have suggested that the original interpretations of these results should be modified. First, Freire (1978), using electron microscopy, found that many spines in visually deprived mice were reduced in size. Since smaller spines would be detected less frequently because of obscuring by the large apical shaft, this result could account for some or all of the observed reduction in spine frequency. Thus, spine size differences may be confused with number differences during development. (These size changes are, of course, an important potential indicant of the strength or other characteristics of the synapses on these spines; see Greenough & Chang, 1984, for a discussion of this point.) The second qualification has to do with interpretation rather than with accuracy of quantification. Schwartz and Rothblat (1980), studying monocular deprivation by lid suture, reported that the frequency of spines in the visual cortex opposite the deprived eye of rats was increased to normal levels if the light level was sufficient to offset the attenuation of retinal illumination associated with the closed eyelid. This result suggests that absolute light level, rather than patterned vision, may be the determinant of spine frequency. Taken together, these recent findings suggest that interpretations of the classical spine frequency results as indicating pattern vision-dependent synapse formation may have been premature.

Visual deprivation effects upon the second light microscopic measure mentioned above, dendritic length, have also been described. Coleman and Riesen (1968) reported that dark-reared cats had, on average, reduced dendritic tree dimensions in several visual cortex populations.

A reduction in the average number of synapses per neuron after visual deprivation has been confirmed using electron microscopic techniques. Cragg (1975) counted the number of synapses and the number of neuronal cell bodies per unit of visual cortex tissue volume in dark-reared and light-reared cats and reported that there were fewer synapses per neuron in the dark-reared group. Thus, although changes in the morphology of individual synapses may also occur (Greenough & Chang, 1984), there is little question that fewer synapses are formed (or retained) by neurons in the visual systems of visually deprived cats and rodents. This issue is discussed in more detail in a later section.

Quantitatively more dramatic light microscopic effects of experience manipulation have been seen in the complex environment paradigm. It seems somewhat surprising that a severe manipulation such as dark-rearing would not have more profound effects upon neuronal dendritic fields than the absence or the presence of toys and conspecifics. The environmental complexity studies have used carefully matched littermate designs, whereas this is rare in visual deprivation experiments, and this could result in enhanced detectability of experience effects, particularly given the small numbers of subjects in most experiments. Alternatively, it may be that the brain is more susceptible to manipulations which come closer to the environment "expected" by the genotype of the species. Finally, it is possible that effects of experience upon the *pattern* of synaptic connections are more dramatic than effects upon their numbers in visually deprived animals (see pp. 80–83).

The pioneering work of Rosenzweig, Bennett, and Diamond (e.g., 1972) and their colleagues showed that certain cerebral cortical regions, particularly the occipital (or visual) cortex, were thicker and heavier in animals reared in complex environments than in animals reared socially or in isolation. A number of related metabolic changes were also reported. Another report from their laboratory (Holloway, 1966) indicated, although there were technical difficulties, that neurons from animals reared in complex environments had more extensive dendritic fields than those in animals reared in isolation. Our laboratory followed up on this work. We found that (1) rats reared in complex environments had neurons with more extensive dendritic fields in several visual cortex cell populations when compared to isolated rats; (2) to a lesser extent, similar effects were also evident in auditory cortex; and (3) these effects were not manifest in all cerebral cortical regions, which suggested that general hormonal or metabolic mediators of these effects were unlikely

(Greenough & Volkmar, 1973; Greenough, Volkmar, & Juraska, 1973; Volkmar & Greenough, 1972). Typical results for the visual cortex are presented in Figure 2. We also found that these effects were not limited to primary sensory cortical regions. Granule cells of the dentate gyrus of the hippocampal formation of developing (but not adult) rats were similarly affected by environmental complexity (Fiala, Joyce, & Greenough, 1978), as were the cerebellar Purkinje cells of monkeys reared under similar environmental conditions (Floeter & Greenough, 1979). The con-

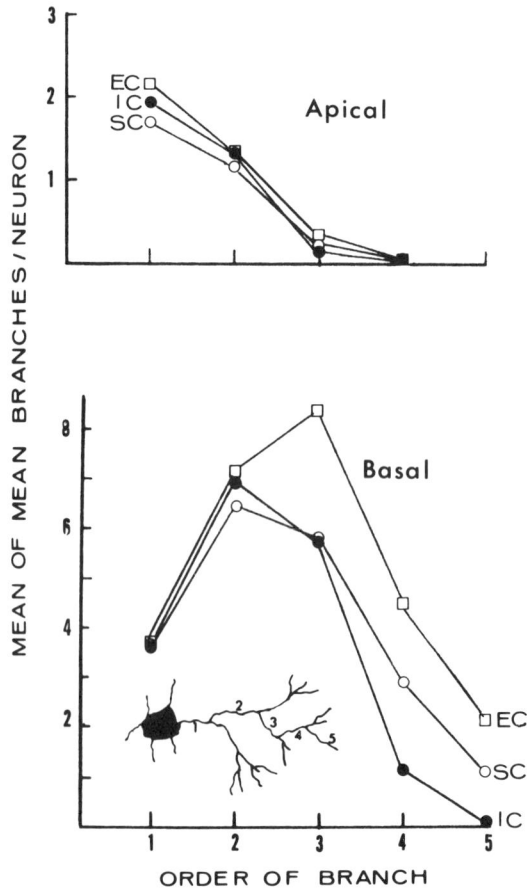

Figure 2. The average number of branches at each order away from the cell body and from the apical dendrite was calculated for a sample of pyramidal neurons from layer 5 of the occipital cortex of rats reared in the experimental environments. EC rats differed significantly from the other groups in basal orders 3, 4, and 5, and from SCs in apical orders 1 and 2. Similar effects were seen in other pyramidal neuron populations and in stellate neurons from layer 4.

clusion from these studies is that the dendritic spaces for synapses, and presumably the number of synapses per neuron, reflects the circumstances under which the organism has developed. We will return below to the mechanisms whereby these effects may arise.

ADULT BRAIN PLASTICITY

Rosenzweig, Bennett, and Diamond (1972) were again the first to demonstrate that environment-dependent differences in brain structure were not limited to traditionally conceived periods of development. They found that cortical weight and thickness in adult rats were increased by exposure to a complex environment. Following up on this, Uylings, Kuypers, and Veltman (1978) and Juraska, Greenough, Elliott, Mack, and Berkowitz (1980) found differences in total dendritic length per neuron in the visual cortex of young adult rats subjected to isolation or environmentally complex housing. These effects were smaller than those seen in developing rats, with some neuron types unaffected, but otherwise comparable. One might argue that these young adult rats were still "developing." However, Green, Greenough, and Schlumpf (1983) have reported similar effects in middle-aged (450-day-old) rats. These results suggested that what had been considered developmental plasticity was not limited to a circumscribed developmental period. It seemed possible that processes involved in brain development were also used by the adult brain and that these processes were involved in the storage of environmentally originating information. Since adult information storage includes material learned and remembered, it seemed possible that these developmental brain information storage mechanisms might be involved in adult learning and memory. One test of this hypothesis was to determine whether similar changes were observed as a consequence of training on traditional learning tasks in adult animals.

We have completed two experiments of this type and are in the midst of a third. The two completed experiments involved the Hebb-Williams maze, a maze with moveable barriers which allows a large number of different maze problems to be presented. In the first experiment rats were taught to solve a series of daily maze problems for water reward, one new problem each day, for 25 days (Greenough, Juraska, & Volkmar, 1979). Controls in this experiment were removed from their cages and given a drink of water in the experimenter's hand. When an animal achieved errorless performance on the new daily maze problem, it was given additional retention trials on previous problems until a total of 25 trials had been run. Results of histological examination of the occipital cortex shown in Table 1 revealed that the trained animals had more dendrite per neuron in the upper region of apical dendrites of

Table 1. Mean Apical Dendritic Intersections with Sets of Six Rings

Rings	Layer IV pyramidal		Layer V pyramidal	
	T	H	T	H
1–6	17.6	17.8	16.3	16.4
7–12	15.7	14.7	11.1	11.0
13–18	8.4	7.5 $p < 0.003$	8.6	7.8 $p < 0.02$
19–24	5.6	5.2	7.0	4.5 $p < 0.00001$
25+	0.7	0.8	0.5	0.2 $p < 0.02$

Note. Effect of adult maze training upon apical dendrite of visual cortex pyramidal neurons in layers IV and V: *Dendritic intersections* refers to a method for quantifying dendritic field size by counting intersections between dendrites and a superimposed set of rings of increasing diameter centered on the cell body. Ring intersection numbers are closely proportionate to dendritic length. Training effects were localized to the upper regions in both neuron populations; within that region, trained animals' (T) values exceed nontrained counterparts' (H) by about 25%. (Data from Greenough, Juraska, & Volkmar, 1979.)

layer IV and V pyramidal neurons. No differences were detected in basilar dendrites of these neurons or on layer IV stellate neurons.

This result indicated increased dendritic space for synapses (and presumably more synapses) in some neurons of animals subjected to a training experience. However, as in all experiments investigating brain changes resulting from experience, it is not possible to ascribe the changes to the neuronal processing and/or storage of information (or memory) from the experience. The changes could arise, for example, from sensory or motor activity or from some hormonal or metabolic activity associated with the training experience. While probably no single experiment can rule out all of these potential causes of training-produced brain differences (see Greenough & Maier, 1972), a within-subject design allows one to evaluate the relative contribution of general factors such as metabolic and motor activity. To provide for a within-subject design, Chang and Greenough (1982) surgically separated the cerebral cortical hemispheres by severing the corpus callosum, the large axon bundle which unites them. An opaque contact lens was developed which, when inserted in the eye, prevented visual input from the training from reaching the brain through that eye. (Recall that, in the rat, visual input from a given eye is largely transmitted to the opposite brain hemisphere.) Three groups of these "split-brain" rats were studied. For one group (trained-nontrained or TN), the contact lens was inserted in the same (randomly determined at the outset) eye each day, such that visual input from the maze training experience (essentially similar to that described above) was directed to only one hemisphere. In a second group (trained-trained or TT) the contact lens was alternated between left and right eye on successive days, such that both hemispheres received equivalent visual input from maze training. A control group (nontrained-nontrained or NN) received no maze training, and was subdivided into a group with the same daily lens position and a group with alternating lens positions. For all groups the contact lenses were removed immediately following the 4–5 hour daily training period. There was no effect of fixed versus alternating lens position within the control group upon our measurements of visual cortex nerve cell dendrites, indicating that occluder insertion alone did not significantly affect brain structure. Overall results for the upper apical dendrite region of layer V visual cortex pyramidal cells that we had previously found to be affected by training are shown in Figure 3. For the TN group this dendritic region was more extensive in the hemisphere opposite the nonoccluded eye, indicating an effect of training specific to the hemisphere receiving visual input and hence unlikely to be the exclusive result of general hormonal, metabolic, or motor activity factors. Results for the other two groups replicated the original maze findings: In Figure 3 it can be seen that the

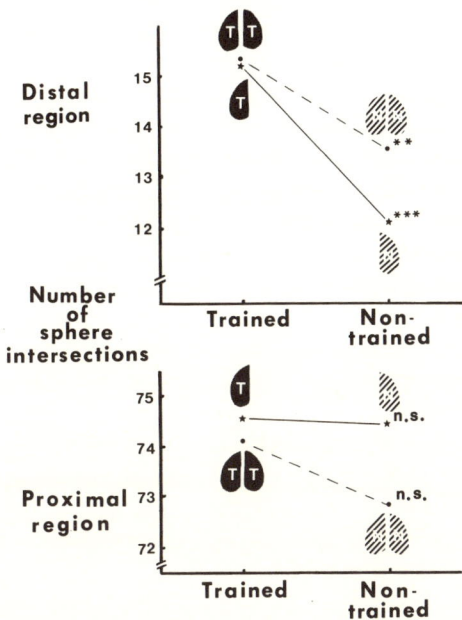

Figure 3. Number of intersections within (proximal) and beyond (distal) 250 micrometers of the cell body (spheres 1–12, 13–30 respectively). For statistical comparison, litter was treated as the replication factor, neurons as repeated measures nested within litter x training condition. TT versus NN and TN(T) versus TN(N) were analyzed separately. **, p0.025; ***, p0.01; n.s., statistically nonsignificant. This experiment used a three dimensional data analysis resulting in sphere rather than ring intersections.

amount of dendrite for the combined hemispheres of the alternately trained group (TT) is more extensive than in the combined hemispheres of the NN group. This experiment indicates that maze training brings about relative increases in dendritic space available for synaptic connections in the brain region receiving visual input from the training experience.

An experiment still in progress examines this sort of specificity in another training paradigm. Larson and Greenough (1981) trained rats to overcome their intrinsic handedness preference in a task requiring the rats to reach into a narrow tube for bits of chocolate chip cookies. A region of contralateral motor cortex thought to subserve handedness preference (Peterson, 1951; Peterson & Devine, 1963) was identified by electrically stimulating the surface of the brain to find the region where forepaw reaching movements were most easily elicited. When neurons in this region opposite the trained forepaw were compared with those in the homologous contralateral motor cortex, a difference in dendritic material quite similar to that seen in the preceding experiments was observed. This experiment again indicates that structural effects are specific to (or maximal in) the hemisphere associated with the training experience. We are currently examining neurons in another part of motor cortex to determine whether the structural effects are confined to the region which, by our electrical stimulation criterion, is most closely associated with the learned movement.

Taken together, these results suggest that the adult brain retains the capacity to form new synaptic connections and that connections are formed as a consequence of exposure to a learning situation. Although it is very tempting to speculate that such new connections actually represent memory, or some aspect of it, in the brain, the actual role of such connections in brain function is not yet clear. In future experiments we hope to determine whether the areas structurally altered after learning are electrophysiologically and metabolically active during performance of the learned task. If the connections are appropriate to some aspect of memory of the training experience, it is of critical interest to determine how appropriate connections might arise as a result of neural activity. In this regard, an intriguing hypothesis has arisen as a result of developmental research described below.

A MECHANISM OF PLASTICITY: SELECTIVE PRESERVATION OF SYNAPSES

Although the data described above suggest that brain plasticity is present throughout life, it is nevertheless apparent that dramatic effects of brief experience manipulations are seen during critical or sensitive periods of rapid anatomical, physiological, and behavioral organization of adult patterns. Examination of the nature of the changes occurring during these periods may help to reveal the basic mechanism by which experience exerts its effects. We propose that this mechanism may involve initial synapse overproduction followed by (or concurrent with) a period of competition among fiber populations for survival, resulting in a strengthening and preservation of appropriate synapses and a loss of inappropriate ones. This selective preservation hypothesis incorporates the ideas of previous workers such as Hebb (1949) and Changeux and Danchin (1976) in postulating an experience-related change in synaptic strength and survivability. In addition, it appears that the temporal characteristics of this period within a particular anatomical system may be related to the relative degree of plasticity exhibited by that system.

A difficulty in the study of the developmental dynamics of a population of synapses is the complexity of the nervous system, in which many different populations of fiber terminals may intermingle and synapse on the same target cells. The problem then becomes one of identifying the relevant populations in order to study their numbers and distributions. This problem has been overcome with greatest success in studies of the visual systems of primates and carnivores, especially in terms of development of binocularity, in which the fibers from individual eyes, which terminate in the same brain region, can be identified physiologically and anatomically. In the sections which follow the development of visual binocularity in primates will be used to illustrate: (1) the fact that a critical or sensitive period exists for the effects of environmental stimulation; (2) the anatomical correlates of this period, that is, an overproduction and subsequent loss of connections which appears to arise from (3) competitive interactions between fiber populations from the two eyes; and (4) a difference in the type of development which may be related to the degree of responsiveness to environmental manipulations. Other examples of these same types of processes occurring in the central nervous system will also be discussed. Because a recent review (Purves & Lichtman, 1980) has described similar developmental patterns in the peripheral autonomic nervous system and at the neuromuscular junction, these examples will not be presented here.

THE PRIMATE GENICULOSTRIATE VISUAL SYSTEM

In most primates the optic nerves from each eye decussate at the optic chiasm and send half of their fibers to each dorsal lateral geniculate nucleus (lgn) in the thalamus. Within the lgn the fibers from each eye are segregated into six laminae. Laminae 1, 4, and 6 receive contralateral eye fibers and laminae 2, 3, and 5 receive ipsilateral eye fibers (Brouwer & Zeeman, 1926; Matthews, Cowan, & Powell, 1960; Polyak, 1957). Cells in the lgn send fibers to visual, or striate, cortex (area 17 of Brodmann, 1905), where they terminate mainly in sublaminae IV A and IV C. An additional separation is evident within sublamina IV C in which parvocellular lgn input (from laminae 3–6) terminates in the lower part of the sublamina (IV Cβ) and magnocellular input (laminae 1 and 2) terminates in the upper portion (IV Cα) (Hubel & Wiesel, 1972; Lund, 1973; Lund & Boothe, 1975). Geniculate input to the cortex is also segregated by eye into ocular dominance columns approximately 400 micrometers wide (see Figure 4) which can be demonstrated physiologically (Hubel & Wiesel, 1968; 1977; LeVay, Hubel, & Wiesel, 1975) or anatomically (Wiesel, Hubel, & Lam, 1974) by transneuronal transport of radioactive amino acids and sugars injected into the eye (Grafstein, 1971). This high degree of separation of binocular inputs in the lgn and in the cortex arises during development from an initially overlapping, homogeneous pattern and therefore provides an excellent system in which to study the development of connectivity patterns and their dependence on environmental input.

Development of the Geniculostriate System

Fibers from the retinal ganglion cells of both eyes initially overlap completely in the lgn and do not begin to show signs of segregation into laminae until approximately three months before birth. This process continues until approximately two months before birth, when an adult-like pattern is attained (Hendrickson & Rakic, 1977; Rakic, 1976, 1977). Descriptive Golgi studies of lgn neurons during the postnatal period suggest that an initial overproduction of spines is followed by a postnatal loss until, by four weeks postnatal, adult levels are reached in the magnocellular laminae. The development of the parvocellular laminae appears to take longer (Garey & Saini, 1981).

Lamination of IV A and IV C is apparent in the visual cortex at six weeks before birth, and ocular dominance column separation begins at approximately three weeks before birth (Rakic, 1977). This process continues, producing an adult-like pattern by three weeks postnatal (Hubel, Wiesel, & LeVay, 1977) and culminating in the final adult pattern with well-defined columnar boundaries at approximately six weeks postnatal (LeVay, Wiesel, & Hubel, 1980)(see Figure 4). Studies of the developing human visual system suggest that the same types of developmental changes may occur in the lgn (Hickey, 1977; Hickey & Guillery, 1981; Hitchcock & Hickey, 1980) and visual cortex (Huttenlocher, de Courten, Garey, & Van Der Loos, 1982; Takashima, Chan, Becker, & Armstrong, 1980).

The same general pattern of development has also been reported to occur in the cat lgn (Shatz & DiBerardino, 1980) and visual cortex (LeVay, Stryker, & Shatz, 1978; Shatz & Stryker, 1978; Shatz, Lindstrom, & Wiesel, 1977). These studies have also shown that axon terminal fields of single lgn neurons are initially spread over a wide territory of cortex and, during development, some of the terminals are lost and those that remain become confined to individual ocular dominance columns. This apparently represents the cellular basis of the system-wide changes revealed physiologically and by transneuronal transport during ocular dominance column formation (see above).

Effects of Experience Manipulation during Development

The importance of binocular vision to development of ocular dominance columns was first demonstrated in the cat by Wiesel and Hubel (1963, 1965), who found that monocular lid suture soon after birth resulted in a physiological bias in cortex favoring the nondeprived eye. It appeared that the columns receiving input from the nondeprived eye expanded at the expense of those receiving input from the deprived eye (Figure 4). These same types of changes have also been demonstrated in monkeys subjected to monocular lid suture (Hubel, Wiesel, & LeVay, 1977; LeVay, Wiesel, & Hubel, 1980). Anatomical changes in the sizes of the ocular dominance columns as a result of altered visual input were confirmed by transneuronal transport studies (Hubel, Wiesel, & LeVay, 1977; LeVay, Wiesel, & Hubel, 1980; Shatz & Stryker, 1978). These studies demonstrated the critical role played by experience in the development of the visual system. In order to determine the period of maximal susceptibility to environmental manipulation and the nature of the columnar development, Hubel, Wiesel, and LeVay (1977; LeVay, Wiesel, & Hubel, 1980) performed monocular lid suture experiments on monkeys at various times after birth. The experimenters found that, while segregation of fibers in the normal animal appears to be complete by six weeks, monocular lid suture beginning as late as ten weeks still produced a modest expansion of the nondeprived eye columns. This suggests that some form of plasticity outlasts the appearance of adult-like columns and that the process of segregation can be reversed as well as simply stopped prematurely. To examine this latter possibility further,

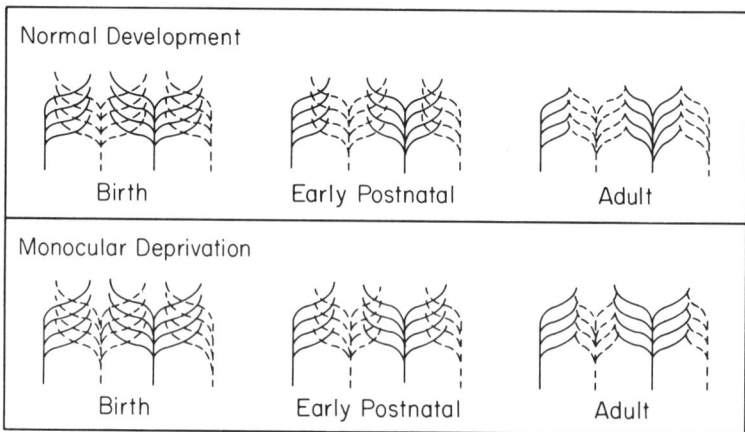

Figure 4. A schematic representation of the development of cortical ocular dominance columns in normal animals or in animals subjected to monocular deprivation during development. The panels on the left represent the large amount of overlap of axon terminal fields from the two eyes around the time of birth. The panels on the right illustrate the result of the competitive interactions which occur during development. In normal development (top), these interactions result in segregation and nearly equal reduction in the axon terminals from each eye. After monocular deprivation, however, terminals from the nondeprived eye are able to maintain a larger area of innervation at the expense of the terminals from the deprived eye.

three monkeys were subjected to reverse lid suture (opening of the deprived eye and closing of the nondeprived eye) at 3 weeks, 6 weeks, or 1 year of age. Deprived columns in the 3-week animal reexpanded, and even overcompensated, following reverse suture. In the 6-week animal the columns reexpanded to normal size, while in the 1-year animal no reexpansion took place. Interestingly, this reexpansion only occurred in sublamina IV Cβ, which receives parvocellular input. Sublamina IV Cα, the recipient zone of magnocellular input, showed no recovery. It therefore appears that fibers from the two main components of the lgn, while showing similar gross developmental patterns, exhibit different responses to environmental manipulation.

This sublaminar difference can also be seen during normal development when the system is examined at a finer structural level. Boothe, Greenough, Lund, and Wrege (1979) measured the frequency of spines and the length of dendrites of spiny stellate cells in sublaminae IV Cα and IV Cβ during normal development. Dendritic spines have been reported to receive 80% of the geniculate input (Garey & Powell, 1971), so these populations of spines probably represent the major target for lgn fibers. Spine numbers in these laminae increased steadily until eight weeks postnatal. After this time spine numbers in IV Cα remained constant, while in IV Cβ they rapidly decreased in numbers. Thus the more prolonged plasticity of the parvocellular system is associated with greater, or more apparent, synapse overproduction. These laminar differences in development were also apparent in counts of somal synapses (Mates & Lund, 1979). Synapses in IV Cα increased in number until five weeks postnatal and then leveled off, whereas in IV Cβ synapses numbers increased until 12 weeks and then decreased. This same general pattern of synapse overproduction has been found throughout primate visual cortex (O'Kusky & Colonnier, 1982) and in the lgn of the cat (Cragg, 1975, Winfield, 1981). The patterns of ocular dominance column development in cat visual cortex, and their responses to deprivation, have also been found to be similar to those in the primate (LeVay, Stryker, & Shatz, 1978; Shatz & Stryker, 1978).

The signals responsible for the onset and the completion of the period of maximal susceptibility to monocular deprivation are unknown (although involvement of the neurotransmitter norepinephrine has been proposed by Kasamatsu and Pettigrew, 1979), but the mechanisms by which column size changes may be produced have been studied. The size of cell bodies in the lgn appear to reflect the size of their axon terminal fields in visual cortex, so that after monocular deprivation lgn laminae innervated by the deprived eye contain smaller cells than the nondeprived laminae (Wiesel & Hubel, 1963). Guillery and Stelzner (1970) observed that, in the monocularly deprived cat, cell shrinkage occurred in the binocular, but not the monocular, segment of the lgn, which seemed to indicate that the cell shrinkage occurred as a result of a competitive disadvantage rather than as a simple deprivation effect. Guillery (1972) tested this idea by lesioning a small portion of a cat's retina, thereby producing an artificial monocular segment. This segment responded to monocular deprivation in the same way as the normal monocular segment, and cells in this segment remained physiologically responsive to stimuli presented to the deprived eye (Sherman, Guillery, Kaas, & Sanderson, 1974). These results suggest that competition between inputs from the two eyes plays an integral role in the modification of ocular dominance columns which result from monocular deprivation.

DEVELOPMENTAL PATTERNS IN OTHER NEURONAL SYSTEMS

Ocular dominance column development is a well-defined example of a general process which may be operating (although more slowly) throughout life. A large number of connections (axon terminals, spines, or dendrites) are produced early in development. Then, as a result of some type of competitive process, some of these connections are preserved, while others are lost. These same trends have been reported in many other places in the central nervous system. In the rat, olfactory bulb granule cell dendrites grow to maximal size between postnatal Days 14 and 21, then shrink (regress) to Day 60 (Brunjes, Schwark, & Greenough, 1982). This loss appears to result in a narrowing of the dendritic field width rather than a simple generalized loss throughout the field, which may indicate the presence of some type of postsynaptic competition.

This process does not appear to be limited merely to developing sensory or motor systems. Hsu, Carter, and Greenough (1978) found that the development of the hamster preoptic area, a sexually dimorphic basal forebrain region, involved postnatal loss of dendrites. In this area, neurons appear to be losing some of their first order (from the cell body) dendrites at the same time that branches are being added to other dendrites.

Other examples of synaptic target overproduction have been studied in the rat medial superior olivary nucleus, in which neurons exhibit an experience-dependent loss of dendrites after postnatal Day 14 (Feng & Rogowski, 1980; Rogowski & Feng, 1981). In rat cerebellum, Purkinje cells have more dendritic material (and spines) at postnatal Day 20 than at Day 250 (Weiss & Pysh, 1978).

Fibers terminating in, and exiting from, prefrontal cortex of the monkey exhibit many of the same patterns of development seen in the geniculostriate system. Goldman and Nauta (1977) noted that callosal fibers terminating in this area of cortex became organized into roughly columnar fields after being distributed widely early in development. It has been shown recently (Goldman-Rakic & Schwartz, 1982) that many of the areas between these columns are occupied by terminals of "associational" fibers from other areas of cortex. This raises the possibility that some sort of competition, similar to that seen in visual cortex, may be occurring between these fiber populations during development. Huttenlocher (1979) has shown that in human frontal cortex the number of synapses at one to two years of age is approximately 50% greater than adult levels. Some of the fibers originating from cells in prefrontal cortex terminate in the corpus striatum. Goldman-Rakic (1981) has found that these terminals are initially spread homogeneously throughout the striatum. At approximately two months before the birth the fibers begin to segregate, and by one month before birth they assume an adult-like annular projection pattern.

Postnatal segregation of fiber populations has also been identified in the cat visual collosal system (Innocenti, Fiore, & Caminiti, 1977; Innocenti, 1981) and in the superior colliculus of rodents (e.g., Land & Lund, 1979). The callosal fibers and efferent fibers from the rat somatosensory cortex exhibit a segregation of terminal fields which appears to involve the complete elimination of axon collaterals (Ivy & Killackey, 1981, 1982; Killackey & Ivy, 1981).

SELECTIVE PRESERVATION AS A CONTINUING PROCESS IN BRAIN FUNCTION

The overproduction of synapses seen in many developing neural systems suggests that some sort of selection process occurs in the production of an adult pattern of connections. Changeux and Danchin (1976) have described the genetic advantages and other characteristics of selective systems. This type of developmental process would involve an overproduction of connections and some type of stabilization process, perhaps at the level of the synapse, through which certain connections are preserved. In the developing geniculostriate system described above overproduction can be seen in the numbers of spines and somal synapses as well as in the distribution of the fiber terminals. Segregation of the terminals into ocular dominance columns at the same time that synapses are being lost apparently represents the operation of the selection process. Swindale (1980) has recently formulated a mathematical model of ocular dominance column formation based upon excitatory and inhibitory interactions among synapses from the two eyes. Successive computer iterations of the model produce a columnar pattern very similar to that seen in the primate cortex, suggesting that relatively simple interactions among synapses might lead to a complex biological pattern. An alternative selection process might occur through preservation of synapses the activity of which is coincident with firing of the postsynaptic cell concomitant with a loss or weakening of inappropriate ones (Greenough & Chang, 1984; Hebb, 1949; Singer, 1979).

Since a selective system could function with only a slight overproduction of connections, it is necessary to ask why a drastic overproduction often occurs. The geniculostriate system again provides examples of drastic overproduction (in layer IV Cβ) and slight overproduction (in layer IV Cα). It appears that synaptic overproduction is associated with greater plasticity (as shown by reverse eyelid suture experiments) and may be characteristic of a rapidly organized system which is very responsive to environmental input. Although little is yet known about the developmental differences between parvocellular and magnocellular layers of the lgn in primates, the physiological dichotomy of these systems (into X and Y systems) presents a unique opportunity to study different developmental patterns within a single sensory system. It may be, for example, that parvocellular input to the cortex serves as a sort of framework upon which later neuronal circuitry is elaborated. A period of rapid development would, in this case, be very desirable.

Evidence for the operation of selective systems in human neural development suggests that the developmental age at which synapse selection occurs is different for different brain systems. Overproduction of synapses in the human visual cortex reaches a peak at 8–12 months after birth (Huttenlocher *et al.*, 1982) whereas in frontal cortex synapse overproduction peaks at 1–2 years (Huttenlocher, 1979). Both the timing and the characteristics of synapse production in these areas correspond to the development of the behaviors with which they are involved. The visual system becomes organized rather early and quite rapidly (the peak of synapse overproduction is quite circumscribed). The frontal cortex develops somewhat later and more gradually, as might be expected given the extended acquisition of complex behavior with which this area is involved. Huttenlocher's (1979) estimate of the age at which stable (though presumably dynamic) adult synapse density levels are achieved in the human frontal cortex is roughly comparable to that at which Inhelder and Piaget (1958) have proposed that the formal operations stage of adult-like thinking is reached, about 13 to 14 years of age.

Finally, an attractive explanation for the results of the adult plasticity studies described above is that the selection process continues throughout life. The evidence for experience-dependent growth in adults discussed above, as well as recent studies suggesting continuing neuron production well into adulthood (Bayer, 1982; Bayer, Yackel, & Puri, 1982; Kaplan & Hinds, 1977; Kaplan, 1981) have forced a revision of the idea of an anatomically static adult nervous system. Such a continuing synapse turnover process may conceivably be involved in the types of brain reorganization required for adult learning and memory. These ongoing processes, which are becoming increasingly apparent as experimental techniques for studying them become more sophisticated, probably underlie the continuous process of behavioral change which is evident throughout life. Thus, apparent discontinuities in synapse formation over the lifespan of an animal may reflect differences in rates of continuing production and selective preservation, rather than different processes at different stages of maturation.

REFERENCES

Bayer, S. A. Changes in the total number of dentate granule cells in juvenile and adult rats: A correlated volumetric and ³H-thymidine autoradiographic study. *Experimental Brain Research*, 1982, *46*, 315–323.

Bayer, S. A., Yackel, J. W., & Puri, P. S. Neurons in the rat dentate gyrus granular layer substantially increase during juvenile and adult life. *Science*, 1982, *216*, 890–892.

Boothe, R. G., Greenough, W. T., Lund, J. S., & Wrege, K. A quantitative investigation of spine and dendrite development of neurons in visual cortex (area 17) of *macaca nemestrina* monkeys. *Journal of Comparative Neurology*, 1979, *186*, 473–490.

Brodmann, K. Beitrage zur histologischen Lokalization der Grosshirnrinde Dritte Mitteilung: Die Rinderfelder der niederen Affen. *J. Psychol. Neurol.* (Leipzig), 1905, *9*, 177–226.

Brouwer, B., & Zeeman, W. P. C. The projection of the retina in the primary optic neuron in monkeys. *Brain*, 1926, *49*, 11–35.

Brunjes, P. C., Schwark, H. D., & Greenough, W. T. Olfactory granule cell development in normal and hyperthyroid rats. *Developmental Brain Research*, 1982, *5*, 149–159.

Chang, F.-L., F., & Greenough, W. T. Lateralized effects of monocular training on dendritic branching in adult split-brain rats. *Brain Research*, 1982, *232*, 283–292.

Changeux, J.-P., & Danchin, A. Selective stabilisation of developing synapses as a mechanism for the specification of neuronal networks. *Nature*, 1976, *264*, 705–712.

Coleman, P. D., & Riesen, A. H. Environmental effects on cortical dendritic fields. I. Rearing in the dark. *Journal of Anatomy (London)*, 1968, *102*, 363–374.

Cragg, B. G. The development of synapses in the visual system of the cat. *Journal of Comparative Neurology*, 1975, *160*, 147–166.

Feng, A. S., & Rogowski, B. A. Effects of monaural and binaural occlusion on the morphology of neurons in the medial superior olivary nucleus of the rat. *Brain Research*, 1980, *189*, 530–534.

Fiala, B. A., Joyce, J. N., & Greenough, W. T. Environmental complexity modulates growth of granule cell dendrites in developing but not adult hippocampus of rats. *Experimental Neurology*, 1978, *59*, 372–383.

Fifkova, E. The effect of unilateral deprivation in visual centers in rats. *Journal of Comparative Neurology*, 1970, *140*, 431–438.

Floeter, M. K., & Greenough, W. T. Cerebellar plasticity: Modification of Purkinje cell structure by differential rearing in monkeys. *Science*, 1979, *206*, 227–229.

Freire, M. Effects of dark rearing on dendritic spines in layer IV of the mouse visual cortex. A quantitative electron microscopical study. *Journal of Anatomy*, 1978, *126*, 193–201.

Garey, L. J., & Powell, T. P. S. An experimental study of the termination of the lateral geniculo-cortical pathway in the cat and monkey. *Proceedings of the Royal Society of London*, 1971, *B179*, 41–63.

Garey, L. J., & Saini, K. D. Golgi studies of the normal development of neurons in the lateral geniculate nucleus of the monkey. *Experimental Brain Research*, 1981, *44*, 117–128.

Goldman, P. S., & Nauta, W. J. H. An intricately patterned prefrontocaudate projection in the rhesus monkey. *Journal of Comparative Neurology*, 1977, *171*, 369–386.

Goldman-Rakic, P. S. Prenatal formation of cortical input and development of cytoarchitectonic compartments in the neostriatum of the rhesus monkey. *Journal of Neuroscience*, 1981, *1*, 721–735.

Goldman-Rakic, P. S., & Schwartz, M. L. Interdigitation of contralateral and ipsilateral columnar projections to frontal association cortex in primates. *Science*, 1982, *216*, 755–757.

Grafstein, B. Transneuronal transfer of radioactivity in the central nervous system. *Science*, 1971, *172*, 177–179.

Green, E. J., Greenough, W. T., & Schlumpf, B. E. The effects of complex or isolated environments on cortical dendrites of middle-aged rats. *Brain Research*, 1983, *264*, 233–240.

Greenough, W. T., & Chang, F.-L., F. Synaptic structural correlates of information storage in mammalian nervous systems. In C. W. Cotman (Ed.), *Neuronal plasticity* (2nd ed.). New York: Guilford Press, 1984.

Greenough, W. T., & Maier, S. F. Molecular changes during learning: Behavioral strategy—a comment on Gaito and Bonnet. *Psychological Bulletin, 1972, 78,* 480–482.

Greenough, W. T., & Volkmar, F. R. Pattern of dendritic branching in occipital cortex of rats reared in complex environments. *Experimental Neurology, 1973, 40,* 491–504.

Greenough, W. T., Volkmar, F. R., & Juraska, J. M. Effects of rearing complexity on dendritic branching on frontolateral and temporal cortex of the rat. *Experimental Neurology, 1973, 41,* 371–378.

Greenough, W. T., Juraska, J. M., & Volkmar, F. R. Maze training effects on dendritic branching in occipital cortex of adult rats. *Behavioral and Neural Biology, 1979, 26,* 287–297.

Guillery, R. W. Binocular competition in the control of geniculate cell growth. *Journal of Comparative Neurology, 1972, 144,* 117–130.

Guillery, R. W., & Stelzner, D. J. The differential effects of unilateral lid closure upon the monocular and binocular segments of the dorsal lateral geniculate nucleus in the cat. *Journal of Comparative Neurology, 1970, 139,* 413–422.

Hebb, D. O. *The organization of behavior.* New York: Wiley, 1949.

Hendrickson, A., & Rakic, P. Histogenesis and synaptogenesis in the dorsal lateral geniculate nucleus (LGd) of the fetal monkey brain. *Anatomical Record, 1977, 187,* 602.

Hickey, T. L. Postnatal development of the human lateral geniculate nucleus: Relationship to a critical period for the visual system. *Science, 1977, 198,* 836–838.

Hickey, T. L., & Guillery, R. W. A study of Golgi preparations from the human lateral geniculate nucleus. *Journal of Comparative Neurology, 1981, 200,* 545–577.

Hitchcock, P. F., & Hickey, T. L. Prenatal development of the human lateral geniculate nucleus. *Journal of Comparative Neurology, 1980, 194,* 395–411.

Holloway, R. L. Dendritic branching: Some preliminary results of training and complexity in the rat visual cortex. *Brain Research, 1966, 2,* 393–396.

Hsu, C.-H., Carter, C. S., & Greenough, W. T. Postnatal development of hamster preoptic area: A Golgi study. Society for Neurosciences Abstract 339, vol. 4, 1978.

Hubel, D. H., & Wiesel, T. N. Receptive fields and functional architecture of the monkey striate cortex. *Journal of Physiology, 1968, 195,* 215–243.

Hubel, D. H., & Wiesel, T. N. Laminar and columnar distribution of geniculo-cortical fibers in the macaque monkey. *Journal of Comparative Neurology, 1972, 146,* 421–450.

Hubel, D. H., & Wiesel, T. N. Functional architecture of macaque monkey visual cortex. *Proceedings of the Royal Society of London, 1977, B198,* 1–59.

Hubel, D. H., Wiesel, T. N., & LeVay, S. Plasticity of ocular dominance columns in monkey striate cortex. *Philosophical Transactions of the Royal Society of London, 1977, B278,* 377–409.

Huttenlocher, P. R. Synaptic density in human frontal cortex—developmental changes and effects of aging. *Brain Research, 1979, 163,* 195–205.

Huttenlocher, P. R., de Courten, C., Garey, L. J., & Van Der Loos, H. Synaptogenesis in human visual cortex—evidence for synapse elimination during normal development. *Neuroscience Letters, 1982, 33,* 247–252.

Inhelder, B., & Piaget, J. *The growth of logical thinking from childhood to adolescence* (trans. A. Parsons & S. Milgram). New York: Basic Books, 1958.

Innocenti, G. M. Growth and reshaping of axons in the establishment of visual callosal connections. *Science, 1981, 212,* 824–827.

Innocenti, G. M., Fiore, L., & Caminiti, R. Exuberant projections into the corpus callosum from the visual cortex of newborn cats. *Neuroscience Letters, 1977, 4,* 237–242.

Ivy, G. O., & Killackey, H. P. The ontogeny of the distribution of callosal projection neurons in the rat parietal cortex. *Journal of Comparative Neurology*, 1981, *195*, 367–389.

Ivy, G. O., & Killackey, H. P. Ontogenetic changes in the projections of neocortical neurons. *Journal of Neuroscience*, 1982, *2*, 735–743.

Juraska, J. M., Greenough, W. T., Elliott, C., Mack, K. J., & Berkowitz, R. Plasticity in adult rat visual cortex: An examination of several cell populations after differential rearing. *Behavioral Neural Biology*, 1980, *29*, 157–167.

Kaplan, M. S. Neurogenesis in the three-month-old rat visual cortex. *Journal of Comparative Neurology*, 1981, *195*, 323–338.

Kaplan, M. S., & Hinds, J. W. Neurogenesis in the adult rat: Electron microscopic analysis of light radioautographs. *Science*, 1977, *197*, 1092–1094.

Kasamatsu, T., & Pettigrew, J. D. Preservation of binocularity after monocular deprivation in the striate cortex of kittens treated with 6-hydroxydopamine. *Journal of Comparative Neurology*, 1979, *185*, 139–162.

Killackey, H. P., & Ivy, G. O. Ontogenetic changes in the targets of layer Va barrel field neurons. Society for Neuroscience Abstract 23.6, vol. 7, 1981.

Land, P. W., & Lund, R. D. Development of the rat's uncrossed retinotectal pathway and its relation to plasticity studies. *Science*, 1979, *205*, 698–700.

Larson, J. R., & Greenough, W. T. Effects of handedness training on dendritic branching of neurons in forelimb area of rat motor cortex. Society for Neuroscience Abstract 23.7, vol. 7, 1981.

LeVay, S., Hubel, D. H., & Wiesel, T. N. The pattern of ocular dominance columns in the macaque visual cortex revealed by a reduced silver stain. *Journal of Comparative Neurology*, 1975, *159*, 559–576.

LeVay, S., Stryker, M. P., & Shatz, C. J. Ocular dominance columns and their development in layer IV of the cat's visual cortex: A quantitative study. *Journal of Comparative Neurology*, 1978, *179*, 223–244.

LeVay, S. Wiesel, T. N., & Hubel, D. H. The development of ocular dominance columns in normal and visually deprived monkeys. *Journal of Comparative Neurology*, 1980, *191*, 1–51.

Lund, J. S. Organization of neurons in the visual cortex, area 17, of the monkey (Macaca mulatta). *Journal of Comparative Neurology*, 1973, *147*, 455–495.

Lund, J. S., & Boothe, R. G. Interlaminar connections and pyramidal neuron organization in the visual cortex, area 17, of the macaque monkey. *Journal of Comparative Neurology*, 1975, *159*, 305–334.

Mates, S., & Lund, J. S. Development of somal synapses in visual cortex (area 17) of the macaque monkey. Society for Neuroscience Abstract 2679, vol. 5, 1979.

Matthews, M. R., Cowan, W. M., & Powell, T. P. S. Transneuronal degeneration in the lateral geniculate nucleus of the macaque monkey. *Journal of Anatomy*, 1960, *94*, 145–169.

Mos, L. P. Light rearing effects on factors of mouse emotionality and endocrine organ weight. *Physiological Psychology*, 1976, *4*, 503–510.

O'Kusky, J., & Collonier, M. Postnatal changes in the number of neurons and synapses in the visual cortex (Area 17) of the macaque monkey: A stereological analysis in normal and monocularly deprived animals. *Journal of Comparative Neurology*, 1982, *210*, 291–306.

Peterson, G. M. Transfers of handedness in the rat from forced practice. *Journal of Comparative and Physiological Psychology*, 1951, *44*, 184–190.

Peterson, G. M., & Devine, J. J. Transfer of handedness in the rat resulting from small cortical lesions after limited forced practice. *Journal of Comparative and Physiological Psychology*, 1963, *56*, 752–756.

Polyak, S. L. *The vertebrate visual system.* Chicago: University of Chicago Press, 1957.

Purves, D., & Lichtman, J. W. Elimination of synapses in the developing nervous system. *Science, 1980, 210,* 153–157.

Rakic, P. Prenatal genesis of connections subserving ocular dominance in the rhesus monkey. *Nature, 1976, 261,* 467–471.

Rakic, P. Prenatal development of the visual system in rhesus monkey. *Philosophical Transactions of the Royal Society of London, 1977, B278,* 245–260.

Rogowski, B. A., & Feng, A. S. Normal postnatal development of medial superior olivary neurons in the albino rat: A Golgi and Nissl study. *Journal of Comparative Neurology, 1981, 196,* 85–97.

Rosenzweig, M. R., Bennett, E. L., & Diamond, M. C. Chemical and anatomical plasticity of brain: replications and extensions. In J. Gaito (Ed.), *Macromolecules and behavior* (2nd ed.). New York: Appleton-Century-Crofts, 1972, pp. 205–277.

Rothblat, L. A., & Schwartz, M. L. The effect of monocular deprivation on dendritic spines in visual cortex of young and adult albino rats: Evidence for a sensitive period. *Brain Research, 1979, 161,* 156–161.

Schwartz, M. L., & Rothblat, L. A. Long-lasting behavioral and dendritic spine deficits in the monocularly deprived albino rat. *Experimental Neurology, 1980, 168,* 136–146.

Shatz, C. J., & DiBerardino, A. C. Prenatal development of the retinogeniculate pathway in the cat. Society for Neuroscience Abstract 160.3, vol. 6, 1980.

Shatz, C. J. & Stryker, M. P. Ocular dominance in layer IV of the cat's visual cortex and the effects of monocular deprivation. *Journal of Physiology (London), 1978, 281,* 267–283.

Shatz, C. J., Lindstrom, S., & Wiesel, T. N. The distribution of afferents representing the right and left eyes in the cat's visual cortex. *Brain Research, 1977, 131,* 103–116.

Sherman, S. M., Guillery, R. W., Kaas, J. H., & Sanderson, K. J. Behavioral, electrophysiological and morphological studies of binocular competition in the development of the geniculo-cortical pathway of cats. *Journal of Comparative Neurology, 1974, 158,* 1–18.

Singer, W. Neuronal mechanisms in experience dependent modification of visual cortex function. *Progress in Brain Research, 1979, 51,* 457–478.

Swindale, N. V. A model for the formation of ocular dominance stripes. *Proceedings of the Royal Society of London, 1980, B208,* 243–264.

Takashima, S., Chan, F., Becker, L. E., & Armstrong, D. L. Morphology of the developing visual cortex of the human infant. A quantitative and qualitative Golgi study. *Journal of Neuropathology and Experimental Neurology, 1980, 39,* 487–501.

Uylings, H. B. M., Kuypers, K., & Veltman, W. A. M. Environmental influences on the neocortex in later life. In M. A. Corner *et al.* (Eds.), *Maturation of the nervous system: Progress in brain research (Vol. 48).* Amsterdam: Elsevier/North Holland, 1978, pp. 261–272.

Valverde, F. Rate and extent of recovery from dark-rearing in the visual cortex of the mouse. *Brain Research, 1971, 33,* 1–11.

Volkmar, F. R., & Greenough, W. T. Rearing complexity affects branching of dendrites in the visual cortex of the rat. *Science, 1972, 176,* 1445–1447.

Weiss, G. M., & Pysh, J. J. Evidence for the loss of Purkinje cell dendrites during late development: A morphometric Golgi analysis in the mouse. *Brain Research, 1978, 154,* 219–230.

Wiesel, T. N., & Hubel, D. H. Single-cell responses in striate cortex of kittens deprived of vision in one eye. *Journal of Neurophysiology, 1963, 26,* 1003–1017.

Wiesel, T. N., & Hubel, D. H. Comparison of the effects of unilateral and bilateral eye closure on cortical unit responses in kittens. *Journal of Neurophysiology, 1965, 28,* 1029–1040.

Wiesel, T. N., Hubel, D. H., & Lam, D. M.-K. Autoradiographic demonstration of ocular dominance columns in the monkey striate cortex by means of transneuronal transport. *Brain Research*, 1974, *79*, 273–279.

Winfield, D. A. The postnatal development of synapses in the visual cortex of the cat and the effects of eyelid closure. *Brain Research*, 1981, *206*, 166–171.

PART II

STUDIES OF DISCONTINUITIES AND DEVELOPMENTAL TRANSITIONS

CHAPTER 5

Detecting Developmental Discontinuities

METHODS AND MEASUREMENT

Kurt W. Fischer, Sandra L. Pipp, and Daniel Bullock

In a sense, discontinuities are commonplace. In cognitive-developmental psychology, discontinuity typically refers to qualitative change, and behavior undergoes qualitative change routinely. Whenever people learn something new, some part of their behavior shows qualitative change.

According to the definition of discontinuity as qualitative change, then, people show frequent discontinuities in their behavior. Yet scholars do not usually count such everyday changes as discontinuities. When an individual learns a technique for walking in snowshoes or a way of conceptualizing discontinuity, there seems to be little useful purpose in referring to the resulting behavioral changes as discontinuities.

What is it that differentiates these ostensibly minor qualitative changes from changes that scientists are more willing to accept as discontinuities? Indeed, if researchers happened upon a real discontinuity, how would they know they had found it?

One of the most obvious answers would seem to be that for a qualitative change to count as a discontinuity, it must be substantial. In

Kurt W. Fischer and Daniel Bullock • Department of Psychology, University of Denver, University Park, Denver, Colorado 80208.   Sandra L. Pipp • Department of Psychiatry, University of Colorado School of Medicine, Denver, Colorado 80262. Preparation of this article was supported by a grant from the Carnegie Corporation of New York and a grant from the Foundation for Child Development. The statements made and views expressed are solely the responsibility of the authors.

95

developmental theories, for example, the equation of developmental stages or levels with qualitative change is misleading. What is apparently meant is that each stage or level involves a *large* qualitative change—a discontinuity. Otherwise, the small qualitative changes that occur in children's behavior every day would all reflect new developmental stages, and the number of stages in development would be virtually infinite.

Magnitude itself is not enough, however. In some developmental research, very large changes occur but at glacially slow rates. It is common, for example, for reasoning characteristic of a certain stage to appear at a low frequency and then to increase slowly over 10 or 20 years (e.g., see Colby, Kohlberg, Gibbs, & Lieberman, 1983; Fischer, 1983). The amount of qualitative change in such a case may be substantial, but because of the slow rate of development it hardly seems justified to refer to such a change as a discontinuity.

What seems to be required for something to count as a discontinuity is that the qualitative change be both large and rapid. In other words, a developmental discontinuity is essentially a *spurt in development.*

If discontinuities such as developmental stages or levels indeed are large, rapid qualitative changes in behavior, then the obvious way to detect them is to use methods for measuring the amount and speed of qualitative change. That is, a yardstick and a clock for qualitative change are needed.

A number of developmental methods and measurement techniques are available that provide such yardsticks and clocks, although researchers have generally neglected to use them (e.g., see Bart & Krus, 1973; Coombs & Smith, 1973; Krus, 1977; Wohlwill, 1973). These methods all use a developmental scale that provides a relatively continuous measure of the qualitative changes predicted to show discontinuity, and they measure the speed of change along the scale. With these methods, then, a discontinuity is demonstrated when the scale shows a large, rapid degree of change.

To use these methods to detect discontinuities, however, it is not enough simply to apply them in any study. Investigators must measure behavior under environmental conditions in which a discontinuity is likely to occur. Failure to attend to this seemingly obvious requirement appears to be one of the most common failings of developmental research. Apparently, developmental discontinuities appear and disappear as a function of the environmental conditions under which behavior is assessed, as we will demonstrate later. The ideal study would appear to include both conditions under which a discontinuity is expected to occur and conditions under which no discontinuity is expected.

To avoid confusions about the meaning of key terms, it is important to note that by this analysis several traditional dichotomies must be recast. First, discontinuity and continuity are not necessarily opposed. Developmental change can be continuous in the sense that it can occur along a finely graded scale of qualitative change, and at the same time it can be discontinuous in the sense that the scale can demonstrate large, rapid change at some point or interval. The one sense in which discontinuity and continuity can still be considered opposites is that if a spurt marks a discontinuity, then a slow increase can be said to mark a continuity.

Second, quantitative and qualitative change are not opposed within this analysis. The quantity of qualitative change can be measured by developmental scales, and indeed a qualitative change must be large on such a scale if it is to count as a discontinuity. Similarly, one index of a discontinuity will be rapid, large change in measures traditionally considered quantitative as well as qualitative. For example, a spurt in vocabulary size (a quantitative change) may be one index of a discontinuity.

To illustrate how such developmental discontinuities can be detected, we will consider several available measurement techniques that provide effectively continuous scales of qualitative change, as well as some measures of quantitative change. We will demonstrate how appropriate methods for using those techniques can both provide evidence for developmental discontinuities and test specification of the conditions under which those discontinuities will be detected. Research to date suggests that developmental spurts will be detected primarily when people are performing at or near their best capabilities.

TWO METHODS FOR DETECTING DISCONTINUITIES

In the study of psychological development, at least two methods are available for detecting discontinuities—one based on scalogram analysis and one based on the method of multiple tasks.

Scalogram Analysis

The method of scalogram analysis uses the Guttman (1944) scale, or scalogram. A set of tasks is ordered in a sequence such that if people pass one task in the sequence, they will pass all earlier steps, and if they fail a task, they will fail all later steps, as illustrated in Table 1. Conventional scalograms are often constructed *post hoc* rather than *a priori*. A large number of tasks are administered, and only those that form a clear scale are included in the scalogram (e.g., Hooper, Sipple, Goldman, & Swinton, 1979; Kofsky, 1966). It is also possible to build what can be

Table 1. The Scalogram Method

Scale step	Tasks					
	A	B	C	D	E	F
0	−	−	−	−	−	−
1	+	−	−	−	−	−
2	+	+	−	−	−	−
3	+	+	+	−	−	−
4	+	+	+	+	−	−
5	+	+	+	+	+	−
6	+	+	+	+	+	+

Note. Correct performance of a task is indicated by a +.

called a *strong scalogram,* in which tasks are designed *a priori* to test a predicted sequence (Fischer & Bullock, 1981).

Scalograms can be used to assess any linear ordering of tasks, but they are especially useful for testing developmental sequences. In such sequences the highest step a person passes correlates with age or some other measure of developmental maturity.

The scalogram method provides a powerful test of developmental sequences with either cross-sectional or longitudinal designs, because every subject performs all tasks and consequently his or her behavior can be categorized in terms of the predicted profiles shown in Table 1. In a longitudinal design, each subject should show one of the predicted profiles at each assessment, every profile should appear in some subjects, and the highest step in the profile should increase as the person grows more mature. In a cross-sectional design, all subjects should show one of the profiles, every profile should be demonstrated by some subjects, and subjects' highest step should increase with older or more mature groups.

Spurts at Specific Ages. When a developmental scalogram has many steps, it provides exactly the kind of virtually continuous scale that is required to test for developmental spurts. Although scalograms have not been used often in developmental research, some of the studies that have employed them have found evidence for spurts at certain steps. Generally these steps have coincided with predictions from cognitive-developmental theories about where new developmental levels or stages should emerge.

In one such study, Corrigan (1977, 1983) tested the development of a number of skills between 1 and 2 years of age, including search (what Piaget, 1936/1954, called *object permanence*) and speech. A number of scientists have posited developmental discontinuities at roughly 12 and

21 months of age (e.g., Fischer, 1982; Kagan, 1982; McCall, Eichorn, & Hogarty, 1977; Uzgiris, 1976; Zelazo & Leonard, 1983). Corrigan's findings showed spurts for both of these postulated discontinuities.

Her measure of the development of search combined tasks from a number of previous studies to produce a 21-step developmental scale, with a separate task to assess each step. In both longitudinal and cross-sectional samples she found strong evidence for a spurt of approximately 10 steps at around 1 year of age. Figure 1 presents the results for the cross-sectional sample. This spurt reflected the following qualitative change in children's behavior: They developed from solution of visible displacements (finding an object only when they had actually seen it being hidden under a specific cloth) to solution of invisible displacements (finding it when they had not seen the movements of the object itself but merely those of the experimenter's hand, in which the object was hidden until the experimenter surreptitiously placed it under a cloth). Other skills that appeared to spurt at approximately the same age included the first use of single words.

Toward the end of the second year, Corrigan found another spurt, in speech production. Children showed dramatic spurts in both vocabulary size and mean length of utterance (Figure 2). Such spurts in speech have also been documented by other investigators (e.g., Bates, 1979; Bloom, 1973). Note that the measures of speech did not involve scalograms, but they did employ continuous scales measuring what are conventionally categorized as quantitative changes. What is required to test for a spurt is not a scalogram *per se* or a measure of qualitative change, but any scale that can vary continuously across a wide range of scores.

Corrigan's study went beyond documenting the occurrence of a

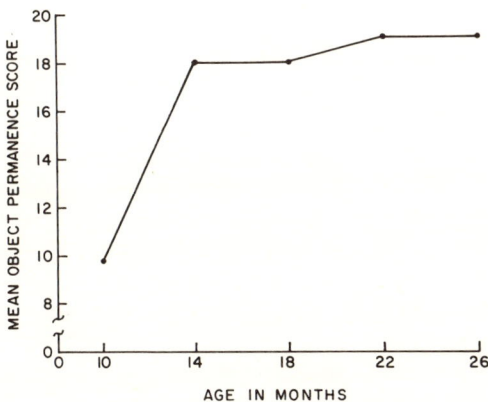

Figure 1. A spurt in search performance at 1 year of age. (From Corrigan, 1976, and Nelson, 1979.)

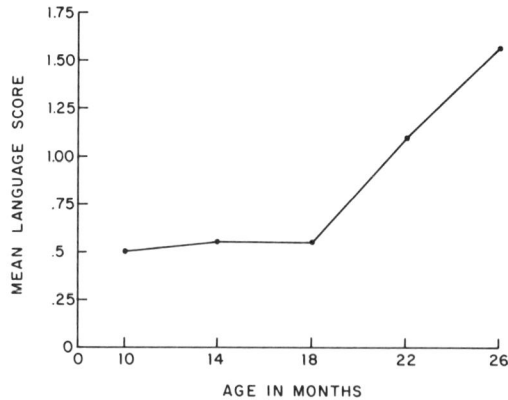

Figure 2. A spurt in speech late in the second year. (From Corrigan, 1976, and Nelson, 1979.)

spurt in speech at the end of the second year. With assessments of skills in a number of domains, she found that the age of the spurt in speech correlated closely with the ages if qualitative changes in various other skills, including search, seriation of nesting cups, and the use of toys as if they were independent agents in pretend play. For most of these other skills, her measures did not provide continuous scales that could assess directly for developmental spurts. Nevertheless, the correlated qualitative changes generally reflected abilitites linked to what Piaget (1936/1954) referred to as *representation.* Her findings thus support the hypothesis that a new developmental level involving representation emerges at this age (Fischer & Corrigan, 1981).

Clusters of Spurts. For a cognitive-developmental discontinuity of any generality, such a cluster of spurts should occur. The composition of the skills in the cluster will vary with the type of discontinuity producing the cluster (Fischer & Bullock, 1981). For a set of spurts to count as an age-related cluster, all the skills should occur within a relatively short age interval.

At times, definitions of stage seem to imply that children enter a new stage instantaneously, as if, for example, Piaget's concrete operations emerged on a child's seventh birthday (see Pinard & Laurendeau, 1969). All that is necessary to show a spurt, however, is that the change be rapid relative to the amount of change that occurred earlier. That is, the change need not occur at a point on the age scale but must take place over a relatively short age interval.

Similarly, it is not reasonable to expect spurts in different content domains or for different children to occur at exactly the same age. Various environmental variables, including task and context, have been shown repeatedly to have pervasive effects on assessments of development (Biggs & Collis, 1982; Fischer, 1980). With what seem to be minor variations on the same task, children often show dramatic changes in developmental level (e.g., Hand, 1981a; Jackson, Campos, & Fischer, 1978). Also, different children inevitably have had different degrees of experience with any content domain. Consequently, the age at which a spurt occurs will vary substantially with different tasks and contexts as well as across children. Developmental levels or stages should be expected to produce not tightly synchronous spurts but instead a cluster of spurts in a restricted age region, as shown in Figure 3.

Assessment of Spurts Independent of Age. Indeed, the scalogram method allows testing for discontinuities independently of age. The speed of a spurt can be measured in terms of the distribution of scores along the scale itself. This type of assessment is particularly useful when individuals are expected to show wide variations in the age at which they demonstrate a discontinuity.

With this method, the dimension of age is replaced by the steps of the developmental scale, and the frequency of subjects at each step is analyzed. A discontinuity is reflected by the bunching of subjects at certain steps in the sequence. Consider, for example, a 5-step scalogram

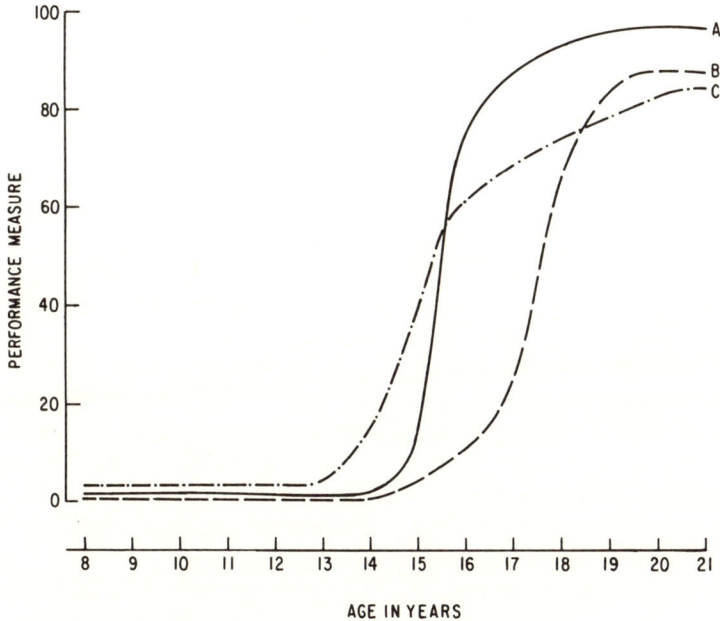

Figure 3. A hypothetical cluster of spurts in three domains.

sequence in which, by hypothesis, Steps 1 to 4 are successively harder tasks at one developmental level and Step 5 is the first task requiring the next developmental level. If individuals are followed as they develop through this sequence, they will show a spurt for the first level, moving quickly through Steps 1 to 4. But if there is a discontinuity between Steps 4 and 5, they will be unable to move onto the next level, and therefore they will remain for a long time at Step 4. When the discontinuity for the next level occurs, they will move to Step 5. If there are further steps at this level, they will spurt through them until they reach the highest step at the new level. Even if these individuals demonstrate their spurts at different ages, they will all spurt through the steps for one level and remain at the highest step of that level until the next level emerges. Thus, in a longitudinal design the spurt can be detected by measuring the length of time that individuals remain at each step.

In a cross-sectional design, the discontinuity will be evident in the distribution of subjects across steps. If subjects are categorized in terms of the highest step passed, a developmental spurt will produce the bunching of subjects at the later steps for each level. The frequency of subjects at the early steps for a level will be low because individuals move quickly through these steps, while the frequency at the later steps for a level will be high because individuals are unable to move quickly to the steps characteristic of the next level. Figure 4 shows such a distribution of subjects for a 9-step sequence assessing three developmental levels.

This method can also be used to detect clusters of spurts across different task domains. What is required is some theory or metric for indicating which steps in different scales represent the points where the same discontinuity will occur. A cluster will be evidenced by the occurrence of bunching at comparable steps across domains. With the emergence of representation toward the end of the second year, for example, Corrigan's results (1977, 1983) suggest that a cluster of spurts will occur at comparable scale points for speech, search, pretend play, and seriation.

Assumptions of the Method. The age-independent analysis just presented rests on two assumptions that are in fact made in all the above scalogram tests for developmental discontinuities. (Similar assumptions are also made in the other methods described below.)

The first assumption is that the distances between steps in the scale vary independently of the predicted developmental discontinuity. Ideally, all steps will be equidistant from adjacent steps, but equal-interval scales are hardly ever obtainable in psychological research, and they are not necessary in this case. All that is required to test for discontinuity is

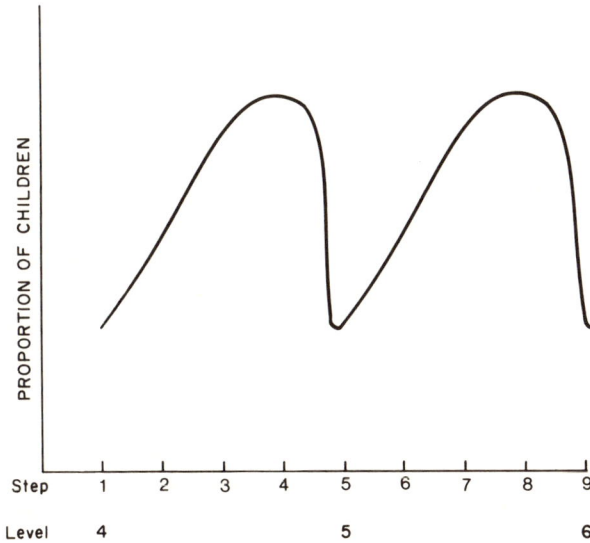

Figure 4. The bunching of subjects at two points on a scalogram. This example is an idealization of what is predicted for the case where discontinuities occur at Steps 1, 5, and 9.

that the distance between steps not be confounded with the points of discontinuity.

Assessment of the distances between steps on a psychological scale can be difficult in general, and it is especially complex for a developmental scale. What is the meaning of distance, and how should it be measured? One obvious measure is developmental distance, the time it takes to develop from one step to the next. Developmental distance, however, cannot be readily used to check for confounding because it is the variable in which a discontinuity is predicted. At the steps on the scale where the spurt occurs, distances between steps are expected to be temporally closer than at other points on the scale.

To meet the first assumption then, some measure of distance is required that assesses an attribute of the scale other than the type of developmental distance reflected in the spurt. Fortunately, a straightforward method is available for obtaining such a measure. For a single scale the researcher can specify both the environmental conditions under which the spurt will occur and those under which it will not occur. The distance between steps under the nondiscontinuity conditions will serve as a test of the first assumption. For example, research in our laboratory indicates that developmental spurts appear in individuals' optimal performance but not in their behavior under typical testing conditions, which do not elicit optimal performance (Fischer, Hand, & Russell, 1983; Fischer & Pipp, 1984). Under typical conditions, the steps on a scale seem to be roughly equidistant, but under conditions eliciting optimal performance, developmental spurts appear. (The environmental conditions for spurts will be discussed further below.)

A second assumption underlying the scalogram method for detecting discontinuities is that extraneous characteristics of the subjects in the sample do not vary systematically with the discontinuity. The most obvious potentially confounding characteristic is age. If the distribution of subjects' ages is not uniform but has large numbers of subjects or observations at certain ages and few at other ages, then age alone may produce bunching of subjects at certain steps in the scale. If a large number of 24-month-olds are tested, for example, but few children between 18 and 23 months, then it is not informative to find apparent bunching at 24 months or at a step that commonly emerges in the interval between 18 and 24 months.

A simple procedure is available to eliminate this potential problem. Subjects can be sampled so that their ages are evenly distributed. In a study of 1- to 3-year-olds, for example, two subjects can be tested for every month of age from 1 year 0 months to 3 years 0 months. Then variations in age are strictly controlled, and much more confidence can be placed in the reality of any spurts obtained.

Method of Multiple Tasks

Scalogram analysis employs an ordered sequence, in which each step develops before all later steps and after all earlier ones. The second method for detecting discontinuities, called the method of multiple tasks, uses a different kind of continuous scale. For each level of the scale, a number of tasks are administered that require the capacity or other quality that defines the level, but the tasks need not be ordered into a microsequence within that level. An investigator might devise, for instance, 10 tasks for each of 3 levels to be tested, with no predictions about the ordering of the 10 tasks within each level.

The result for each subject is a score for number of tasks passed at each level. If the tasks are valid and reliable measures of the predicted levels, the data provided by such a measure will have some general similarities to those from a scalogram: The number of tasks each person passes at a lower level will be greater than or equal to the number passed at the next higher level. No individuals will pass any task at a higher level unless they have also passed some tasks at every lower level. And no individuals will fail all tasks at a lower level unless they also fail all tasks at every higher level. If such a scale in fact measures development, then the number of tasks passed at each level and the highest level at which any tasks are passed will correlate with age or some other measure of developmental maturity.

With this scale, a discontinuity for a given level will be evidenced when an individual's performance spurts abruptly from passing zero or few tasks to passing many tasks. As with the scalogram, this type of scale can be used to test developmental hypotheses with either cross-sectional or longitudinal designs, since each subject's data can be analyzed for the requisite developmental patterns.

Environmental Conditions for Discontinuities. One of the central questions for developmental science is, what are the environmental conditions under which discontinuities appear? In developmental research, most studies have found slow and gradual developmental change (see Fischer, 1983), and only a few have found evidence for spurts (e.g., Corrigan, 1983; Emde, Gaensbauer, & Harmon, 1976; Fischer & Pipp, 1984; Tabor & Kendler, 1981). We believe that a major reason for the scarcity of evidence for spurts is that most studies have neither used methods designed to detect spurts nor specified how environmental conditions will affect the nature of developmental change.

Hundreds of studies have demonstrated that environmental variations change the Piagetian stages that children show (see reviews by Fischer, 1980; Flavell, 1971; Gelman, 1978). But few studies have gone beyond such demonstrations to investigate systematically the relation between stage and environmental conditions. Several studies suggest that such investigations might well illuminate the nature of developmental stages or levels (e.g., Day & Stone, 1982; O'Brien & Overton, 1982).

In a study in progress on the development of arithmetic concepts, we have used the method of multiple tasks to test a hypothesis about the environmental conditions for discontinuities (Fischer, Hand, & Russell, 1983; Kenny & Fischer, in preparation). According to the hypothesis, when individuals develop a new cognitive level, their abilities do show discontinuous change. The spurt is evident, however, only in *optimal performance*. With familiar materials, the opportunity for practice, and the provision of environmental support for high-level skills, performance becomes optimal, and development shows a spurt as a new level emerges (Fischer & Pipp, 1984).

Testing this hypothesis was straightforward with the method of multiple tasks. A set of tasks was devised for each developmental level, and subjects performed them under four environmental conditions varying in the degree to which optimal performance was promoted. Only the two most extreme conditions will be discussed here. The condition providing the least support for optimal performance, called the *spontaneous* condition, was similar to the typical testing conditions in most cognitive-developmental research. Students were simply given several arithmetic problems for each level, without any opportunity to practice or any help from the experimenter.

The condition providing the most support for optimal performance was called the *practice and support* condition. After students had completed the problems in the spontaneous condition, they were shown an example of a good answer for each problem, and they were encouraged to practice the problems on their own during the interim before the next testing session. Two weeks later they were again shown a good answer to each problem, the answer was taken away, and they provided their own answer. Thus they had been able to practice the problem for two weeks, and they were provided with the support of seeing a good answer to the problem.

The prediction was that for the spontaneous condition students would show little or no evidence of a spurt with the emergence of a new developmental level. For the practice-and-support condition, on the other hand, they would show a striking spurt in performance for each new level.

The developmental levels were predicted by skill theory, a general framework for analyzing and predicting the course of psychological development in any domain (Fischer, 1980). Two of the levels tested were *Level 7 single abstractions,* which are predicted to emerge at approximately 10 to 13 years of age in American middle-class children, and *Level 8 abstract mappings,* predicted to emerge at approximately 14 to 17 years. A spurt reflecting Level 7 was therefore predicted for the interval between 10 and 13 years, and a spurt reflecting Level 8 for the interval between 14 and 17 years.

In the study, the tasks for Level 7 single abstractions were six problems requiring the definition of one of the arithmetic operations (addition, subtraction, or multiplication) and the application of that definition to a concrete computational problem, such as $7 + 3 = 10$ for addition. (Division was omitted from the Level 7 tasks because elementary-school children are generally incapable of dealing with division.) The tasks for Level 8 abstract mappings were eight problems requiring explanation of the relation between two similar arithmetic operations (addition–subtraction, addition–multiplication, multiplication–division, or division–subtraction) and the application of that explanation to two concrete computational problems (such as $7 + 3 = 10$ and $10 - 3 = 7$ for addition and subtraction).

Subjects were middle-class Anglo students in Denver. Thus far,

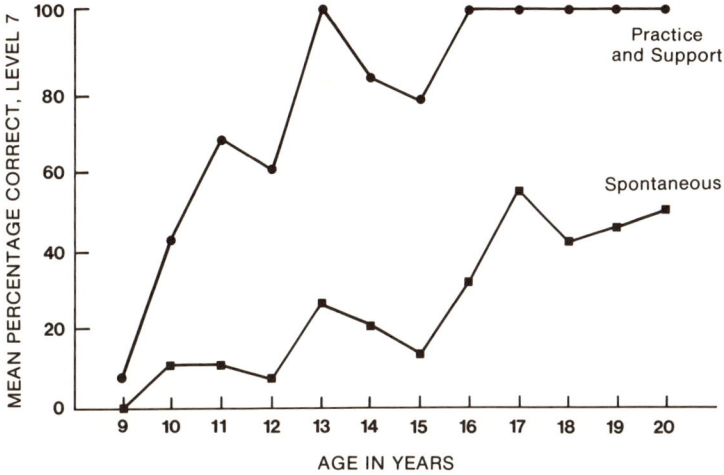

Figure 5. Changes with age in the percentage of Level 7 arithmetic problems solved under two conditions. Eight subjects, four of each sex, were tested at each age. The same subjects performed the tasks under both conditions.

four males and four females have been tested for every year of age from 9 to 20.

The results to date strongly support the predictions. For the spontaneous condition, performance showed a slow, gradual increase over a period of many years, as illustrated in Figures 5 and 6. Spontaneous performance of Level 7 tasks, depicted in the lower curve in Figure 5, increase gradually between 9 and 17 years of age, showing only one relatively small, statistically significant spurt at 15 to 17 years. Spontaneous performance of Level 8 tasks, shown in the lower curve in Figure 6, increased gradually between 15 and 20 years, showing no statistically significant spurt at all.

The pattern was markedly different for the practice-and-support condition. An abrupt spurt was evident for each of the two levels in the predicted age intervals. Performance on the Level 7 tasks, shown in the upper curve in Figure 5, spurted at 10 to 11 years to over 60% correct. Performance on the Level 8 tasks, shown in the upper curve in Figure 6, was near zero until 15 years of age and spurted sharply to over 80% correct at 16 years and after.

With tasks for a given level, an additional prediction can be made about spurts. When the initial emergence of the level does not result in near-perfect performance, then a *second spurt* is predicted upon the emergence of the next developmental level. The new capacity at the next

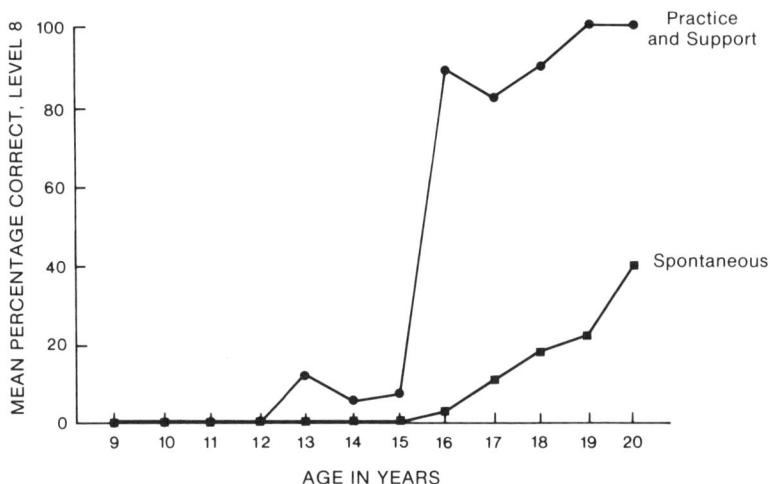

Figure 6. Changes with age in the percentage of Level 8 arithmetic problems solved under two conditions. Eight subjects, four of each sex, were tested at each age. The same subjects performed the tasks under both conditions.

level produces this second discontinuity because the skills for the earlier level undergo consolidation and differentiation when they are re-organized at the new level. In general, development of a later level will lead to a spurt in performance at an earlier level whenever such consolidation and differentiation occurs (Fischer, 1980; Flavell, 1971). For instance, the concept of subtraction, which can be understood with a Level 7 single abstraction, will be differentiated and consolidated when the adolescent is capable of comparing subtraction with addition, which requires a Level 8 abstract mapping.

The results of the arithmetic study support the second-spurt hypothesis. For Level 7 tasks, the spurt at the emergence of Level 7 was to approximately 60% correct. Consequently, an additional spurt could be predicted when Level 8 developed. This second spurt did indeed occur beginning at 13 years, as shown in the upper curve of Figure 5. The spurt at 15 to 17 years for the spontaneous condition, shown in the lower curve of Figure 5, also seems to reflect this second discontinuity.

A similar second spurt can be predicted for Level 8 tasks. According to skill theory, another developmental level, Level 9 abstract systems, emerges at approximately 18 to 22 years of age. Consequently, a second spurt in performance on Level 8 tasks can be expected at that age interval so long as performance at earlier ages remains low enough to allow such a spurt. With the emergence of Level 8 at age 15, performance on Level 8 tasks in the practice-and-support condition (upper curve in Figure 6) spurted to approximately 85% correct, which allowed only small room for improvement. Nevertheless, performance did seem to spurt to 100% correct at 19 years of age. The interpretation of this small increase as a second spurt is supported by what appears to be the beginning of a spurt for the spontaneous condition at 20 years, shown in the lower curve in Figure 6. According to this hypothesis, a further increase in spontaneous performance would be expected at 21 to 22 years.

These results suggest that discontinuities may indeed be common in optimal performance. The reason that researchers have discovered few developmental spurts may well be that they have seldom investigated the upper limits of performance. Also, most studies have not used continuous scales for assessing behavior. With either the method of multiple tasks or the scalogram method, investigators should find clearly discontinuous development under conditions that promote optimal performance and mostly continuous development under conditions of modal or average performance.

Assessment of Spurts Independent of Age. In the arithmetic study, performance for each level spurted at a particular age interval. Discontinuities need not occur in the same interval for all subjects, however. Just as with the scalogram method, the method of multiple tasks allows testing for a discontinuity independently of age, because the scale provided by the method can substitute for age.

If a cognitive level develops discontinuously, as hypothesized for the levels in the arithmetic study, then the distribution of performances for the tasks assessing that level will be bimodal. Some subjects will be capable of building the necessary skills at the level, and they will pass many or most of the problems. Others will not yet be capable of building the skills, and they will fail all or most of the problems. In a plot of the frequency of subjects passing each of the possible numbers of tasks, then, a large proportion of subjects will fail all or most of the tasks, and a large proportion of subjects will pass many or most of the tasks.

The results of the arithmetic study illustrate this type of analysis for discontinuity, as shown for Level 8 tasks in Figure 7. In the spontaneous condition, shown by the white bars, most subjects failed most tasks, and many passed a moderate number of tasks. The distribution was therefore unimodal and heavily skewed toward zero. On the other hand, in the practice-and-support condition, shown by the black bars, most subjects either passed most tasks or failed most tasks. The distribution was therefore strongly bimodal, demonstrating a developmental discontinuity with respect to the multiple-task scale.

In summary, the two methods proposed—the scalogram and the method of multiple tasks—allow strong tests for developmental discontinuities when the proper research designs are used. In both cases,

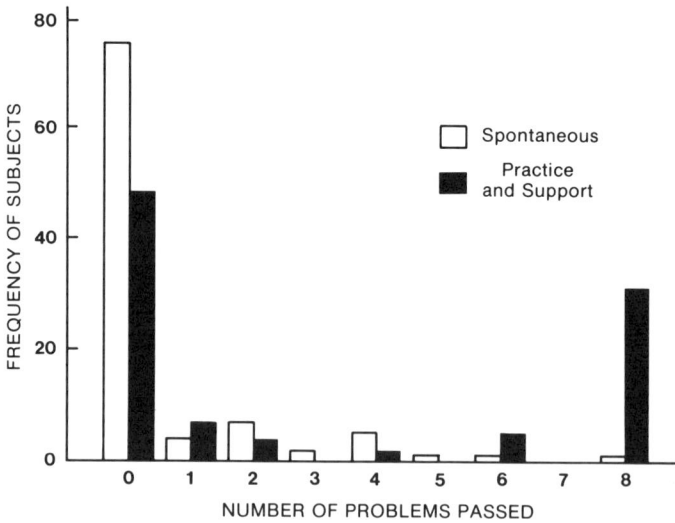

Figure 7. Distribution of subjects solving Level 8 arithmetic problems under two conditions. Eight subjects, four of each sex, were tested at each age. The same subjects performed the tasks under both conditions.

discontinuities can be detected in terms of age-related changes or in terms of the distribution of subjects along the scale itself. Discontinuities do not seem to appear under all environmental conditions. Research suggests that discontinuities in cognitive development may be most likely when the conditions encourage optimal performance. Methods other than the two proposed can also be used to test for discontinuities so long as they provide a continuous scale for assessing the behavior of interest.

SECOND-ORDER DISCONTINUITIES

All the analyses discussed thus far have tested for what might be called a *first-order discontinuity,* a sudden spurt in performance on a single developmental scale. There is another general type of discontinuity, a *second-order discontinuity*—an abrupt change in the relation between performance for two or more domains or conditions. In terms of measurement and data analysis, second-order discontinuities are clearly different from first-order discontinuities, but it remains to be determined empirically whether the two types involve distinct developmental processes.

One example of a second-order discontinuity involves a change in the relation between behaviors in different domains. At one period of development, for instance, performance across domains may correlate highly, but at a later period the correlation may abruptly drop. Another example of a second-order discontinuity involves a change in the relation between behaviors under different conditions. The developmental levels of behaviors under two conditions may be virtually identical at one period of development, but at a later point this concordance may suddenly disappear.

There is good reason to believe that second-order developmental discontinuities do occur. For example, cognitive-developmental levels of the kind tested in our arithmetic study are predicted to affect not only the rate of change of skills in individual domains (first-order discontinuities), but also the relations between skills across domains (second-order discontinuities).

The best method for detecting second-order discontinuities appears to involve the simultaneous use of continuous developmental scales in several domains or under several conditions. A discontinuity will be evidenced by an abrupt change in the relation between the performances for the several domains or conditions. In previous research, at least two types of second-order discontinuities have been identified by the use of two different kinds of methods. The correlation method has been used to uncover discontinuities in relations among domains, and the concordance method has been used to find discontinuities in the relations among conditions.

Correlation Method: Relations among Domains

In the correlation method, performances in a number of domains are measured, and changes in the correlations among the scales for those domains are examined. A second-order discontinuity occurs when the relation among rank orders on several scales changes abruptly.

Drops in Correlations at Developmental Transitions.   Robert McCall and his colleagues used this method to search for what they called *transitions* in performance on infant intelligence tests (McCall, 1983; McCall, Hogarty, & Hurlbut, 1972; McCall, Eichorn. & Hogarty, 1977). They reasoned that if at certain points in development infants' capabilities are undergoing major reorganizations, those changes should be evident in patterns of performance on intelligence tests, which assess performances in a variety of domains. To search for the transitions, they examined data from the Fels and Berkeley longitudinal studies, in which infants were tested repeatedly during infancy.

The performance of infants in each age group was first factor-analyzed with one of the simplest possible techniques—principal components analysis without rotations. The first factor for each age group was extracted, representing the factor around which most of the variations in performance fit. Any major changes in general intellectual capacity would presumably be evident in changes in these first factors across age groups.

Initially the researchers examined whether the factors changed in content from age to age—that is, whether they showed qualitative changes. There did seem to be major changes in content at certain ages, but the significance of these changes remained debatable because of a problem with the tests used. As is typical in intelligence tests, the test items changed from age to age, so that it was sometimes difficult to separate item changes from genuine changes in the infants' behaviors. If the items had been kept constant, the method of multiple tasks could have been used to test for a first-order discontinuity, a spurt in performance on the relevant tasks at certain ages.

To overcome this problem, the researchers examined changes in the correlations between factors at successive ages. With a broad sample of items, variations in content could be assumed to be insignificant for this analysis. A developmental transition would presumably be evident in abrupt changes in the correlations at certain ages.

The results did show four such second-order discontinuities—at approximately 2 months, 7 months, 13 months, and 20 months. In the early weeks of life, there was virtually no correlation from one testing session to the next, but at 2 to 4 months the first stability appeared, as the correlations rose dramatically. The three later discontinuities were of a different sort. There was a sudden drop in the correlations at those ages, followed by a sudden return to approximately the previous level of stability. That is, children showed less consistency in the rank-orders of their performance from one age to the next during these periods. Seibert, Hogan, and Mundy (in press) have done a related analysis with retarded children and found discontinuities at the mental ages (not chronological ages) of approximately 2, 7, and 13 months.

Relation between Spurts and Transitions. It seems significant that the ages of the third and fourth discontinuities are virtually the same as those at which Corrigan (1977, 1983) found spurts in her study of language and cognitive development in late infancy, described earlier. Also, the ages of the first two discontinuities are similar to those at which Emde, Gaensbauer, and Harmon (1976) found spurts in biobehavioral developments. Other evidence also suggests developmental spurts at approximately these four ages (see Fischer, 1982, for a summary).

McCall and his colleagues have not addressed the issue of how their findings of sudden drops in correlations might relate to these spurts, but there is a straightforward way in which such spurts (first-order discontinuities) will produce drops in correlations (second-order discontinuities). When a new developmental level is emerging, spurts are distributed across an age interval, as illustrated in Figure 3. Different children show the spurts at somewhat different ages, and within the same child the exact age of a spurt varies across domains. Also, as explained earlier, spurts appear to occur not only for tasks that require the new level but also for those that can be solved at the previous level. Consequently, performance on many different tasks shows spurts, but at differing ages.

The results of these variations in spurts is increased inconsistency in the developmental level of behaviors. While some performances jump ahead to the new level, others remain at the lower level until a later age. The inconsistency during the period of frequent spurts naturally produces a drop in the correlation, which is essentially a measure of consistency. After enough time has passed for the spurts to spread across a wide range of domains, children will show much more consistency in performance across domains, and the correlation will rise again. This same phenomenon is well documented for periods when large spurts in physical growth occur. With the onset of puberty, for instance, the variance of physical growth increases and various correlations among physical-growth measures drop temporarily, including the correlation of present size with adult size (Tanner, 1962).

Thus, the studies by McCall and his colleagues illustrate the use of the correlation method to uncover one type of second-order discontinuity. Given the type of data they analyzed, their discovery of developmental discontinuities is particularly impressive. Indeed, as McCall himself (1983; McCall et al., 1977) notes, the odds were biased against findings such discontinuities, because items on intelligence tests are typically chosen to produce slow, gradual increases in intelligence scores with age. With studies specifically designed to assess for spurts, the evidence for drops in correlations with the emergence of a new level would presumably be even stronger.

Concordance Method: Relations among Conditions

In the second method for detecting second-order discontinuities, the concordance method, a continuous scale is used to assess performance in the same subjects under two or more environmental conditions. A second-order discontinuity occurs when there is an abrupt change in the relation between performance in the different conditions.

The scalogram is particularly well suited for detecting such second-order discontinuities, because it provides a fine-grained assessment of developmental change. With such a measure the researcher can easily determine the exact point on the scale at which the relation between performances changes.

Elicited-Imitation and Spontaneous Conditions. In a series of scalogram studies in our laboratory we have uncovered a second-order discontinuity of this type. Performance was assessed under two types of conditions: elicited imitation and spontaneous behavior. At a certain point in development, there was a dramatic change in the relation between performance under the two types of conditions—a sudden drop in concordance.

Several of these studies used the developmental sequence shown in Table 2, which describes the development of agency and the social role of doctor (Watson & Fischer, 1977, 1980). The children were first tested in the elicited-imitation condition, in which the experimenter demonstrated a story for each step in the sequence and the child was asked to act our or describe that story. This procedure provided strong environmental support for every step assessed. After the elicited-imitation conditions, two spontaneous conditions were administered: free play and best story. In free play the experimenter asked the child to make up stories like those that had been demonstrated and then left the room. When the experimenter came back, he asked the child to show him the best story he or she could. The stories in the two spontaneous conditions could be easily scored on the same developmental scale as the elicited-imitation condition, but they provided much less environmental support for optimal performance.

When there was a difference between conditions, elicited imitation produced more advanced performance on the developmental sequence than the two spontaneous conditions, which in turn produced nearly identical performances. The difference was not uniform throughout the sequence, however. At one point in the sequence, there was an abrupt change in the relation between elicited imitation and the spontaneous conditions—a second-order discontinuity. For the first four steps in the sequence, nearly all children demonstrated the same highest step in all three conditions, although a few children showed lower steps in the spontaneous conditions. Beginning at Step 5, the pattern changed abruptly: There was a sharp drop in the percentage of children showing the same highest step across conditions, so that most children produced lower performance in the spontaneous conditions that in elicited imitation.

Figure 8 presents the results for the relation between two of the conditions—elicited imitation and free play. Each child was first categorized according to his or her highest step in elicited imitation. Then for each elicited-imitation step, the proportion of children who showed that same highest step in free play was plotted. The proportion dropped

Table 2. A Developmental Sequence of Social Role Playing

Cognitive level/step	Role-playing skill	Modeled behaviors
Sensory-motor systems		
1	Self as agent: A person pretends to carry out one or more behaviors, not necessarily fitting a social role.	Experimenter pretends to drink from an empty cup.
Single representations		
2	Active other agent: An agent performs one or more behaviors, not necessarily fitting a social role.	Experimenter pretends that a doll is talking, walking, eating, and washing, as if it were actually carrying out the actions itself.
3	Active substitute agent: An object substituting for an agent performs one or more behaviors, not necessarily fitting a social role.	Experimenter pretends that a block is walking, eating, talking, and going to sleep, as if it were a person or a doll.
4	Behavioral role: An agent performs several behaviors fitting a social role (in this case, doctor).	Experimenter pretends that doctor-doll uses a thermometer and an otolaryngoscope on another doll.
Representational mappings		
5	Social role: One agent behaving according to one social role (doctor) relates to a second agent behaving according to a complementary social role (patient).	Experimenter pretends that a patient doll is sick and a doctor doll examines her and gives her some medicine. The patient doll makes appropriate responses during the examination.
6	Social role with three agents: Two complementary roles (patient and nurse), instead of just one, are simultaneously related to the first role (doctor).	Experimenter pretends that a doctor doll examines a sick patient doll and is aided by a nurse doll. Both patient and nurse respond appropriately.
Representational systems		
7	Intersection of social roles for two agents: Two separate, agent-complement role relations are coordinated so that one agent is	Experimenter pretends that a doctor doll examines a sick patient doll and also acts as a father to the patient, who is his daugh-

(continued)

Table 2. (*Continued*)

Cognitive level/step	Role-playing skill	Modeled behaviors
	in two roles simultaneously and relates to another agent in two complementary roles (doctor-father with patient-daughter).	ter. The patient doll responds appropriately as both patient and daughter.
8	Intersection of social roles for three agents: Three separate agent-complement role relations are coordinated so that one agent in two or three roles simultaneously relates to two other agents each in the relevant complementary roles (doctor-father-husband with patient-daughter and mother of patient-wife).	Experimenter pretends that one doll is a doctor, father, and husband relating to two other dolls. The second doll is a sick patient and the first doll's daughter. The third doll is the patient's mother and the first doll's wife.

Note. The stories and exact scoring criteria are available from the authors on request. From "Development of Social Roles in Elicited and Spontaneous Behavior During the Second Year" by M. W. Watson and K. W. Fischer. *Developmental Psychology*, 1980, *16*, 483–194. Copyright 1980 by the American Psychological Association.

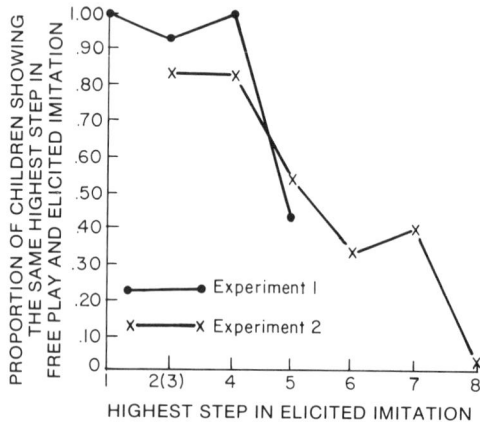

Figure 8. A second-order discontinuity in pretend play: Change in the proportion of children playing spontaneously at their highest step. In Experiment 1, subjects were 40 children ranging in age from 1½ to 4 years. In Experiment 2, subjects were 68 children ranging from 1½ to 7½ years of age.

precipitously at Step 5, and it continued to drop throughout the rest of the sequence.

In other words, performance in the several conditions showed little or no gap in the early steps of the sequence, but a substantial gap appeared at Step 5 which grew even larger at later steps. This effect was replicated in another study, assessing pretend play involving nice and mean social interactions rather than social roles (Hand, 1981a).

Tiers in Development. This second-order discontinuity appears to reflect a fundamental developmental process, involving what are called *tiers* in skill theory (Fischer, 1980). A tier is a major reorganization in behavior, larger than the changes reflected by most developmental levels. In a tier, a new type of behavioral unit emerges and gradually develops to specified levels of greater complexity. For example, the emergence of single representations at about 2 years of age marks the beginning of the representational tier. The next several developmental levels involve the combination of representations in increasingly complex relations within the representational tier. Then, at approximately 10 years of age, another new tier begins with the emergence of single abstractions. As with representations, there is a process of combination of abstractions to form increasingly complex relations within the abstract tier. Thus, each tier begins with a simple form of skill and at succeeding levels builds up increasingly complex and difficult structures growing from that simple form.

One prediction from the tier hypothesis is that skills at the later levels within a tier will be more difficult to sustain than skills at the earlier levels (Hand, 1981b). Single representations, for example, will be easier to sustain than representational mappings or systems, and single abstractions will be easier to sustain than abstract mappings or systems. This differential difficulty will be most evident under spontaneous environmental conditions, which provide little environmental support for optimal performance. It will be least evident under conditions that provide high support for optimal performance. That is, the gap between the levels of performance under spontaneous and supportive conditions will grow larger at higher levels within a tier.

This pattern is exactly what is shown in Figure 8. Steps 1 to 4 are all at the first level of representations, and Step 5 marks the beginning of the second level, representational mappings. At Step 5 there is a second-order discontinuity, the sudden appearance of a large gap between elicited imitation and spontaneous performance. Thus, movement to the second level of the representational tier seems to produce this sudden gap. Skills at the first level seem to be easy for the individual to sustain without much environmental support, but beyond the first level they become much more difficult to sustain without support.

According to skill theory, the representational tier is only one of a series of tiers. At about 10 to 11 years, a new tier begins as the child develops the capacity for abstractions. If the tier hypothesis is correct, then development in the abstract tier should show a second-order discontinuity similar to that found for the representational tier.

In a study of the development of the concepts of intention and responsibility in adolescents, Hand and Fischer (1981) found the same type of second-order discontinuity for the abstract tier. At the first developmental level, single abstractions, 80% of the subjects whose highest step was at that level in a supportive condition also showed the same highest step in a spontaneous condition. For the second and third levels of abstractions, however, all subjects showed lower performance in the spontaneous condition than in the supportive one.

Whether or not our interpretation of these data is correct, they clearly demonstrate that the relation between supportive and spontaneous conditions in development deserves further investigation. A focus on detecting second-order discontinuities in such conditions (and probably in other conditions as well) promises to shed new light on the processes of developmental change.

SUMMARY AND CONCLUSIONS

Discontinuities cannot be simply equated with qualitative change because behavior shows small qualitative changes continually. What most theories of developmental process seem to mean by *discontinuity* is qualitative change that is both large and rapid.

Methods and measures are available that provide powerful tools for detecting such developmental discontinuities. Key features of these methods and measures are that (1) they provide at least one continuous scale measuring the amount of qualitative change; (2) they provide a way of measuring speed of change on the scale; and (3) as a practical matter in research, it also appears to be important for the investigator to specify the environmental conditions under which discontinuities occur.

Two methods are especially useful for detecting discontinuities. In the scalogram, an independent task is used to assess every step in a many-step developmental sequence. In the method of multiple tasks, a number of unordered tasks are used to assess a single developmental level or stage. With each of these methods, developmental discontinuities can be straightforwardly detected in two ways; (1) by a spurt in performance at a certain age period or (2) by bunching in the distribution of subjects at certain points along the scale. These methods can be used effectively with either longitudinal or cross-sectional designs.

Certain key assumptions are made under both methods, but studies can be designed to ensure that these assumptions are met. First, it is assumed that the distance between steps is independent of the locations of discontinuity. A straightforward way to deal with this assumption is to include not only an environmental condition under which the discontinuity will occur but also a condition under which it will not occur. Second, it is assumed that variations in characteristics of the subjects sampled other than the one being assessed are independent of the discontinuity. The most obvious such characteristic is age. To deal with this assumption, the investigator can sample subjects evenly along the dimension of age (or of whatever characteristic needs to be controlled).

Because environmental factors produce such large variations in developmental step, the emergence of a level or stage will be evidenced by a cluster of spurts in an age region. All measures will not spurt at exactly the same age.

One hypothesis about the environmental conditions for discontinuity is that spurts will occur under conditions that optimize performance. When performance is not optimal, spurts will seldom occur.

Spurts or drops in performance on a continuous scale reflect first-order discontinuities. Second-order discontinuities involve an abrupt change in the relation between performance in two or more domains or conditions. There are at least two methods for detecting second-order discontinuities: In the correlation method, developmental changes in the consistency of individual rank-orders of performances across domains is analyzed; in the concordance method, changes in the similarities of performance on the same scale under different conditions are examined.

The use of these various methods for detecting and describing developmental discontinuities can substantially enhance the ability of developmental researchers to analyze and explain the processes of developmental change.

Acknowledgments

We should like to thank the following people for their contributions to this chapter: Bennett Bertenthal, Roberta Corrigan, Robert Emde, Helen Hand, Susan Harter, Sheryl Kenny, Marilyn Pelot, George Potts, and Malcolm Watson.

REFERENCES

Bart, W. M., & Krus, D. J. An ordering-theoretic method to determine hierarchies among items. *Educational and Psychological Measurement*, 1973, *33*, 291–300.

Bates, E. *Language and context: Studies in the acquisition of pragmatics.* New York: Academic Press, 1976.

Bertenthal, B. I. The significance of developmental sequences for investigating the what and how of development. In K. W. Fischer (Ed.), *Cognitive development.* San Francisco: Jossey-Bass New Directions for Child Development, No. 12, 1981.

Biggs, J., & Collis, K. *A system for evaluating learning outcomes: The SOLO taxonomy.* New York: Academic Press, 1982.

Bloom, L. *One word at a time: The use of single word utterances before syntax.* The Hague: Mouton, 1973.

Colby, A., Kohlberg, L., Gibbs, J., & Lieberman, M. A longitudinal study of moral judgment. *Monographs of the Society for Research in Child Development,* 1983, *48,* (1, Serial No. 200).

Coombs, C. H., & Smith, J. E. K. On the detection of structure in attitudes and developmental processes. *Psychological Review,* 1973, *80,* 337–351.

Corrigan, R. Patterns of individual communication and cognitive development (Doctoral dissertation, University of Denver, 1976). *Dissertation Abstracts International,* 1977, *37,* (10), 5393B. (University Microfilms No. 77–7400).

Corrigan, R. The development of representational skills. In K. W. Fischer (Ed.), *Levels and transitions in children's development.* San Francisco: Jossey-Bass New Directions for Child Development, No. 21, 1983.

Day, M. C., & Stone, C. A. Developmental and individual differences in the use of the control-of-variables strategy. *Journal of Educational Psychology,* 1982, *74,* 749–760.

Emde, R., Gaensbauer, T., & Harmon, R. Emotional expression in infancy: A biobehavioral study. *Psychological Issues,* 1976, *10,* No. 37. New York: International Universities Press.

Fischer, K. W. A theory of cognitive development: The control and construction of hierarchies of skills. *Psychological Review,* 1980, *87,* 477–531.

Fischer, K. W. Human cognitive development in the first 4 years. *The Behavioral and Brain Sciences,* 1982, *5,* 282–283.

Fischer, K. W. Illuminating the processes of moral development: A commentary. In A. Colby, L. Kohlberg, J. Gibbs, & M. Lieberman, A longitudinal study of moral judgment. *Monographs of the Society for Research in Child Development,* 1983, *48* (1, Serial No. 200).

Fischer, K. W., & Bullock, D. Patterns of data: Sequence, synchrony, and constraint in cognitive development. In K. W. Fischer (Ed.), *Cognitive development.* San Francisco: Jossey-Bass New Directions for Child Development, No. 12, 1981.

Fischer, K. W., & Corrigan, R. A skill approach to language development. In R. E. Stark (Ed.), *Language behavior in infancy and early childhood.* Amsterdam: Elsevier, 1981.

Fischer, K. W., & Pipp, S. L. Processes of cognitive development: Optimal level and skill acquisition. In R. J. Sternberg (Ed.), *Mechanisms of cognitive development.* New York: Freeman, 1984.

Fischer, K. W., Hand, H. H., & Russell, S. The development of abstractions in adolescence and adulthood. In M. Commons, R. Richards, & C. Armon (Eds.), *Beyond formal operations.* New York: Praeger, 1983.

Flavell, J. H. Stage-related properties of cognitive development. *Cognitive Psychology,* 1971, *2,* 421–453.

Gelman, R. Cognitive development. *Annual Review of Psychology,* 1978, *29,* 297–332.

Guttman, L. A basis for scaling qualitative data. *American Sociological Review,* 1944, *9,* 139–150.

Hand, H. H. The development of concepts of social interaction: Children's understanding of *nice* and *mean.* (Unpublished doctoral dissertation, University of Denver, 1981.) *Dissertation Abstracts International,* in press. (a)

Hand, H. H. The relation between developmental level and spontaneous behavior: The importance of sampling contexts. In K. W. Fischer (Ed.), *Cognitive development*. San Francisco: Jossey-Bass New Directions for Child Development, No. 12, 1981. (b)

Hand, H. H., & Fischer, K. W. *The development of concepts of intentionality and responsibility in adolescence*. Paper presented at the Sixth Biennial Meeting of the International Society for the Study of Behavioral Development, Toronto, Canada, August 1981.

Hooper, F. H., Sipple, T. S., Goldman, J. A., & Swinton, S. S. A cross-sectional investigation of children's classificatory abilities. *Genetic Psychology Monographs*, 1979, *99*, 41–89.

Jackson, E., Campos, J. J., & Fischer, K. W. The question of decalage between object permanence and person permanence. *Developmental Psychology*, 1978, *14*, 1–10.

Kagan, J. *Psychological research on the human infant: An evaluative summary*. New York: W. T. Grant Foundation, 1982.

Kenny, S. L., & Fischer, K. W. Optimal levels in the development of abstractions in arithmetic. Manuscript in preparation.

Kofsky, E. A scalogram of classificatory development. *Child Development*, 1966, *37*, 191–204.

Krus, D. J. Order analysis: An inferential model of dimensional analysis and scaling. *Educational and Psychological Measurement*, 1977, *37*, 587–601.

McCall, R. B. Exploring developmental transitions in mental performance. In K. W. Fischer (Ed.), *Levels and transitions in children's development*. San Francisco: Jossey-Bass New Directions for Child Development, No. 21, 1983.

McCall, R. B., Hogarty, P. S., & Hurlbut, N. Transitions in infant sensorimotor development and the prediction of childhood IQ. *American Psychologist*, 1972, *27*, 728–748.

McCall, R. B., Eichorn, D. H., & Hogarty, P. S. Transitions in early mental development. *Monographs of the Society for Research in Child Development*, 1977, *42*, (3, Serial No. 171).

Nelson, K. The role of language in infant development. In M. Bornstein & W. Kessen (Eds.), *Psychological development from infancy*. Hillsdale, N.J.: Lawrence Erlbaum, 1979.

O'Brien, D. P., & Overton, W. F. Conditional reasoning and the competence-performance issue: A developmental analysis of a training task. *Journal of Experimental Child Psychology*, 1982, *34*, 274–290.

Piaget, J. *The construction of reality in the child*. (trans. M. Cook). New York: Basic Books, 1954. (Originally published 1937.)

Pinard, A., & Laurendeau, M. "Stage" in Piaget's cognitive-developmental theory: Exegesis of a concept. In D. Elkind & J. H. Flavell (Eds.), *Studies in cognitive development*. New York: Oxford University Press, 1969.

Seibert, J. M., Hogan, A. E., & Mundy, P. C. Mental age and cognitive stage in very young handicapped children. *Intelligence*, in press.

Tabor, L. E., & Kendler, T. S. Testing for developmental continuity or discontinuity: Class inclusion and reversal shifts. *Developmental Review*, 1981, *1*, 330–343.

Tanner, J. M., *Growth at adolescence* (2nd ed.). Oxford: Blackwell Scientific Publications, 1962.

Uzgiris, I. C. Organization of sensorimotor intelligence. In M. Lewis (Ed.), *Origins of intelligence: Infancy and early childhood*. New York: Plenum Press, 1976.

Watson, M. W., & Fischer, K. W. A developmental sequence of agent use in late infancy. *Child Development*, 1977, *48*, 828–836.

Watson, M. W., & Fischer, K. W. Development of social roles in elicited and spontaneous behavior during the preschool years. *Developmental Psychology*, 1980, *16*, 483–494.

Wohlwill, J. F. *The study of behavioral development*. New York: Academic Press, 1973.

Zelazo, P. R., & Leonard, E. The dawn of active thought. In K. W. Fischer (Ed.), *Levels and transitions in children's development*. San Francisco: Jossey-Bass New Directions for Child Development, No. 21, 1983.

CHAPTER 6

Developmental Continuities and Discontinuities in a Form of Familial Dyslexia

Bruce F. Pennington, Shelley D. Smith,
Linda L. McCabe, William J. Kimberling, and
Herbert A. Lubs

This paper is concerned with theoretical and methodological issues which have arisen in our ongoing studies of a particular form of familial dyslexia. This form of dyslexia appears to be related to a locus on Chromosome 15 that we have identified through linkage analysis (Smith, Kimberling, Pennington, & Lubs, 1983). This form of dyslexia allows us to look for developmental continuities and discontinuities at four levels of analysis, as well as to consider the interaction among these levels across development. These levels are I—genetic, II—neurological, III—cognitive, and IV—the surface phenotype of the reading and spelling problems themselves (Pennington, McCabe, Smith, Kimberling, & Lubs, in preparation; Smith, Pennington, McCabe, Kimberling, & Lubs, in preparation).

Bruce F. Pennington • Department of Psychiatry, University of Colorado, Denver, Colorado 80262.   **Shelley D. Smith and William J. Kimberling** • Boys Town Institute for Communication Disorders in Children, Omaha, Nebraska 68131.   **Linda L. McCabe** • Department of Psychology, University of Colorado, Denver, Colorado 80202. **Herbert A. Lubs** • Mailman Center for Child Development, University of Miami Medical Center, Miami, Florida 33101. This research was supported by NICHD grant RO1-MD-12741.

TWO KINDS OF CONTINUITY IN GENETIC FAMILY STUDIES

We would like to use our study as a forum for discussing several issues and problems regarding developmental continuity and discontinuity that arise in a genetic family study of a behavioral trait but which have wider applicability to other developmental studies. The developmental psychologist usually thinks of continuity in terms of an individual or group of individuals across age and within a life span. However, in behavioral genetics we address two other, somewhat different kinds of continuity. These are (1) continuity across generations, which, if the trait has a common etiology, can be analogous to developmental or longitudinal continuity, and (2) *lateral* continuity across affected individuals, all of a given age. The term *lateral* is used here to emphasize the contrast with both developmental and cross-generational continuity, both of which involve continuity of a behavioral characteristic across time. Lateral continuity refers instead to the continuity or consistency of a behavioral characteristic across individuals, all of whom are at the same point in developmental time. It is, concerned, therefore, with the extent of individual differences or phenotypic variability, to put it in genetic terms. We will first discuss cross-generational continuity and its possible relationship with the more familiar concept of developmental continuity. We will then discuss lateral continuity.

A developmentalist doing work in behavior genetics is faced with dilemmas in the case of each kind of continuity. In terms of cross-generational continuity, such an investigator must demonstrate that the behavioral trait in question shows such continuity across generations; otherwise, it is not convincing that the trait is inherited. However, because of his own limited life span, he cannot have access to members of each generation in a family at the same developmental period. Since it is likely that the trait or phenotype in question *changes* with development and is modified by different life experiences, it is difficult to demonstrate that members of different generations have (or have had) the same behavioral trait—that the trait "runs true." For example, both hyperactivity and stuttering show evidence of heritability but also show marked developmental changes in phenotype.

In the case of hyperactivity, two adoption studies (Cantwell, 1972; Morrison & Stewart, 1973) have reported a significantly increased incidence of retrospectively reported hyperactivity in childhood among natural parents of hyperactive probands as compared with adoptive parents of hyperactive probands. Such a finding provides evidence for the heritability of hyperactivity. Moreover, since a low level of retrospectively reported hyperactivity in childhood was found for both adoptive parents of hyperactive children and for parents of normal control children, there is an implication that the parenting environment may be less significant than heritable factors in the development of hyperactivity. However, a number of interpretative issues remain in both studies, including the problem of developmental discontinuity in the phenotype of an affected individual across the life span. For instance, both of these studies found an increased incidence of other, adult psychiatric diagnoses in the biological parents of the hyperactive probands, including alcoholism in both mothers and fathers, hysteria in mothers, and sociopathy in fathers. Another adoption study which found a significant association between hyperactivity in adopted children and alcoholism in biological parents is that of Cadoret and Gath (1978). These findings suggest that there is a significant developmental discontinuity in the phenotypic expression of hyperactivity, such that the putative hyperactivity genotype may manifest itself in qualitatively different maladaptive behaviors in childhood and adulthood. Such a developmental discontinuity would make it difficult to do good family or linkage studies, since, aside from history, it would be difficult to decide which of the adults in a family should be counted as affected. Moreover, since there is evidence for heritability in some of these adult psychiatric disorders (e.g., alcoholism and sociopathy), it may be that the genetic factors involved in these disorders produce a phenocopy of hyperactivity in childhood. Finally, the association between parental (specifically maternal) alcoholism and hyperactivity in the offspring might be due to the effects of alcohol on the fetus, rather than a genetic effect *per se*. Because of all of these problems, as well as difficulties in defining the phenotype in children, genetic studies of hyperactivity face considerable challenges. Identifying subtypes with particular response patterns to stimulant medication could be a useful strategy, however.

In the case of stuttering, Kidd (1977) and colleagues have conducted a systematic investigation of its inheritance, involving the first-degree relatives of 395 adult stutters (see Pennington & Smith, 1983, for a review of this research). Stuttering also shows a significant developmental discontinuity, in that many stutterers show spontaneous recovery by adulthood. Therefore, these investigators' criterion for deciding whether individual relatives carried the trait was whether they had *ever* stuttered for a significant period. This definition of the phenotype allowed them to test several genetic models for the inheritance of stuttering and to rule out simple autosomal dominant or recessive, or X-linked modes of inheritance. They did find a sex difference of 4:1 males to females, however. Their data were compatible with both polygenic and single gene inheritance when environmental modification and differential thresholds for sex are considered. A remaining puzzle in this work is how to account for individual differences in both the severity or persistence of stuttering, since the current data do not suggest a relationship between genotype (in either the polygenic or modified single gene models) and these two important aspects of the phenotype. A possibility that these investigators are considering is that of a two-locus system, which one gene influencing stuttering and the other influencing recovery.

These two examples show how the issue of developmental discontinuity can complicate a genetic study of a behavioral trait, making it difficult to demonstrate cross-generational continuity, which is a prerequisite for considering the trait to be familial and therefore of possible interest for genetic investigation.

It is also important to make clearer at this point the logical relationship between the more familiar concept of developmental continuity and that of cross-generational continuity. The two are logically independent in that each can occur without the other. A nongenetic behavioral trait could persist across the lifespan and thus show developmental continuity; it would not necessarily show cross-generational continuity, however. A presumably genetic trait such as stuttering shows a significant developmental discontinuity, as discussed above, but nonetheless shows cross-generational continuity, in that across generations affected individuals begin to stutter at around the same age.

However, the two kinds of continuity are logically connected in the special case of a genetically based condition in which the *same* phenotype is found among family members of *different* ages and generations. Such a finding obviously provides evidence of cross-generational continuity. What is more subtle is that such a finding provides strongly suggestive evidence of developmental continuity, at least across the portion of the lifespan bounded by the youngest and oldest affected individuals. That is, given the same etiology and the same phenotype, it is reasonable to assume the youngest affected will continue to show that phenotype when he reaches the age of the oldest affected. Thus, in this special case, cross-sectional data provide information about developmental continuity that ordinarily could be obtained only from a longitudinal study. We have belabored this point because, as we will see below, this special case applies to the phenotype analyses in this form of familial dyslexia: There is evidence of a significant degree of developmental continuity in both the cognitive and surface phenotypes from the youngest to oldest affected subjects.

In terms of lateral continuity, the investigator is faced with the likelihood of phenotypic variability due to uncontrolled genetic and environmental variance. This phenotypic variability stands in direct opposition to the important requirement that the phenotype be rigorously defined at some level of analysis so that it can be determined *which* relatives of the proband have the same disorder and which do not.

Phenotypic variability makes it quite likely that some individuals with the abnormal genotype will exhibit a normal phenotype and vice versa.

Sex differences in the behavioral trait in question are a special case of lateral discontinuity. Let us take the case of a behavioral trait the inheritance of which has been shown to be consistent with major gene inheritance, which is true of the type of familial dyslexia discussed in this paper and stuttering, discussed above. Besides the abnormal genotype, these behavioral traits would also be expected to be influenced by the individual's overall speech and language abilities. These are, in turn, likely to be determined polygenically in interaction with diverse environmental influences. Therefore there can be a polygenic background which helps to determine an individual's overall verbal abilities. An abnormal major gene for either fluency or reading ability would quite likely interact with this polygenic background. If there are sex differences in this polygenic background, as there appears to be for speech and language abilities, then, as a group, males and females with the same abnormal genotype (for either stuttering or reading disability) would be expected to show phenotypic differences, even though the abnormal gene in question is on an autosome and not X-linked. Figure 1 depicts sex differences in the polygenic background for verbal abilities and how those would result in a preponderance of males falling below the threshold value of verbal skills necessary for normal reading development. If males are already at greater risk for reading disability (or stuttering), then the imposition of an abnormal gene should result in a greater degree of severity in males, as well as a somewhat greater number of males being affected. In contrast, females would be somewhat shielded against the effects of such an abnormal gene. The sex differences in the stuttering research discussed above, as well as the sex differences in our own study, are consistent with this hypothesis.

In our study, we have found a sex ratio of males to females of 1.89 to 1 in this presumably autosomal dominant form of familial dyslexia, which appears to reflect the kind of gene–genome interaction described

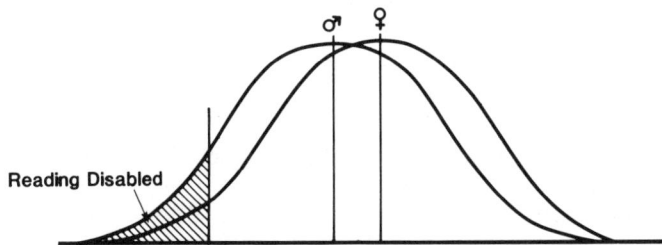

Figure 1. Distribution of verbal abilities by sex.

above. We have also found that affected females generally showed less severe reading and spelling problems than affected males. Additionally, in using the various definitions of the phenotype discussed below, we have found that the false negatives (i.e., adults with a clear history of reading disability in childhood who are normal according to the definition of the phenotype) are invariably females. Thus, for these recovered females there has been a developmental discontinuity, which suggests the role genetic factors can play in developmental changes across the lifespan (Plomin, 1983).

A final consideration with respect to both cross-generational and lateral continuity is that each may vary across the three different levels of phenotype analysis considered here: neurological, cognitive, and surface. Differential environmental experiences, including remedial teaching, may impact differently on each level, thereby creating *intraindividual* phenotypic discontinuity. For example, an individual with the abnormal genotype may be similar to other affected subjects neurophysiologically but may appear normal on tests at both the cognitive and surface levels.

In what follows, we will review the methods used and the evidence we have found for these kinds of continuity and discontinuity at the genetic, cognitive, and surface levels. Important methodological issues will be discussed as they arise. This will be followed by a review of what is known about the neurological and neurophysiological basis of dyslexia and our proposed strategy for identifying the neurophysiological deficit in this population.

GENETIC METHODS AND RESULTS (LEVEL I): THE USE OF LINKAGE ANALYSIS

Quite a number of studies spanning diverse behavioral genetic methodologies (including pedigree and twin studies, as well as more sophisticated segregation analyses) have strongly suggested both that dyslexia can be transmitted genetically and that the genetic mode of transmission is heterogeneous (see Finucci, 1978; Herschel, 1978; and Pennington & Smith, 1983, for reviews of genetic studies of dyslexia). There is evidence for polygenic inheritance, autosomal dominant inheritance, and, in females, autosomal recessive inheritance. Thus the existing literature suggests etiologic heterogeneity not only for dyslexia as a whole (since it is generally accepted that some dyslexias are non-familial), but also *within* the class of familial dyslexia. We may describe this as another type of discontinuity (i.e., *etiologic discontinuity*) which must be dealt with in the investigation of dyslexia. Given this genetic heterogeneity, it would be important to identify a subsample of affected families whose pedigrees are all consistent with a particular type of inheritance and then attempt to ascertain more directly what the genetic mechanism is in that particular genetic subsample.

This is the strategy that was followed in the present project. Specifically, the assumption was made that one form of familial dyslexia is inherited in an autosomal dominant fashion, since the literature suggested that such a genetic subtype existed and since the hypothesis of autosomal dominance could be directly tested by a genetic linkage analysis.

Genetic linkage is the deviation from Mendel's law of independent assortment of genes and results when two genes are located close together on the same chromosome such that crossing over between them during meiosis is reduced. Thus *linkage* here refers quite literally to two genes' being physically linked because they are close together on the same chromosome. Two genes which are not linked are constantly reshuffled. The phenomenon of linkage provides a method of localizing a gene, that is, by seeing if it is linked to a gene the locus of which is known. Linkage is measured by the "lod score," or logarithm of odds of likelihood that a given deviation from random assortment of two genes could be found fortuitously. Classical linkage analysis involved two traits both known to be genetic. More recently, linkage analysis has been applied to behavioral traits with evidence of heritability such as schizophrenia and alcoholism (e.g., Elston, Kringlen, & Namboodiri, 1973; Winokur, Tanna, Elston, & Go, 1976). A linkage analysis is only appropriate for a trait expected to be due to major gene inheritance; a polygenic trait obviously would not give evidence of linkage.

In the Genetics of Specific Dyslexia Project (Smith, Kimberling, Pennington, & Lubs, 1983), eight three-generation kindreds with pedigrees consistent with autosomal dominant inheritance of dyslexia were selected for study (Figure 2). The linkage analysis employed 23 genotyping markers (i.e., variant gene products detectable by biochemical tests) and Q and C banding chromosomal heteromorphisms (i.e., different patterns of light and dark bands on chromosomes which are visualizable through a light microscope and which are transmitted genetically). These markers and heteromorphisms are each specific for particular portions of particular chromosomes and as a group potentially cover all 22 pairs of autosomes in the human genome (see Figure 3). The results of the linkage analysis were that all loci tested gave lod scores less than 1.0 (i.e., $p > .10$) except for Chromosome 15 short-arm banding heteromorphisms, which gave a lod score greater than 3.0 (i.e., $p < .001$). Thus, the linkage analysis provided suggestive evidence for an autosomal dominant locus on Chromosome 15 responsible for one subtype

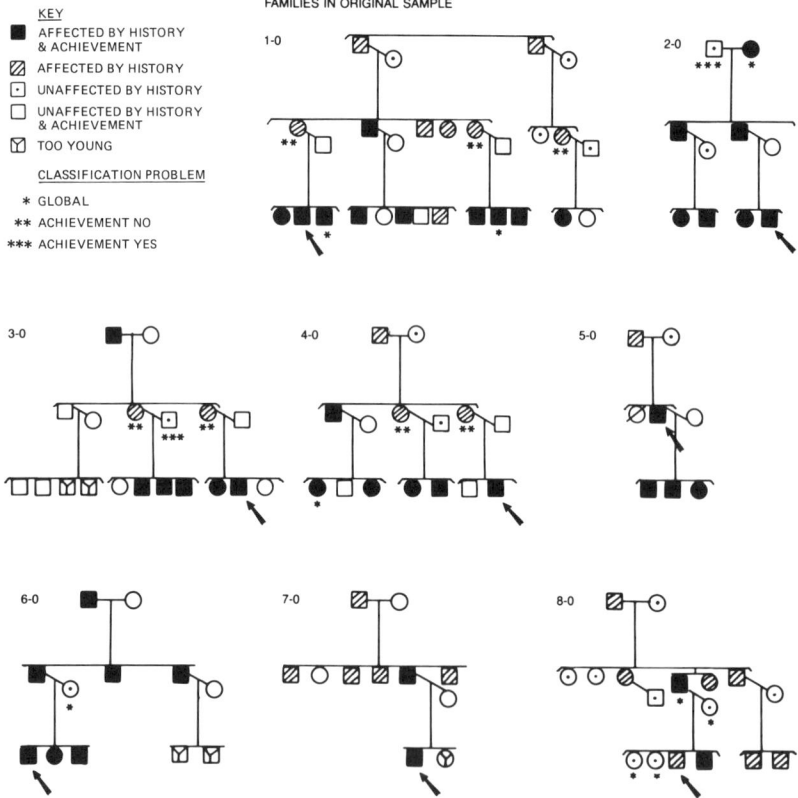

Figure 2. Pedigrees of dyslexic kindreds.

of familial dyslexia. This result provides support for Hallgren's (1950) theory of autosomal dominant transmission.

Thus this sample of kindreds gives evidence of genetic homogeneity, or continuity at the etiologic level. It is obviously important to replicate this finding in a second sample of kindreds, and this work is in progress. It is also important to point out that the finding of a positive linkage did *not* depend on finding an abnormal marker which was present in all affected subjects. Rather, using chromosomal markers (i.e., light and dark banding patterns) which show several variations in the normal population, we found an association between a pattern of markers and dyslexia *within* kindreds, whereas the particular pattern found varied *across* kindreds. Thus it is quite unlikely that the markers themselves are involved in the transmission of dyslexia; rather, they appear

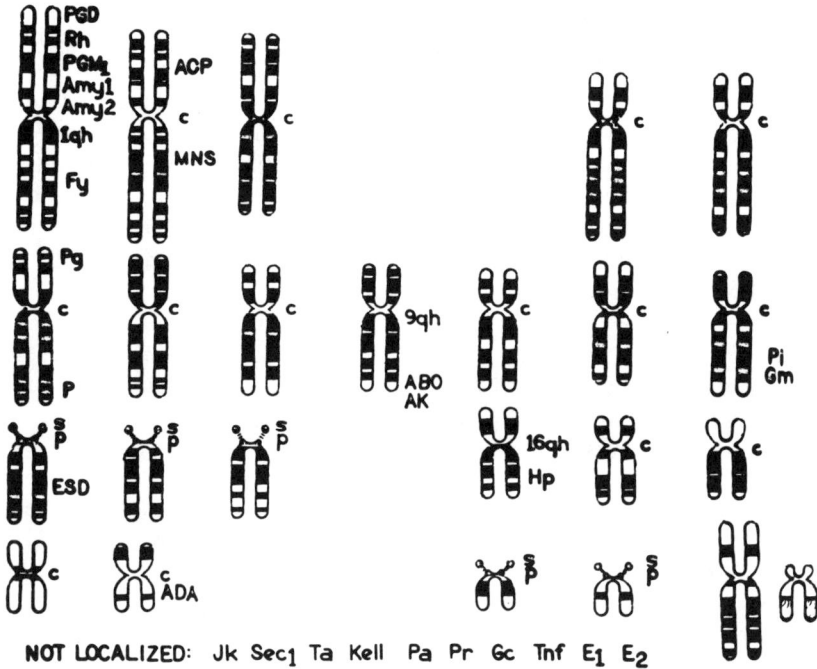

Figure 3. Linkage analysis with dyslexia.

to be linked to a yet unidentified gene or group of genes on Chromosome 15 involved in the transmission of this form of familial dyslexia.

Genetic continuity is a profound thing for a developmentalist to contemplate, involved as he usually is in the search for much more ephemeral continuities. A gene or closely linked group of genes is truly continuous across generations, except for the rare events of either a spontaneous mutation or a crossover and recombination within the closely linked group of genes. In the case of this kind of continuity, one is prompted to realize that the appropriate unit of analysis is not the individual or groups of individuals all of a given age, but kindreds.

Since we do not have data regarding Level II—the neurological level—we will next discuss Level III and return later to a discussion of possible neurological deficits in developmental dyslexia.

COGNITIVE METHODS AND RESULTS (LEVEL III)

Our initial investigation of the cognitive phenotype was motivated by a desire to see whether our familial dyslexics matched any of the dyslexia subtypes described by previous investigators and, if so, whether there was consistency of subtype within and across families. The work on subtypes done by Mattis, French, and Rapin (1975) seemed most promising to us at that time. Therefore, we composed a subtypes battery (see Appendix) which includes, with some modifications. the tests used by Mattis and colleagues (1975) to identify subtypes of dyslexics. This battery also includes the Wechsler Intelligence Scales. Mattis' battery identifies distinct neuropsychological syndromes associated with three different subtypes of dyslexia, including a visuospatial subtype and two language-related subtypes. Camp's modification (Camp & McCabe, 1978) of Boder's (1973) test, was also given, which identifies subtypes through patterns of spelling errors; these results will be discussed in the following section since they pertain to the surface phenotype.

This battery was given to 43 affected and 25 unaffected subjects, which is a subset of the total number of subjects in the sample ($N = 63$ affected and 41 unaffected). These were the subjects available for this more in-depth testing; there were no systematic biases in the selection of this subset.

The analysis of the cognitive phenotype is presented in detail in Smith, Pennington, McCabe, Kimberling, and Lubs (in preparation); the main results and methodological issues are discussed here.

Few affected children or adults fell into one of the three Mattis *et al.* (1975) subtypes. Most were classified as normal. Moreover, a chi-square analysis showed that those affected were not significantly more likely to fall into any of the Mattis subtypes than were those unaffected. This held true when children were analyzed separately; thus the overall failure of the Mattis subtypes to describe this population is not attributable to the large number of adult dyslexics in our sample. It is of some interest that for those few affected who did fall into a Mattis subtype, all fell into the Language Disorder subtype. There were few diagnosis main effects in an analysis of variance of all the tests in the Mattis battery, suggesting that these dyslexics have fewer neuropsychological deficits than most of the subtypes identified clinically. The only investigators who appear to have identified a subtype similar to this one are Hughes and Denckla (1978), who proposed a dyslexia-pure subtype; the Colorado Family Reading Study (Decker & DeFries, 1981), who found a more specifically reading disabled subtype; and Satz and Morris (1980), who identified an "unexpected" subtype. When age differences were covaried out, significant diagnosis main effects were found only for a test of word retrieval (Boston Naming Test), auditory discrimination of words in a noise background (Goldman–Fristoe–Woodcock Test of Auditory Discrimination: Noise), and auditory verbal short-term memory (Wechsler Digit Span). These results suggest a subtle language-processing deficit involving either lexical retrieval, phonetic awareness and segmentation skills or phonological memory, or some combination of these.

As for Verbal, Performance, and Full scale IQ, the affected and unaffected subjects did not differ when age was covaried out. There was a significant difference in mean Verbal IQ between affected and unaffected before age was covaried out. Since there were a considerably larger number of affected as opposed to unaffected children and adolescents, this result may indicate that the language-processing deficit is more pronounced in younger dyslexics and lessens in adulthood.

Cross-Generational Continuity

Evidence for this kind of continuity is provided by the analysis of covariance which shows that poorer performance on these few tests is independent of age and therefore holds across generations. Of course, this is not strong evidence for continuity since a significant difference in means may have little power at the level of individual diagnosis. In order to consider individual differences, a multiple discriminant analysis was also performed, using a history of early reading and spelling problems as the independent criterion for diagnosis of an individual as affected or unaffected. (History information was obtained in a detailed interview with families.) The multiple discriminant function correctly identified 90.5% of the affected and 73.9% of the unaffected, for an overall correct classification rate of 84.6%. The variables in the discriminant function, in order, were Auditory Discrimination: Noise; Freedom from Distractibility (FD) factor score; Perceptual Organization (PO) factor score; Grooved Pegboard—nondominant hand; and the Benton Visual Retention Test. (The two factor scores, FD and PO, are simply the means of subtests which have clustered together on previous factor analytic studies of the Wechsler IQ Scales. FD equals the mean of Digit Span and Arithmetic; PO equals the mean of Block Design and Object Assembly.) The high correct classification rate for affected subjects provides reasonably strong evidence for cross-generational continuity in the cognitive phenotype. In further describing what this cognitive phenotype is, we must point out that the weighting of the PO and Visual Retention variables in the discriminant function was *opposite* in sign to that of the first two. This means that *good* scores on these variables were characteristic of the dyslexics.

This latter result is quite interesting neuropsychologically because it provides evidence for a suggestion made by Symmes and Rappaport (1972) and by Denckla (1979); namely, that in specific dyslexics and their families there is evidence for good or superior right-hemisphere spatial abilities. This raises the intriguing possibility that the genetic mechanism which causes the presumably left-hemisphere problems with reading and spelling also causes enhanced right-hemisphere development. Since the posterior brain areas which are primarily involved in reading and spelling in the left hemisphere are roughly homologous to those involved in spatial abilities in the right hemisphere, there is the possibility in this population either that both of these areas are under common genetic control or that maldevelopment in left posterior areas somehow leads to overdevelopment of right posterior areas. Other aspects of the possible neurological basis of dyslexia are discussed in greater detail in a subsequent section.

Lateral Continuity

Even though the results of the discriminant function analysis demonstrate lateral continuity in the cognitive phenotype, there were definitely individual differences among affected subjects, even those of similar age and education level. For instance, the fact that some adult and child dyslexics fall into Mattis's Language Disorder subtype is evidence for a greater degree of language-processing difficulty in these subjects. Whether there is a definite *qualitative* discontinuity in the nature of the cognitive deficit among the affecteds cannot be definitely determined with these tests for reasons discussed below.

Discussion

These results do suggest that an underlying problem exists in either phonological memory, phonetic awareness and segmentation skills, or lexical retrieval, or some combination of these. Evidence for deficits in each of these areas has been found by other investigators of reading disability. Specifically, I. Liberman and associates (Shankweiler & Liberman, 1976; Mann, 1981) have conducted a number of studies which show that poor readers rely *less* on phonological memory than good readers for short-term recall of letters, words, and sentences, whether presented visually or aurally. They interpret this result as reflecting a general deficiency in phonological memory in poor readers. In terms of phonetic awareness and segmentation skills, a number of studies have shown that problems in these areas are highly predictive of problems in early reading development (e.g., Liberman, Shankweiler, Liberman, Fowler, & Fischer, 1977; Wallach & Wallach, 1976, 1979). Finally, in terms of lexical retrieval, Denckla and Rudel (1976a,b) found that dyslexics make more errors and showed longer latencies on a variety of word-retrieval tasks than either normal or nondyslexic MBD controls.

Paralleling these experimental reports are the clinical observations of Denckla (1979), who reports on two subtypes of dyslexia (among her six subtypes) which are characterized by difficulties in one or more of these three areas. Specifically, she discusses an anomic-repetition disorder, characterized by excessive naming errors and a poor ability to repeat sentences and digits in the absence of either articulation or verbal comprehension problems, or a depressed verbal IQ. Her second subtype is a dysphonemic-sequencing disorder, characterized by omissions, substitutions and sequencing errors in sentence and digit repetition, and misunderstanding of complex syntactical constructions. Individuals in the second subtype do not make excessive naming errors, she reports, but the errors they do make involve phonemic or sequential problems. Again there are no difficulties in articulation or verbal IQ. However, Denckla does not present data on how reliable the discrimination between these two subtypes is or how frequently they overlap.

Determining the relative contribution of problems in phonological memory, lexical retrieval, or phonetic awareness and segmentation skills to the cognitive phenotype in our population is complicated by the question of whether these are discrete, separable processes in normal language processing and by the certainty that, even if discrete, these component processes undoubtedly interact. The clinical tests used by us and other dyslexia investigators to identify problems in one or more of these linguistic processes, in fact, confound all three processes. Thus, a problem in phonetic awareness and segmentation skills could affect the efficiency of both auditory verbal short-term memory (measured by sentence and digit span tasks) and word-retrieval (measured by naming tasks) since a phonologic representation appears to be a preferred and efficient mode of holding verbal information in short-term memory and since words to be retrieved from the lexicon (an individual's internal "dictionary") have a phonologic "address", among others. Similarly, a deficit in phonological memory span would limit both phonetic segmentation skills and word-retrieval since both of these processes require a certain amount of space in working memory. Finally, a problem in lexical access or retrieval would make phonemic codes and other linguistic information stored in long-term memory less automatically available for utilization in phonemic segmentation and phonological memory. These possibilities could be disambiguated experimentally but there is little research to date which has attempted to do so.

We are planning experimental cognitive studies to separate these possibilities more clearly.

METHODS AND RESULTS PERTAINING TO THE SURFACE PHENOTYPE (LEVEL IV)

An initial battery (see Appendix) of tests was given to all subjects (N = 63 affected, N = 41 unaffected), and these results have been used to clarify the surface phenotype. This was done in both a quantitative and qualitative way. The former is discussed first.

Subjects were initially classified as affected or unaffected on the basis of history. As can be seen in Figure 4, the affected were, not surprisingly, significantly worse than their unaffected relatives on achievement tests of reading and spelling. Moreover, this figure shows a profile of achievement results which has proved to be almost diagnostic for specific dyslexia, both in this population and in a large clinical sample of learning-disabled individuals. That is, the Peabody Individual Achievement Test (PIAT) Mathematics and General Information scores are at the norm or higher, whereas Reading Recognition (i.e., reading single words aloud) and both recognition and recall tests of spelling (i.e., the PIAT and Wide Range Achievement Test—WRAT—spelling tests respectively) are markedly depressed. In addition, Reading Comprehension is usually better than oral reading or spelling and may even be at the norm. The explanation of this profile is that specific dyslexia involves mainly a deficiency in the use of phonemic and orthographic codes for decoding and encoding words, rather than a more generalized language-processing or conceptual deficit. People with average or better conceptual skills would be expected to learn mathematics and general information and would also be able to rely on several strategies to extract

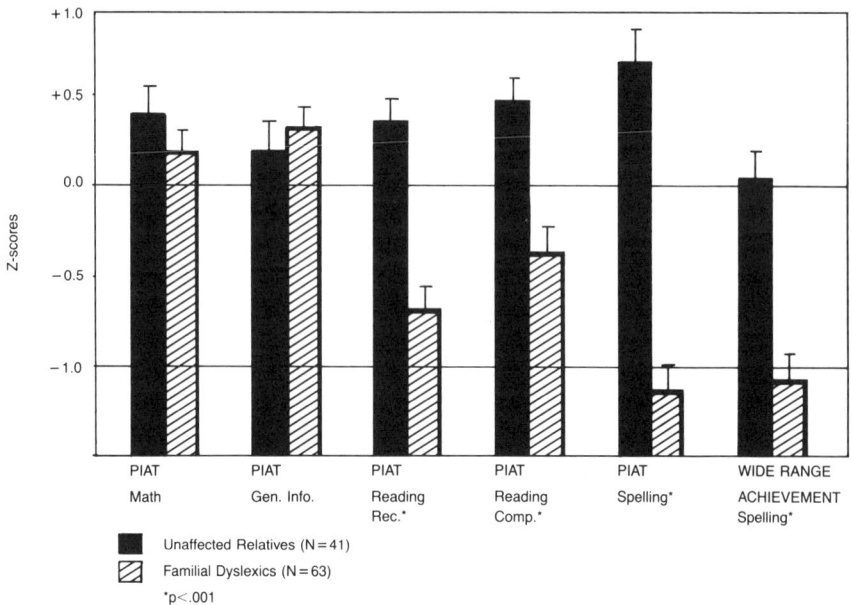

Figure 4. Achievement test scores ($\overline{X}$ + 1 SEM).

meaning from text, even if they were unable to decode completely every word into its full phonetic representation.

We operationalized this profile into a diagnostic algorithm and applied it to all the subjects in the sample. The algorithm is as follows:

1. Mathematics or General Information age Z score is $\geq -.5$ and $\geq$ all the oral reading and spelling scores.
2. The age Z score difference between Mathematics or General Information and any one (or more) of these tests—PIAT, Reading Recognition or Spelling, WRAT Spelling, or Gray Oral Reading—is $\geq +1.0$.
3. The age Z score for Reading Comprehension is $\geq$ the age Z score for either oral reading (either PIAT Reading Recognition or Gray Oral) or spelling (either PIAT or WRAT Spelling).
4. One oral reading or spelling test age Z score must be $\leq .5$, unless the discrepancy found for Criterion 2 is ≥ 2.0.

We found that the algorithm identified 88% of those subjects affected by history and rejected 93% of those unaffected by history, for an overall concordance of 86%. There were eight false negatives (i.e., affected by history but normal on the algorithm), all of whom were adult females. This offers further evidence for a sex difference in expressivity in this form of familial dyslexia, which we have discussed above. The percent for true positives includes eight subjects affected by history who failed to meet the criteria for *specific* dyslexia because both their Math and General Information scores were also depressed, indicating that they had more widespread language and learning disorders. However, for most of these subjects reading and spelling were the weakest areas. These nonspecific dyslexics did not appear restricted to a particular developmental period but were clustered in a few families. They also all fell into the Language Disorder subtype described by Mattis *et al.* (1975).

A more sensitive measure of the surface phenotype is provided by our *qualitative* analyses of the spelling errors made by the affected subjects. This was explored using three different category systems for classifying spelling errors. The first of these was the Denver Reading and Spelling Test (DRST), which is the Camp–McCabe adaption of Boder's (1973) system for identifying subtypes of disabled readers. Boder's typology identified distinct types of dyslexic spellers, including a *dyseidetic* type, who have a poor visual memory for the exact spelling of words, and a *dysphonetic* type, who have a poor ability to generate phonetic equivalent spellings. Boder and others (e.g., Sweeney & Rourke, 1978) have related these two types of disabled spellers to different kinds of hemisphere dysfunction.

Unfortunately, our ability to test Boder's typology in this population was limited because through most of the data collection we had only a form of the DRST which went up to the eighth-grade level. The subsample who were validly tested with the DRST included 24 dyslexics and 4 unaffected subjects, classified as such by *both* history and the achievement algorithm described above. Sixteen of the dyslexics fell into the *dyseidetic* category, none into the *dysphonetic* or *mixed* categories, and eight into the normal category. All four unaffecteds were classified as normal. Thus, in this subsample, the Boder system is not very sensitive (i.e., 67% correct classification rate) in identifying these dyslexics. It also tended to be even less sensitive with older dyslexics. Moreover, when the performance of these subjects was compared on the WRAT spelling error categories described below, the analysis of variance found few significant differences among the three groups. Specifically, those dyslexics classified as *dyseidetic* made significantly fewer phonetic errors on the WRAT Spelling Test. This would appear to contradict their classification as *dyseidetic*, since this term is used by Boder to describe dyslexics who have a poor visual memory for the exact spelling of words but have adequate phonetic skills for generating approximate spellings.

For the other two qualitative analyses of spelling, we used the errors on the WRAT Spelling Test. The range of spelling difficulty was wide enough on this test so that there did not appear to be a problem with floor and ceiling effects. Thus we were able to obtain an adequately large corpus of errors from nearly all subjects, both normal and dyslexic.

The rationale for the actual error categories in each coding system is as follows: The first coding system (Coding System I; see Table 1) dealt mainly with phonetic accuracy. We examined the consonant structure of the whole word as to whether it had all the consonant sounds of the target word in the correct order. If so, it was scored either as PE, phonet-

Table 1. Spelling Error Types: Coding System I

Phonetic equivalent (PE): conchans, angezity
Orthographically illegal phonetic equivalent (OIPE): canshuns, angziaty
Phonetic sequencing error (P-Seq): anxeity, medevial

Addition (A): consciosence, anxitet
Substitution (Sub): conchunch
Omission (O): consciene, anzaty
Bizarre (Biz): chios, angx
Nonphonetic sequencing error (NP-Seq): kichten

Other
Letter reversals: orber, gress
Refusals

ic equivalent, an OIPE, orthographically illegal ponetic equivalent, or a P-Seq, phonetic sequencing error. If the error did not have all the consonant sounds in the correct order, it was scored into one of the other categories, reflecting whether a consonant sound was omitted (O), added (A), substituted (Sub), or missequenced, (NP-Seq). We also scored for bizarre errors and letter reversals. This system did not take into account the variations in orthographic structure of individual words and addressed orthographic rules in only a limited way (i.e., through the orthographically illegal categories). This system is similar to that used by several spelling investigators, including Camp and McCabe (1978) and those in Frith (1980).

The second system (Coding System II; see Table 2) dealt with orthographic regularities and was motivated by the fact that much of spelling development involves more than the learning of simple correspondences between phonemes and graphemes. In fact, the correct spelling of most words depends on orthographic rules and regularities which cannot be predicted from such simple correspondences. These issues are discussed in detail by Gleitman and Rozin (1977) and Shankweiler and Liberman (1976). In our second scoring system, we characterized individual words on the WRAT according to the different kinds of spelling rules and regularities they embodied. The rules and regularities we identified are based on the work of Venezky (1970), Marsh, Friedman, Welch, and Desberg (1980), and others. *Phonetic-sequential* refers to words which can be spelled using simple phonetic-sequential rules (e.g., *majority*). *CVCe* refers to consonant-vowel-consonant-silent *e* words which involve this hierarchical spelling rule (e.g., educ*ate*). The

Table 2. Spelling Error Types: Coding System II

Phonetic sequential	CVCe	High-frequency vowel clusters	Low-frequency vowel clusters	Single consonant alternations
run	educ*ate*	tr*ai*n	prec*ious*	*c*ircle
arm	instit*ute*	sh*ou*t	loqua*cious*	ne*c*essity
majority	exagger*ate*	bel*ie*ve		mu*s*eum
				re*s*ilient

Double consonants	Consonant clusters	Analogy words	Exception words
su*gg*estion	*sh*out	fa*sh*ion	ruin
posse*ss*ion	*ch*arlatan	sugge*st*ion	anxiety
pusi*ll*animous	iride*sc*ence	*phy*sician	conscience
		*cour*teous	

next two categories, *high-frequency* and *low-frequency vowel clusters,* include words which have vowel clusters that are pronounced as one phoneme (e.g., tr*ai*n). The next category is *single-consonant alternations,* that is, words which have consonants pronounced differently from their most frequent pronunciation (e.g., ne*c*essity). *Double (or geminate) consonants* refers to words with doubling of the same consonant (e.g., su*gg*estion). *Consonant clusters* refers to words with a group of two or more consonants which do not cross a syllable boundary and thus are sounded together (e.g., *sh*out). *Analogy words* contain letter groups the motivation of which is morphological and which are irregular in a phonetic sense but for which there is a family resemblance or analogy to other words. Finally, *exception words* contain letter groups with a sound correspondence to few, if any, other English words and are not regular in either a phonetic or analogy sense. Errors were scored for the *part* of the word embodying the particular rule or regularity in question and were divided by the total number of words embodying that rule that the individual misspelled. For *phonetic-sequential* and *exception* words, any misspelling of the word counted as an error.

In order to delineate more clearly the spelling phenotype of subjects who are both specifically and unequivocally dyslexic, we initially omitted both the eight global subjects, the eight females affected by history only, and the four subjects who were normal by history but specifically dyslexic according to the achievement algorithm Thus we compared subjects who were affected according to *both* history and the algorithm with subjects who were unaffected by both criteria.

The results of the analyses of variance for both coding systems were as follows: Significant mean differences were found for the phonetic equivalents and omissions categories in Coding System I and for all the Coding System II categories except double consonants and analogy words. Although both coding systems only considered errors, and thus presumably were unconfounded by a subject's quantitative performance on the WRAT, we nonetheless performed an analysis of covariance on these error categories using WRAT Spelling score as a covariant. We found that affected subjects were significantly different from the unaffecteds on Subs ($p < .05$) and nearly so on PEs ($p = .12$) among the Coding System I types and only on single consonant alternations ($p < .01$) among the Coding System II error types. Thus the significant differences found for omission, consonant cluster, and vowel clusters on the ANOVA appear to be explained by level of spelling ability rather than by a clear qualitative difference in spelling strategy between the two groups.

These results show that these dyslexics have difficulty with *both* simple phonetic correspondences and more subtle orthographic rules. Interestingly, on the two types of spelling pattern which would seem to depend most heavily on visual memory, double consonants and analogy words, the affected subjects do not differ from their unaffected relatives. This reinforces our earlier conclusion that, despite the classification of these individuals as dyseidetic in Boder's system, visual memory problems do not appear to be at the root of their spelling difficulty.

We also did discriminant function analyses on both the Coding System I and Coding System II variables. The Coding System I variables alone correctly classified 71% of the affected and 92% of the unaffected, for an overall correct classification rate of 75%. The Coding System II variables were considerably more powerful in identifying affected subjects, attaining a correct classification rate of 91% for them, 82% for the unaffected, and 88% overall. This 20% increase in sensitivity confirms our hypothesis that when these dyslexics produce a phonetic equivalent it frequently still differs from a normal spelling error in terms of its sensitivity to orthographic rules and regularities.

Cross-Generational Continuity

At a quantitative level, the success of the classification algorithm gives evidence for cross-generational continuity in the surface phenotype and demonstrates that this aspect of the phenotype runs true across generations. This, of course, does not mean that there are not developmental changes in reading and spelling ability in specific dyslexics, both of a quantitative and of a qualitative nature. It means only that the overall *pattern* of basic academic skills remains relatively invariant across ages.

The multiple discriminant analyses of the spelling error types also provide evidence of overall *qualitative* continuity across generations in the spelling phenotype in this form of familial dyslexia. There were some developmental changes in spelling phenotype which were explored using analyses of covariance. Some of the error categories gave evidence of age- and education-related effects and others clearly did not, and this varied somewhat between affected subjects and unaffected. In Coding System I, neither phonetic equivalent nor omission errors showed developmental effects for either group. This indicates that most of the development in the ability to use simple letter–sound correspondences has already occurred before the age of the youngest subjects in this sample (i.e., 8 to 9 years). For omission errors, which were twice as common among the dyslexics, this suggests that the cognitive processing error underlying these errors is fairly constant across development. One can readily imagine the phonological memory problems discussed above causing the dyslexics to make omission errors. Some of the rarer errors, nonphonetic sequencing errors, letter reversals, and orthographically illegal combination errors, almost never occurred in normal spellers and preponderantly occurred among younger dyslexics. Thus these more unusual (and bizarre) errors show a strong developmental effect in dyslexia, amounting to a significant developmental discontinuity.

For Coding System II, the analysis of covariance showed that there were no developmental effects for single-consonant rules and analogy words in either affected or unaffected subjects. In the case of single-consonant rules, this would appear to mean that these rules present a persistent problem for dyslexics across development and that for normal individuals a certain level of mastery of these rules is reached early in spelling development. Analogy words are probably learned by both groups in an all-or-none fashion. Thus, if the individual misspells the word, he is likely to misspell the analogy portion of it. Since our measure considers errors only, it could be insensitive to developmental effects for this kind of word. Developmental effects were found for vowel clusters, consonant clusters, and double consonants in both groups. This would indicate that spelling development continues into adulthood for both groups for these more complex spelling patterns.

Lateral Continuity

Both the quantitative and qualitative data provide fairly strong evidence for lateral continuity as well, because of the high success rate both for the classification algorithm and the discriminant function analyses of spelling errors. The major discontinuity was the finding of eight affected subjects with more global depression on the PIAT. Since the qualitative analyses considered only subjects affected both by history and the algorithm, we do not know how much lateral discontinuity would be found in the omitted subjects.

THE NEUROLOGICAL BASIS OF DEVELOPMENTAL DYSLEXIA (LEVEL II)

We have taken no neurophysiological measures so far on our sample of familial dyslexics and thus know by far the least about them at this level of analysis. Consequently, this section will focus on a brief review of neurological theories of dyslexia and then present our hypotheses regarding the localization of the neurological dysfunction in this particular population.

At the time of its discovery by Hinshelwood (1896) and others at the turn of the century, developmental dyslexia was hypothesized to involve the posterior left hemisphere, especially the angular gyrus. This theory was based on an analogy with acquired alexias in adults. Well-known examples were Déjerine's (1891, 1892) classic cases of patients with alexia with and without agraphia. In the first case, involving alexia with agraphia, the patient could neither read nor write spontaneously but could speak and understand spoken language and, was not, therefore, aphasic in the usual sense. Postmortem examination revealed a lesion in the angular gyrus. It thus appeared that the angular gyrus had a special role in reading and spelling. In particular, it appears that the angular gyrus serves as a sort of symmetrical, intersensory transducer, which coverts the visual image of letters and words into some kind of auditory image, and vice versa. Thus damage to the angular gyrus would impair reading and spelling but leave other language functions intact.

In Déjerne's (1892) second case, involving alexia without agraphia, the patient similarly could not read, but he could write spontaneously. Postmortem neuropathological analysis revealed what was essentially a disconnection syndrome. There was damage to both the left visual cortex and the portion of the corpus callosum known as the splenium, which connects the left and right hemisphere visual areas. Thus the angular gyrus and other language areas were isolated from visual input, and the visual images of words could therefore not be converted into their linguistic equivalents. However, the patient could still write spontaneously because the angular gyrus was intact.

These two basic types of acquired alexias have been confirmed by numerous subsequent investigators. (See Albert, 1979, for a review.) It is of interest that some degree of anomia is a frequent correlate of both forms of acquired alexia.

The syndromes of acquired alexia provide strong evidence for a distinct, neuropsychological system subserving reading and spelling in the adult brain. This is significant because the phenomenon of specific developmental dyslexia, including the results presented here, provides evidence for a similar, distinct neuropsychological system in the developing brain. Additional developmental evidence is provided by the rare syndrome of hyperlexia (Huttenlocher & Huttenlocher, 1973), which is usually an associated symptom in some cases of infantile autism and other serious developmental disorders. Hyperlexic children spontaneously learn to read before age five and have excellent phonic skills, in the absence of either reading comprehension or normal overall language or cognitive functions. Thus hyperlexia is the neuropsychological opposite of dyslexia; both point to a brain system that is distinct even in the preschool period and which is specialized for reading.

Despite the evidence discussed above, the neurological basis of developmental dyslexia remains confusing and controversial. This is a result both of the likely heterogeneity of the disorder and the nearly total lack of reliable neurological data from developmental dyslexics, in contrast to the numerous careful neurological case studies of acquired alexics. There have been two neuropathological studies of youths described as dyslexic (Drake, 1968; Galaburda & Kemper, 1979). Both had clear histories of reading and spelling problems, positive family histories for similar problems, some speech milestone delays, some problems with arithmetic, and a history of seizures. The latter three characteristics make them unlike the dyslexics in the present sample. On autopsy, Drake's case had abnormal gyri in the parietal areas bilaterally, a thinning of related areas of the corpus callosum, and ectopic cells in the underlying white matter. The case reported by Galaburda and Kemper had abnormalities only in the cortex of the left hemisphere, mainly an "area of micorpolygyria involving the posterior part of the planum temporale and Heschl's gyrus and the immediately adjacent parietal operculum" (p. 263). Thus the dysplasia involved an area roughly adjacent and anterior to the angular gyrus and definitely involved part of the classical language areas. The other evidence for a structural neurological abnormality in dyslexia comes from the CT scan study of Hier, LeMay, Rosenberger, and Perlo (1978), who found a reversal of the usual parieto-occipital asymmetry (which normally favors the left hemisphere) in 10 of 24 dyslexics. This reversal of asymmetry was associated with a significantly lower Verbal IQ relative to the dyslexics with a normal asymmetry. Taken together, these neuropathological and neuroradiological results provide evidence for an anatomical abnormality in *some* dyslexics, especially those with evidence of more widespread language-processing disorders.

Since the time of Hinshelwood, there have been numerous other neurological theories of developmental dyslexia. Orton (1937) believed that dyslexia represented a failure in the development of left hemisphere dominance. Witelson (1977) proposed that dyslexics have, to use her title, "two right hemispheres and none left" and that they show a preference for processing most information in the right hemisphere. More recently, some researchers (e.g., Gross, Rothenberg, Schottenfeld, & Drake, 1978) have proposed deficient development of the corpus callosum, which would prevent integration of visuospatial information in the right hemisphere with auditory linguistic information in the left. This theory faces the immediate criticism that most children with congenital absence of the corpus callosum do not have language or reading deficits, not do adults with split brains. The proponents of this view would argue that a weak or inefficient callosum is worse than an absent one, since both hemispheres would still be trying to work together but would be doing it poorly. As can be seen from this cursory summary, all these theories can find at least partial justification from what is known about acquired alexias.

An adequate neurological theory for the basis of dyslexia faces a number of issues:

1. It must address the issue of the extent to which reading is dependent upon the speech and language system in the brain, both developmentally and in maturity. Is silent reading pure visual information processing that does not involve internal speech at some level or another? Developmentally, there is little question that reading is dependent upon the speech and language system. However, the puzzle in specific dyslexia is that it is not associated with a definite dysphasia. Therefore, a related requirement for an adequate theory of developmental dyslexia is that the neurological dysfunction must be restricted enough to produce specific reading and spelling problems. Dysfunction in the angular gyrus would appear to satisfy these requirements.

2. It must somehow explain a major developmental *discontinuity* in specific dyslexia—that it has no well-established developmental precursors in the preschool period. Specific dyslexic children, for the most part, appear to be normal in infancy and toddlerhood and have problems only when they first begin to have to name letters and learn letter–sound correspondences. Theoretically, this could reflect the fact that the brain areas specific to reading and spelling do not mature until around age 4, and thus any deficit in those areas is functionally "silent" in earlier development. Yakovlev and Lecours's (1967) data suggest that many association areas in the cortex myelinate this late or later; this could, then, be true of the angular gyrus and/or other brain regions important for reading.

3. It must deal with the increasingly appreciated fact that, as Allen Gevins (1982) put it, "the thinking brain is complicated" and that reading is a complex functional system in the brain involving the coordination of several separated brain regions.

This last point is illustrated by two more complex studies of the neurophysiology of reading. Using regional cerebral blood flow, Lassen, Ingvar, and Skinhoj (1978) have shown that, in adults, reading aloud involves the coordination of six brain centers: the visual association area, auditory association areas, the frontal eye field, the supplementary motor area, Broca's area, and the mouth area of primary motor cortex. The auditory areas activated during reading aloud appear to include the primary auditory area, Wernicke's area, and the angular gyrus.

Using EEG and evoked potential measures submitted to brain electrical activity mapping (BEAM) methodology, Duffy and colleagues (Duffy, Denckla, Bartels, & Sardini, 1980) identified four similar areas which showed *differences* when they compared normal and dyslexic boys on several nonreading and two reading tasks. These four areas were the supplementary motor area, Broca's area, the left midtemporal area, and the left posterolateral quadrant (i.e., Wernicke's area, parietal association areas, including the angular gyrus, and visual association areas). These results would suggest that the neurophysiological deficit in dyslexia is complex and involves several or all of the parts of the functional system shown by Lassen *et al.* (1978) to be involved in normal reading.

We have planned magnetoencephalographic studies with our population of familial dyslexics in collaboration with Martin Reite and Lynn Snyder. Unlike the EEG or EP, magnetoencephalography permits much more localized recordings of brain activity. We will be recording from primary auditory cortex, Wernicke's area, angular gyrus, and Broca's area during three tasks: simple auditory processing, phoneme discrimination, and reading single words and pseudowords. We expect the familial dyslexics to show neurophysiological differences from controls only during reading. We do not have a firm basis for predicting whether only the angular gyrus will show differences during reading or whether differences will also be observed in the two classical language areas, consistent with the results of Duffy *et al.* (1980) discussed above.

OVERALL DISCUSSION

We have described a particular form of familial dyslexia which shows a considerable degree of cross-generational continuity at three levels of analysis: the genetic, cognitive, and surface levels. Contrary to the possibility mentioned earlier, we did not find that phenotypic variability increased as we moved from the cognitive to the surface level. Instead, we found a similar high level of both cross-generational and lateral continuity at both of these levels of analysis.

As was discussed in the *Introduction*, the finding of cross-generational continuity in the phenotype of individuals of different ages with a common genetic etiology strongly implies developmental continuity across the age range studied. Thus, though we do not have longitudinal data on this sample, it appears very likely that the distinctive cognitive and surface phenotype patterns described above are continuous across development, at least from age 8 until well into adulthood. This, of course, does not mean that reading and spelling skills do not develop in these dyslexics across this age span. There were certainly age differences in *level* of reading and spelling performance across the affected subjects.

Rather, what is continuous is the underlying cognitive deficit and a *qualitative* pattern of reading and spelling performance. The continuity in this qualitative pattern suggests that these dyslexics may be encumbered with a more rigid and less automatic strategy for reading and spelling, whereas their unaffected relatives may benefit from considerable developmental change in their strategies. These results also indicate that this form of familial dyslexia is not a developmental lag which spontaneously recovers.

We do not have data as yet bearing on the neurophysiological level, but the convincing evidence for continuity in the genetic level on the one hand and the cognitive and surface levels on the other hand point to the likelihood that continuity will be found at the neurophysiological level as well, since the neurophysiological level is intermediate to these two other levels.

Although these findings of continuity assure us that we are dealing with a homogeneous and specific form of dyslexia which runs true across generations, we also found evidence for some developmental and lateral discontinuities. An interesting example of the latter category was sex differences in expressivity.

In stressing continuity in this genetic condition, we do not mean to imply that the impact of this putative dyslexia gene is immutably fixed at birth. First of all, as discussed above, the fact that there are no clear precursors of dyslexia in infancy or early childhood may mean that the main effect of this gene occurs considerably *after* birth by acting on postnatal brain development. If so, that would represent a genetically based developmental discontinuity. Second, genetic causation does not imply that environmental interventions are hopeless. Remedial reading instruction definitely helps these dyslexics compensate, and we are currently involved in research to develop better reading remediation techniques, including the use of microcomputers. Thus it is important to emphasize the role that genetic factors can play in developmental change as well as in continuity (e.g., Plomin, 1983).

An important next step would be to attempt to relate variables *across* levels of analysis. This poses numerous serious conceptual problems, because an overall conceptual framework is needed to think about how variables at one level should relate to variables at another. It does appear likely that the correspondences across levels will be complex. Yet this integration across levels is very important in a study such as this because it leads to the discovery of the most important kind of continuity, causal continuity. Until this step can be taken, the continuities demonstrated at each level of analysis are essentially *descriptive*; they do not permit us to make strong inferences about causal relationships or the mechanism involved in this disorder. The long-term goal of this project is to map out completely the mechanism in this disorder, going from the genetic alteration to the surface phenotype. This will doubtless involve adding other levels of analysis than those considered here.

Some preliminary ideas about how to relate variables across levels of analysis are presented in this paragraph. For instance, it would be useful to know how the cognitive variables relate to the surface phenotype variables. One hypothesis discussed above is that the greater frequency of omission and substitution spelling errors made by dyslexics reflects their phonological memory problems. This is being tested in a preliminary way by comparing the correlation of this error type with scores on the Wechsler Digit Span and the Boston Naming Test, respectively. If a significantly higher correlation were found with the former, this would suggest the hypothesized relationship, whereas similar correlations, or one favoring the latter test, would suggest that these kinds of errors are secondary to less specific or other language-processing problems. A much more sophisticated method for examining the causal relationships among levels is that of structural equation modeling which is described in Chapter 16 of this volume. However, the current data set is too small to permit these more sophisticated analyses.

In conclusion, the issue of continuity and discontinuity in studying development is a particular example of a fundamental issue or principle in scientific thought, that of invariance. The history of science is full of examples of phenomena which appeared heterogeneous through the lens of one set of categories but which proved to share a unifying invariance at some different, usually deeper, level of analysis. Thus the extent to which an investigator finds discontinuity in development depends, in part, on what level of analysis he chooses and the extent to which he also investigates other, possibly biological, levels.

Acknowledgments

The Development Psychobiology Research Group provided valuable stimulation and criticism in the development of the ideas in this chapter. In particular, the authors wish to thank Marshall Haith and Robert Plomin for helpful comments on an earlier version.

APPENDIX

Initial Battery of Psychological Tests

Children and Adults
Peabody Individual Achievement Test

Mathematics
Reading Recognition
Reading Comprehension
Spelling
General Information
Wide Range Achievement Test—Spelling
Gray Oral Reading Test
Colorado Perceptual Speed Test

Adults only (Age over 18)
Stress Tests (Finucci)
Upside-Down (2 Subtests)
Mirror Image (2 Subtests)
Backwards (2 Subtests)
Nonsense Passages

Subtypes Battery of Psychological Tests

WISC–R or WAIS

Mattis Subclassification System
Language Disorder
Boston Naming Test
Spreen-Benton Token Test
Spreen-Benton Sentence Repetition Test
Goldman-Fristoe-Woodcock Test of Auditory
Discrimination
Articulatory and Graphomotor Dyscoordination
Goldman-Fristoe-Woodcock Sound Blending
Mattis Graphomotor Test
Grooved Pegboard
Goodglass-Kaplan Oral Expression
Visuo-spatial Perceptual Disorder
Raven's Progressive Matrices
Benton Visual Retention Test
PMA Spatial Rotations Test

Boder Subclassification System
Denver Reading and Spelling Test (Camp & McCabe, 1978)

REFERENCES

Albert, M. L. Alexia. In K. M. Heilman & E. Valenstein (Eds.), *Clinical neuropsychology*. New York: Oxford University Press, 1979.
Boder, E. Developmental dyslexia: A diagnostic approach based on three atypical reading—spelling patterns. *Developmental Medicine and Child Neurology*, 1973, *15*, 683–687.

Cadoret, R. J., & Gath, A. Inheritance of alcoholism in adoptees. *British Journal of Psychiatry*, 1978, *132*, 252–258.

Camp, B., & McCabe, L. The Denver reading and spelling test. Research Manual, unpublished, 1978.

Cantwell, D. P. Psychiatric illness in families of hyperactive children. *Archives of General Psychiatry*, 1972, *27*, 414–418.

Decker, S. N. & DeFries, J. C. Cognitive ability profiles in families of reading disabled children. *Developmental Medicine and Child Neurology*, 1981, *23*, 217–227.

Déjerine, J. J. Sur un cas de la cécité verbal avec agraphie suivi d'autopsie. *Mémoires de la Société de Biologie*, 1891, *3*, 197–201.

Déjerine, J. J. Contribution à l'étude anatomo-pathologique et clinique des différentes variétés de cécité verbale. *Mémoires de la Société de Biologie*, 1892, *4*, 61–90.

Denckla, M. B. Childhood learning disabilities. In K. M. Heilman & E. Valenstein (Eds.), *Clinical neuropsychology*. New York: Oxford University Press, 1979.

Denckla, M. B., & Rudel, R. Naming of pictured objects by dyslexic and other learning disabled children. *Brain and Language*, 1976, *39*, 1–15. (a)

Denckla, M. B., & Rudel, R. Rapid 'automatized' naming (R.A.N.): Dyslexia differentiated from other learning disabilities. *Neuropsychologia*, 1976, *14*, 471–479. (b)

Drake, W. E. Clinical and pathological findings in a child with a developmental learning disability. *Journal of Learning Disabilities*, 1968, *1*(9), 9–25.

Duffy, F. H., Denckla, M. B., Bartels, P. M., & Sardini, G. Dyslexia: Regional differences in brain electrical activity by topographic mapping. *Annals of Neurology*, 1980, *7*, 412–420.

Elston, R. C., Kringlen, E., & Namboodiri, K. K. Possible linkage relationships between certain blood groups and schizophrenia or other psychoses. *Behavior Genetics*, 1973, *3*, 101–106.

Finucci, J. M. Genetic considerations in dyslexia. *Progress in Learning Disabilities*, 1978, *4*, 41–63.

Frith, V. (Ed.). *Cognitive processes in spelling*. New York: Academic Press, 1980.

Galaburda, A. M., & Kemper, T. L. Auditory cytoarchitectonic abnormalities in a case of familial developmental dyslexia. *Annals of Neurology*, 1979, *6*(2), 94–100.

Gevins, A. The thinking brain is complicated. Presented at *The International Neuropsychological Society* meetings, Pittsburgh, February 1982.

Gleitman, L. R., & Rozin, P. The structure and acquisition of reading. I: Relations between orthographies and the structure of language. In A. S. Reber & D. L. Scarborough (Eds.), *Toward a psychology of reading*. Hillsdale, N.J.: Lawrence Erlbaum, 1977.

Gross, K., Rothenberg, S., Schottenfeld, S., & Drake, C. Duration thresholds for letter identification in left and right visual fields for normal and reading-disabled children. *Neuropsychologia*, 1978, *16*, 709–715.

Hallgren, B. Specific dyslexia (congenital word-blindness): A clinical and genetic study. *Acta Psychiatrica et Neurologica Supplement*, 1950, *65*.

Herschel, M. Dyslexia revisited: A review. *Human Genetics*, 1978, *40*, 115–134.

Hier, D. B., Le May, M., Rosenberger, P. B., & Perlo, V. P. Developmental dyslexia: Evidence for a subgroup with reversal of cerebral asymmetry. *Archives of Neurology*, 1978, *35*, 90–92.

Hinshelwood, J. A case of dyslexia: A peculiar form of word-blindness, *Lancet*, 1896, *2*, 1451–1454.

Hughes, J. R., & Denckla, M. B. Outline of a pilot study of electroencephalographic correlates of dyslexia. In A. Benton and D. Pearl (Eds.), *Dyslexia: An appraisal of current knowledge*. New York: Oxford University Press, 1978.

Huttenlocher, P. R., & Huttenlocher, J. A study of children with hyperlexia. *Neurology*, 1973, *23*, 1107–1115.

Kidd, K. K. A genetic perspective on stuttering. *Journal of Fluency Disorders*, 1977, *2*, 259–269.

Lassen, N. A., Ingvar, D. M., & Skinhoj, E. Brain function and blood flow. *Scientific American*, 1978, *239*, 62–71.

Liberman, I. Y., Shankweiler, D., Liberman, A. M., Fowler, C., & Fischer, F. W. Phonetic segmentation and recoding in the beginning reader. In A. S. Reber & D. Scarborough (Eds.), *Toward a psychology of reading*. Hillsdale, N.J.: Lawrence Erlbaum, 1977.

Mann, V. A. Reading skill and language skill. Paper presented to the *Society for Research in Child Development*, Boston, Mass., April 1981.

Marsh, G., Friedman, M., Welch, V., & Desberg, P. The development of strategies in spelling. In U. Frith (Ed.), *Cognitive processes in spelling*. New York: Academic Press, 1980.

Mattis, S., French, J. M. and Rapin, I. Dyslexia in children and young adults: Three independent neuropsychological syndromes. *Developmental Medicine and Child Neurology*, 1975, *17*, 150–163.

Morrison, J. R., & Stewart, M. A. The psychiatric status of the legal families of adopted hyperactive children. *Archives of General Psychiatry*, 1973, *28*, 888–891.

Orton, S. T. Reading, writing and speech problems in children. New York: Norton, 1937.

Pennington, B. F., & Smith, S. D. Genetic influences on learning disabilities and speech and language disorders. *Child Development*, 1983, *54*, 369–387.

Pennington, B. F., McCabe, L. L., Smith, S. D., Kimberling, W. J., & Lubs, M. A. The spelling phenotype in a form of familial dyslexia. In preparation.

Plomin, R. Developmental behavior genetics. *Child Development*, 1983, *54*, 253–260.

Satz, P., & Morris, R. Learning disability subtypes: A review. In F. J. Pirozzolo & M. O. Wittrock (Eds.), *Neuropsychological and cognitive processes in reading*. New York: Academic Press, 1980.

Shankweiler, D., & Liberman, I. Y. Exploring the relations between reading and speech. In R. M. Knights & D. J. Bakker (Eds.) *Neuropsychology of learning disorders: Theoretical approaches*. Baltimore: University Park Press, 1976.

Smith, S. D., Kimberling, W. J., Pennington, B. F., & Lubs, M. A. Specific reading disability: Identification of an inherited form through linkage analysis. *Science*, 1983, *219*, 1345–1347.

Smith, S. D., Pennington, B. F., McCabe, L. L., Kimberling, W. J., & Lubs, M. A. The cognitive phenotype in a form of familial dyslexia. In preparation.

Sweeney, J. E., & Rourke, B. P. Neuropsychological significance of phonetically accurate and phonetically inaccurate spelling errors in younger and older retarded spellers. *Brain and Language*, 1978, *6*, 212–225.

Symmes, J. S., & Rappaport, J. L. Unexpected reading failure. *American Journal of Orthopsychiatry*, 1972, *42*, 82–91.

Venezky, R. L. *The structure of English orthography*. The Hague: Mouton, 1970.

Wallach, M., & Wallach, L. *Teaching all children to read*. Chicago: University of Chicago Press, 1976.

Wallach, M., & Wallach, L. Helping disadvantaged children learn to read by teaching them phoneme identification skills. In L. B. Resnick & P. A. Weaver (Eds.), *Theory and practice of early reading* (vol. 3). Hillsdale, N.J.: Lawrence Erlbaum, 1979.

Winokur, G., Tanna, V., Elston, R., & Go, R. Lack of association of genetic traits with alcoholism C3, Ss and ABO systems. *Journal of Alcohol Studies*, 1976, *37*, 1313–1315.

Witelson, S. Developmental dyslexia: Two right hemispheres and none left. *Science*, 1977, *195*, 309–311.

Yakovlev, P. I., & Lecours, A. R. The myelogenetic cycles of regional maturation of the brain. In A. Minkowski (Ed.), *Regional development of the brain in early life: A symposium*. Philadelphia: Davis, 1967, pp. 3–69.

CHAPTER 7

The Study of Transitions

CONCEPTUAL AND METHODOLOGICAL ISSUES

J. P. Connell and Wyndol Furman

One of the central and most controversial problems in developmental psychology is the question of continuity and discontinuity in development. Numerous empirical studies have examined this question, and many theoretical positions have been proposed (e.g., Brim & Kagan, 1980; Emde, Gaensbauer, & Harmon, 1976; Emmerich, 1968; Greenough, this volume; Kagan, this volume; Mischel, 1968; Rutter, this volume; Sackett, Sameroff, Cairns, & Suomi, 1981; Wohlwill, 1973). Although little consensus has been reached, developmentalists of any persuasion would agree that human beings undergo a series of changes in the course of the life span. Moreover, human lives appear to be characterized by periods of relative stability and periods of marked change or transition. These transitions are thought to be the times when major reorganizations or discontinuities may occur.

Although the nature of transitions appears to be central to the question of continuity and discontinuity, relatively little theoretical or empirical work has directly addressed this topic. This is not to say that investigators have not referred to the idea of transitions. As early as 1908, the anthropologist Arnold Van Gennep described transitions in status as

J. P. Connell • Graduate School of Education and Human Development, University of Rochester, Rochester, New York 14627.   **Wyndol Furman** • Department of Psychology, University of Denver, University Park, Denver, Colorado 80210.
Portions of this research were supported by grants from the W. T. Grant Foundation (J. P. Connell, Principal Investigator), the National Institute of Child Health and Human Development (HB-09613; Susan Harter, Principal Investigator), and the Biomedical Research Grant Fund (BRS-S07RR07138; Wyndol Furman, Principal Investigator).

times of "ceasing" and "becoming." Since then, social scientists have used the concept to refer to shifts between stages of cognitive development, reorganizations of personality structure (e.g., the midlife transition), and major changes in a person's environment or social structure (e.g., retirement). Typically, however, the concept of a transition has not been defined carefully and, in fact, the term has been used in a variety of ways. In some instances, the term *transition* seems to refer to external events, whereas in other instances it seems to refer to internal transformations. Eurich (1981) observed that transitions can be defined by time periods in the life span, by role changes, or by events. Given the great interest in continuity and discontinuity today, it appears to be a propitious time to take a closer look at the concept of transitions.

In the present chapter, we will try to identify some of the major properties of transitions and propose a vocabulary for describing transitions. Once the general concept has been discussed, we will delineate several potential typologies. Finally, we will outline a methodological framework for studying transitions and illustrate its use with two examples of our own research.

THE CONCEPT OF TRANSITIONS

Our review of the literature led us to conclude that a transition can best be defined as the occurrence of relatively greater change in a characteristic or set of characteristics of an individual or of a group of individuals. Such characteristics may be biological, cognitive, personality, or interpersonal. Moreover, the same general conceptualization of transitions seems applicable regardless of whether the characteristic of interest is overt behavior, an underlying attribute, or the structural or organizational properties of the organism. As will be seen subsequently, however, we believe that the transitions of grestest interest involve changes in the biological or psychological processes that organize and reorganize overt behaviors.

In the present conceptualization, three further definitional terms may be differentiated: *transitional event, transitional period,* and *transitional mechanism*. The former two terms are descriptive concepts that refer to the instigation (event) and duration (period) of the transition. *Transitional mechanism* refers to the reorganizational processes that trigger and maintain the relatively greater change in the organism's pattern of functioning. These three terms are discussed in greater detail in subsequent paragraphs.

Transitional Event

The term *transitional events* is used to refer to the events or changes that initiate the period of relatively greater change. These events may be exogeneous life events, such as entering school or a parental divorce, or endogenous events, such as the discovery that one has a terminal disease. Additionally, *nonevents,* such as not dating as an adolescent, can bring on transitions. That is, some events or forms of experience may be necessary for maintaining periods of stability (Gottlieb, 1976).

The present definition of transitional events bears some similarity to the typical conception of life events (Hultsch & Plemons, 1979; Reese & Smyer, 1980). If life events are thought of as objective, external events, however, the two terms do differ. First, we include endogenous events as do some life-span developmental psychologists. More important, we think of life events as occasions for change or reconstitution in an organism's characteristics. If relatively greater change does not occur following the event, that event would not be considered a transitional event in the present conceptualization. This restriction is necessary because transitions have been defined as the occurrence of change *in the individual's or group's pattern of functioning.* For example, a change in school settings may or may not be followed by change in a child's pattern of functioning. Although others may prefer to think of transitions as changes in either the person or environment, we believe that that has the undesirable effect of equating quite different phenomena.

Transitional Period

Whereas the term *transitional events* refers to the initiating factor(s), *transitional periods* are thought of as the times of relatively greater change in an organism's characteristics. This period can be very brief, such as Saul's conversion on the road to Damascus or perhaps falling in love at first sight. More typically, the period of marked change will extend over weeks, months, or perhaps even a year. Such extended periods seem particularly likely when the transitional event has multiple effects or when these effects initiate further changes. For example, the transitional period following the death of a spouse may initiate a long period of readjustment during which other transitional events may occur, such as a reinstitution of dating or a change in living standard. Transitional periods need not occur immediately after the event but instead may be delayed, such as in the case of a delayed grief reaction (Hultsch & Plemons, 1979). On the other hand, the transitional period could begin prior to the event itself. For example, an adolescent who is anticipating leaving home may begin to change prior to actually leaving. In some important sense, one could argue that the anticipation of leaving home is the transitional event in this case. The important point, however, is that an objective event is not a precise marker of the occurrence, length, or timing of a transitional period. It is the tenure of the organism's manifestation of relatively greater change that defines a transitional period.

In the present conceptualization transitional periods end when the pattern of the organism's functioning stabilizes. Such a stabilization may be masked by the occurrence of a second transition. Transitions often result in some enduring change in a characteristic, but it also seems possible for a transition to end with a return to the preexisting state of functioning. Although some investigators have limited transitions to instances of enduring changes (Stewart, 1982), we have included temporary changes because they too can involve a period of marked change.

Transitional Mechanism

Transitional events and transitional periods are essentially descriptive concepts. When the idea of a transition is used to account for observed discontinuities in behavior, it is important to provide some theoretical explanation for this occurrence. In particular, such an explanation would involve identifying the transitional mechanisms responsible for the change or reorganization of the person's characteristics. For example, it is not sufficient simply to describe a change in a child's cognitive performance; instead, one would want to discuss how the failure of the child's schema either to assimilate an experience or to accommodate to it has produced the disequilibrium.

It should also be noted that the mechanisms responsible for the discontinuity in a transition may be the same ones responsible for continuity at other periods. For example, shifts in the pattern and intensity of hormonal secretion during puberty may be seen as a transitional mechanism yielding observed discontinuities in the pattern of adolescents' functioning. However, the secretion of hormones at earlier and later points in development also functions to maintain the continuity of physical growth. Similarly, the breakdown of primitive ego defenses in the course of dynamic psychotherapy may trigger transitional periods, but the reconstitution of more adaptive defenses may produce subsequent stabilization of patterns of functioning.

Thus, transitional mechanisms are to be understood in the context of theoretically specified processes. Transitional periods may ensue when such processes emerge, change in intensity, or are disrupted. Transitional events may mark the onset of changes in these processes but should not be viewed as the cause of the transition. Instead, transitional events should be thought of as activators for the reorganizational capacities of the organism to modify current structures in response to adaptational demands.

THE CONCEPT OF CHANGE

Up until this point, the idea of change has been introduced without any elaboration. Unfortunately, terms such as change, continuity, consistency, and stability have been used in a variety of manners in the developmental literature. Although our intent is not to resolve these thorny definitional issues, it is important to point out the kinds of changes we believe can occur during a period of transition.

As noted earlier, we have defined a transition as the occurrence of relatively greater change in a characteristic or set of characteristics of an individual or a group of individuals. Such characteristics can be categorized into two types—the observed or measured attributes and the underlying theoretical constructs. Statisticians commonly use the terms *manifest variable* (observed characteristic) and *latent variable* (underlying construct) to make this distinction. For example, in research on attachment, the number of instances of physical contact is commonly measured (i.e., a manifest variable). Theoretically, however, such physical contact is thought to be a behavioral manifestation of the underlying construct of proximity-seeking. That is, one overt way through which children seek proximity is by making physical contact.

Some investigators have argued that changes could occur at one level and not the other (Kagan, 1969). Consider the previous example. From early to late infancy, proximity-seeking may continue to be a highly salient function, but the manifestations of proximity-seeking may change. That is, young infants may principally seek proximity by physical contact, whereas older infants may seek proximity through visual or distal contact. Kagan would argue that the manifest variables of physical and distal contact have been changed but that the underlying latent construct of seeking proximity remains the same. As will be discussed shortly, we believe that it may be more accurate to describe this type of change as a change in the pattern of relations *between* manifest and latent variables rather than as a change at the manifest variable level alone (see the section on structural change, pp. 159–160).

In our definitional framework, change at the level of the underlying construct must have measurable consequences; that is, transitions must have empirical referents. Conversely, the empirical referents must be linked together by an underlying psychological or physiological construct for the transition to have theoretical meaning. Having discussed the crucial distinction between manifest and latent variables, we can now proceed to describe three forms of change that can be quantified.

Level Change

Level refers to the absolute level or amount of a variable for an individual or the mean level of a variable for a group of individuals. Since organisms and their environments are in a constant state of flux, level changes can be expected to occur constantly. Hence, transitional changes in level should refer to significant alterations in the rate of change of a variable. This point is illustrated in the hypothetical depiction of changes in body size at the time of puberty (Figure 1). Body size generally increases with age, but during the transition of puberty the rate of change is greater than before or after this period. It also seems possible for the rate of change to be less in a transition. For example, a severe illness during adolescence may decrease or even temporarily reverse a developmental trend to become more self-sufficient. Thus, transitions can be marked by either increases or decreases in the rate of change.

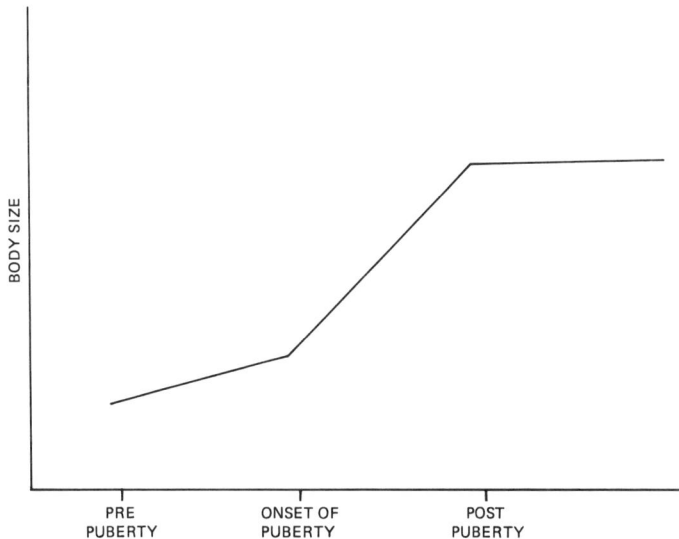

Figure 1. Hypothetical changes in body size around time of puberty.

Structural Change

Several theorists have observed that an underlying variable may have different behavioral manifestations at different developmental points (Baltes & Nesselroade, 1970; Coan, 1966; Kagan, 1969); conversely, any specific behavior's meaning may change with development. We refer to these changes as changes in structure. That is, structural change is change in the degree and pattern of relationship between manifest variables and latent variables; typically, structure is assessed by examining the number and composition of factors. This concept of structural change or consistency has been discussed in the statistical literature in terms of factorial invariance (see Connell & Goldsmith, 1982; Meredith, 1964). Coan (1966) describes three classes of change in structure: (a) factor metamorphosis—the underlying construct retains its substantive interpretation but the manifest variables that load on it change; (b) factor emergence—a new factor is necessary to account for changes in the pattern of correlations among the manifest variables; and (c) factor disintegration—a factor is no longer necessary due to a change in the pattern of correlations.

McCall, Eichorn, and Hogarty's (1977) study of transitions in early mental development provides an illustration of two of these kinds of structural change. These investigators conducted principal components factor analyses of infants' performance on intelligence test items. A primary component (factor) appeared at all ages, but the items that loaded on this component changed markedly at transitions between cognitive stages (i.e. factor metamorphosis). Additionally, several secondary components appeared only during specific cognitive stages (i.e., factor emergence).

Kagan's (1969) concept of phenotypic change can be interpreted in the framework of structural change. He described phenotypic change as change in the overt behavior but continuity in the underlying variables or functions. Our example of a change from physical manifestations of proximity-seeking to distal manifestations of proximity-seeking would be an instance of such. In the present framework this type of change would be considered to be structural change, that is, change in the pattern of relationships between manifest and latent variables, rather than a change in the manifest or phenotypic variable alone.

More generally, structural change is a concept close to the heart of developmentalists interested in cognitive development (e.g., Piaget and his colleagues), personality development (psychoanalytic and ego psychologists), and language development (structuralists such as Chomsky). Commonly such changes have been described as *qualitative* or *discontinuous changes* (Emmerich, 1964; Flavell, 1971). We prefer the term *structural*, however, because terms such as *qualitative* have been used in different ways in other areas of psychological research (e.g., qualitative as opposed to quantitative research methods).

Conceptualizing structural change as a function of and manifestation of alterations in normative developmental processes is also quite common in developmental theory. By placing the idea of structural change in the context of the metatheory of factor analysis (underlying latent constructs expressed in patterns of interrelationships among manifest variables), it may become possible to test hypotheses springing from broad developmental theories in a more sophisticated manner.

Centrality Change

A third form of quantifiable change is change in the centrality or degree and pattern of interrelatedness of variables. A special case of the centrality of a characteristic is referred to by many developmentalists as stability, or the degree of interrelatedness of a variable with itself over time. It is commonly thought of as the degree of consistency in the rank-ordering of individuals over time (Emmerich, 1964; McCall, 1977). Often the relative stability of a characteristic will decrease with the onset of a transition. For example, McCall *et al.* (1977) report that the stability of intelligence scores was greater within a cognitive period than between periods. Similarly, Sells, Roff, Cox, and Mayer (1967) found a decrease in the stability of sociometric status during the transition to junior high. It should be emphasized that these changes in stability need only be relative ones. That is, a variable may be less stable during a transition than before or after, but it still may be somewhat stable. In both of the examples cited, significant, but lower, correlations were observed across the transitional period. Additionally, the change is often only temporary; the stability of a characteristic after a transition may return to or exceed the pretransition stability.

Typically, one would look for lower stability during transitional periods, but it is at least theoretically possible that stability could be increased during a transition. Take, for example, the situation in stock car racing when drivers are required to maintain their positions in the field during the clean up of a crash. Although psychological examples are more difficult to imagine, it seems possible that a stressful transition may lead to a rigidification of the chain of power or social structure in a group of individuals.

Centrality also incorporates the idea of changes in the pattern of relationships among variables. During a transition, some variables may become less related to each other, whereas others may become more related. For example, the correlation between the quality of family relationships and social adjustment decreases markedly with the transition from home to college, while the correlation between the quality of friendships and social adjustment increases (Buhrmester, Shaver, Furman, & Willems, 1982; Furman, Buhrmester, Shaver, & Willems, 1982). Like changes in level or structure, transitional changes in centrality may be either temporary or enduring.

Each of these three forms of change (level, structure, and centrality) can be further differentiated in terms of whether they refer to interindividual or intraindividual change (Block, 1971; Wohlwill, 1973). For example, a population's mean level on a variable may change with a transition or a specific individual's absolute level may change. Neither form of change can be inferred from the other. That is, although the population in general may change, any specific individual may not, and vice-versa. Although interindividual studies are much more common, several investigators have argued that intraindividual studies are particularly important for understanding transitions in development (Baltes & Goulet, 1970; Wohlwill, 1973).

TYPES OF TRANSITIONS

Developmental Transitions

We believe that the present definition of transitions may be useful for studying a broad variety of phenomena. The concepts appear to be appropriate for describing transitions in single biological or psychological characteristics or transitions that involve a wide range of variables. A particularly interesting type of transition is one that we call a *developmental transition*. Here we are referring to transitions such as those that occur between stages in cognitive development, at the onset of puberty, or after a major shift in a person's social environment (e.g., entering school, taking a first job). To be precise, we would define developmental transitions as those transitions that have enduring effects on the organizational characteristics of the person. That is, the underlying structure or processes of the organism undergo change. Thus, changes in overt behavior would not be sufficient evidence that a developmental transition was occurring. Instead, the patterns of behavioral change would have to be a manifestation of hypothesized changes in the underlying properties of the organism.

Additionally, the changes are enduring in the sense that the organism does not return to the pretransitional pattern of adaptation after the transition but instead adopts a new pattern of adaptation. For example, the experience of leaving home may lead someone to develop a new self-image. On the other hand, if the person's self-image or other characteristics do not undergo any lasting changes, it would not be considered a developmental transition by the present definition (although it may be a transition).

The present definition of developmental transitions is just that—a definition. The concept will require some further elaboration and perhaps some basic changes. But we should like to single out these transitions because we think that the study of such transitions is particularly likely to shed light on the central issues of discontinuity and continuity in development. (See Blasi, 1976, for further discussion of the distinction between development and simple change.)

Endogenous and Exogenous Transitional Events

A second major distinction among types of transitions can be made on the basis of whether the transition was triggered by endogenous or exogenous events. Endogenous events are internal events or changes in physiological or psychological processes. Exogenous events are occurrences or changes in the social ecology, such as entering school, leaving home, receiving a draft notice, or the death of a significant other. Of course, in the present definition exogenous events are only transitional events if they trigger transitional mechanisms within the person that lead to changes in the pattern of functioning. Anyone familiar with contemporary interactional perspectives (Magnusson & Endler, 1977; Sameroff & Chandler, 1975) would recognize that endogenous and exogenous events are inextricably intertwined. However, when the role of either can be identified, the description of the transition becomes more precise. For example, in the cognitive-developmental literature, Strauss (1972) has differentiated between external or adaptational disequilibrium and internal or organizational disequilibrium. Aside from this area of work, however, relatively little attention has been given to the differences in the characteristics and effects of these two types of transitional events.

Typologies of Life Events

Several life-span developmental psychologists have proposed taxonomical systems for categorizing life events (Baltes, Reese, & Lipsett, 1980; Brim & Ryff, 1980; Danish, Smyer, & Nowak, 1980; Hultsch & Plemons, 1979; Modell, Furstenberg, & Hershberg, 1976; Reese & Smyer, 1980). Although it is not possible to review all of these dimensional systems or typologies here, a description of one such system may prove to be valuable. In particular, Danish *et al.* (1980) proposed a list of six structural characteristics of life events: (a) the timing, (b) duration, (c) sequencing, (d) cohort specificity, (e) contextual purity, and (f) probability of event occurrence. The *timing* of an event refers to whether it occurs when the person or culture expects it to happen. For example, leaving home at the completion of high school would be "on-time," whereas leaving before or much later would be "off-time." *Event-sequencing* refers to whether the event occurred in the expected order. For example, having a child before getting married would be out of sequence in terms of our culture's expectations. The *duration* of a event refers to the length of time it is experienced. Length of time experienced includes the anticipatory period, the event itself, and the postevent influences. Thus, this characteristic is essentially the same as our concept of length of transitional period. *Cohort specificity* refers to the variation in meaning and effect that an event has on different cohorts. For example, Danish *et al.* point out that a woman's becoming a top executive in 1920 was functionally different from the same occurrence in 1980. *Contextual purity* refers to the extent to which an event is interrelated with other, concurrent events. Contextually pure events occur in relatively stable, uneventful times, whereas contextually impure events occur concurrently with other major events or have effects that range beyond the domain in which they occur. For example, entrance into junior high school is contextually impure because it occurs around the onset of puberty and the beginning of romantic relationships. The sixth characteristic, *probability of occurrence*, refers to the likelihood that the event is experienced at some time by large or small proportions of the population. Other authors have used the terms *normative* and *nonnormative* or *idiographic* to describe this differentiation (Baltes *et al.*, 1980).

Reese and Smyer observed that relatively little work has been done on the interrelationships of the different characteristics or their effects on the experience of the event. We expect, however, that these different properties will affect the nature of the transition, particularly in terms of the effects on adaptation. For example, transitions brought on by off-time, out-of-sequence, or low-probability events may be more difficult to adjust to or may put the individual at a higher risk for developing maladaptive patterns of functioning (Glaser & Strauss, 1971; Neugarten, 1970). In each of these cases, the individual is less likely to have a social support network on which to rely; in fact, some of these events may not even be sanctioned by the society (e.g., having a child prior to marriage). Similarly, if the period of transition after the event is particularly long, it may indicate that the person is having difficulty making the requisite changes. On the other hand, long anticipatory periods are thought to facilitate the process of transition because the person is provided with an opportunity to learn what changes are expected and rehearse different patterns of adaptation (Kimmel, 1974). Of course, none of these characteristics will perfectly predict the adaptive consequences of the event. Although inheriting a million dollars from a distant uncle may be entirely nonnormative, most of us would be more than willing to undergo the transition that is likely to follow.

Like our present distinction between endogenously and exogenously induced transitions, these typologies have focused on the transitional event *per se.* Moreover, the majority of the classification schemes have been based on intuition or observation rather than theory. There is a great need for systems, particularly theoretical ones, for differentiating among various transitional periods or transitional mechanisms.

METHODOLOGICAL CONSIDERATIONS

A central tenet of the present definition of transitions is that they refer to the occurrence of relatively greater change. One implication of this point is that simple change from one age to another would not be sufficient evidence that a transition is occurring. Instead, the relative magnitude of change in a hypothesized transition would have to be shown to be greater than the relative magnitude of change when a transition is not underway.

Design Strategies

One strategy for examining the relative magnitude of change would be to collect information on some individual or group at three points in time, such as during a transitional period and twice after the period. For example, if the entry into elementary school was thought to initiate a transition, one could examine various characteristics of children near the beginning of the first, second, and third grades. If the expected transition had occurred, one should find greater change between the first two time points than between the latter two points. A similar set of considerations would apply if information were obtained at two pretransitional points and once during the transition.

An alternative strategy is to compare the rates of change in a transitional group and a stable group. For example, one could collect longitudinal data on one group of children during preschool and first grade and on another group during the second and third grade. Greater change should occur in the first group. In this example the effects of the transition were determined by comparing the characteristics of children before and after a hypothesized transitional event (entry into the first grade).

One could also assess the effects of a transition by comparing the amount of change between two time points in a transitional period with the amount between two comparable points in a stable period. For example, the degree of change from the beginning to the end of the first grade should be greater than that from the beginning to the end of the second grade.

These design strategies can provide valuable information about transitions and their impact. At the same time, it would be ideal to demonstrate that the change with a transition was greater than either before or after the transition. Such a demonstration would require obtaining information at a minimum of five points in time on one group (two pretransitional, one transitional, and two posttransitional times) or comparing two time points for three groups (a pretransitional stable group, a transitional group, and a posttransitional stable group). Needless to say, such longitudinal designs can be impractical or expensive. Many interesting comparisons can be made with cross-sectional data, particularly when cohort effects seem unlikely. Perhaps initial, exploratory studies of a transition should be kept methodologically simple, whereas the more complex designs may be appropriate for confirmatory studies.

Data Analytic Strategies

Several analytical tools presently exist for examining each form of change discussed in the previous section. For example, changes in level can be assessed by traditional analysis of variance techniques or in some instances by time-series analysis. Factor analysis, principal component analysis, or cluster analysis can be used to look at changes in the underlying structure.

Changes in centrality can be examined with correlational techniques. One could examine the correlations among repeated administrations of the same measure(s) (i.e., stability assessments), or more generally, one could compare the patterns of correlations among variables within and between different time points.

A relatively new procedure, structural equation modeling, may also prove valuable (see Chapter 16; Goldberger, 1973; Jöreskog & Sörbom, 1978). This comprehensive data-analytic strategy integrates the hypothesis-testing capacities of three previously separate procedures—factor analysis, path analysis, and analysis of variance. Not only does this technique permit one to examine all forms of change, but one can examine them simultaneously. Although it is beyond the scope of this chapter to present the technical details of this strategy, a description of its hypothesis-testing capacity and an application of the strategy to the study of attachment can be found elsewhere (Connell & Goldsmith, 1982).

Finally, both interindividual and intraindividual changes in level, structure, or centrality can be studied, but they require different sets of comparisons. Descriptions of the analytic strategies necessary for the study of intraindividual change can be found elsewhere (Block, 1971; Catell, 1966; Nesselroade, in press; Baltes, Reese, & Nesselroade, 1977).

Methodology and Theory

Although one should not necessarily expect a transition to be characterized by all forms of change, social scientists should consider the different aspects of transitional change. To date, studies of transitions have principally focused on changes in level or stability of specific manifest variables such as IQ scores. Studies of structural change are particularly needed. What are the sources of continuities and discontinuities in the patterning of behavior thought to reflect such broad theoretical constructs as attachment, intelligence, self-esteem, or social support? A richer picture of transitional periods and the processes underlying the instigation and resolution of these periods will be potentiated by developing hypotheses regarding these different forms and types of change.

TWO EXAMPLES OF THE STUDY OF TRANSITIONS

These methodological considerations and the conceptual framework underlying them are illustrated in two examples of our research. Connell and his colleagues have been concerned with the effects of changes in the educational setting on children's self-related cognitions and affects (Connell, 1981; Connell & Tero, 1982; Harter & Connell, in press). This research has focused on two specific changes: (a) the shift from primary elementary school to upper elementary school and (b) the shift from elementary school to junior high. The first change in setting occurs between the third and fourth grade and involves a shift from a relatively affect-oriented educational atmosphere to a more achievement-oriented atmosphere. The second one occurs between the sixth and seventh grade and involves a number of alterations, including changes in peer groups and educational structure.

Although these studies have incorporated a range of measures, the present description will focus principally on transitions in children's unknown perceptions of control. This construct was defined as the degree to which children think they understand the reasons for their academic successes and failures. Children with high unknown perceptions of control are puzzled by their school performance and unable to explain why they may have succeeded or failed. In contrast, children with low unknown perceptions of control believe they know the reasons for success or failure. Generally, unknown perceptions should decrease with experience in school, but changes in the educational setting were expected to disrupt this developmental trend.

This study employed a combined cross-sectional and longitudinal design. A sample of third- through seventh-grade children were administered questionnaires in the spring of the school year and once again a year and a half later. To assess changes in level, three longitudinal groups were compared: (a) a young transitional group (third to fifth grade), (b) a stable group (fourth to sixth grade), and (c) an older transitional group (sixth to eighth grade). As expected, the mean level of unknown perceptions of control in the stable group decreased over the year-and-a-half period; that is, the additional year and a half of experience in a relatively stable environment had led the children to say that they knew more about the reasons for academic success and failure. In contrast, the mean level in the two transitional groups did not change over time.[1] Apparently, the change in school settings had counteracted the effects of the year and a half of additional experience. This pattern of results illustrates the point that transitions may alter the developmental curve of a variable but that the changes in level are not necessarily greater during a transition than during stable periods.

The changes in educational setting also affected the stability of children's perceptions of control. As expected, these scores were less stable in the two transitional groups than in the stable group (mean $r = .27$ versus $r = .45$, respectively).

To examine changes in aspects of centrality other than stability, the pattern of correlations between unknown perceptions of control and other variables was examined in cross-sectional groups of children who had recently undergone a change in school settings (fourth- and seventh-graders) and in groups who were at the end of a stable period (third- and sixth-graders). Overall, greater unknown perceptions of control were associated with poorer achievement. These correlations were stronger, however, in the stable groups than in the transitional groups (see Table 1). Children with negative feelings about academic achievement tended to have higher unknown perceptions of control. In contrast to the findings with achievement, however, these correlations were stronger in the transitional groups than in the stable groups. These findings suggest that the determinants of perceptions of unknown control may be different in stable and transitional periods. In stable periods

---

[1] In this and the following example, the data were also analyzed by structural modeling techniques and similar results were obtained.

Table 1. Centrality Correlations of Unknown Control

	Grade			
Centrality relationships	3rd (Young stable)	4th (Young transitional)	6th (Old stable)	7th (Old transitional)
Unknown control–achievement	−.58[a]	−.20	−.36[a]	−.25
Unknown control–competence affect	−.32	−.45[a]	−.05	−.33[a]

Note. Higher scores on unknown control indicate less knowledge about the reasons for action (e.g., greater unknown); higher scores on competence affect indicate more positive feelings toward achievement. Therefore, a negative correlation between unknown control and achievement or competence affect indicates that the greater the unknown control, the lower the level of achievement and/or the more negative the competence affect. (Adapted from Connell & Tero, 1982.)
[a]$p < .05$.

these perceptions appear to be linked to children's ability to think through the reasons of success and failure. In transitional periods, however, this variable may be related to children's affective reactions to school and may reflect their ability to cope with the transition.

The second example comes from a project on adolescents' adjustment to the transition from home to college (Buhrmester *et al.*, 1982; Furman *et al.*, 1982; Shaver, Furman, Buhrmester, & Willems, 1981). In this project, over 200 incoming students completed questionnaires during the summer prior to their arrival at college and again during each of the three academic quarters of their freshman year. The questionnaires assessed numerous aspects of students' social lives and contained several measures of adjustment, including state and trait measures of loneliness, psychological and physical health symptoms, ratings of the difficulty of the transition, and overall satisfaction with social and academic life.

The changes in state and trait loneliness provide a nice illustration of the effects of the transition from home to college. First, the possibility of structural changes was examined by conducting principal component analyses of the items on the loneliness scales at each of the four measurement points. Two primary factors, trait loneliness and state loneliness, emerged in all analyses; moreover, individual items loaded on the same factors at all time points. These findings indicate that the two loneliness constructs had not changed structurally as a result of the transition.

As expected, the transition had striking effects on the mean level of state loneliness. The ratings increased markedly from the summer to the fall and then subsided to their previous level by winter; the level remained stable between the winter and spring. The mean level of trait loneliness was expected to remain consistent across the year. Interestingly, however, the ratings of trait loneliness *decreased* between the summer and fall and then remained stable. These findings suggest that a contrast or adaptation-level effect may have been present. That is, the experience of greater state loneliness in the fall may have led the students to decide that their general level of loneliness over the last few years (i.e., trait loneliness) may not have been so high as they had previously thought.

The transition was also characterized by a decrease in the stability of state loneliness (see Figure 2). The correlation between summer and fall ratings of state loneliness was rather minimal, but the correlations among the three academic quarters at college were substantial. In contrast, ratings of trait loneliness remained highly stable throughout.

The transition also affected the centrality relationships of state and trait loneliness (see Figure 2). First, the positive correlation between state and trait loneliness decreased from summer to fall (.58 to .40) and then increased as the year proceeded (winter $r = .48$; spring $r = .60$). Apparently, trait measures are more predictive of current states during stable periods than during transitional periods. Additionally, the relationships between the loneliness measures and various characteristics of the students' social networks were examined (see Table 2). During the summer, trait loneliness was highly negatively correlated with the students' degree of satisfaction with various aspects of their social life; that is, those with high trait loneliness scores were less satisfied. These correlations decreased, however, during the fall transition. In many instances, the correlations rose again in magnitude as the year proceeded. Ratings of state loneliness showed the opposite pattern of relationships. Generally, the correlations of state loneliness with adjustment were highest during the fall quarter. Although correlational, these results

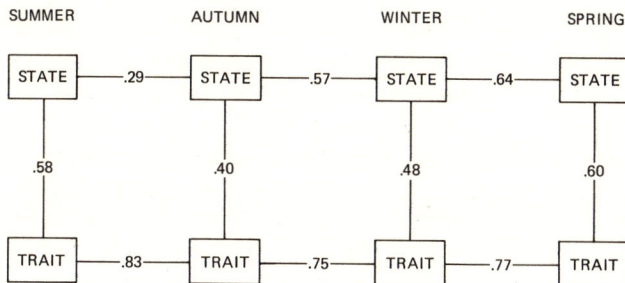

Figure 2. Interactions of state and trait loneliness.

Table 2. Changes in the Centrality of Trait and State Loneliness

	Time period			
	Summer	Fall	Winter	Spring
Trait loneliness				
Satisfaction with quality of romantic life	−.44	−.22	−.10	−.06
Dating frequency	−.29	−.15	−.06	−.09
Satisfaction with quality of friendships	−.42	−.22	−.27	−.38
Satisfaction with number of friends	−.37	−.19	−.18	−.32
Satisfaction with friendship clique	−.49	−.11	−.22	−.24
Perceived popularity	−.37	−.26	−.30	−.32
Overall satisfaction with social network	−.60	−.25	−.33	−.42
State loneliness				
Satisfaction with quality of romantic life	−.32	−.54	−.24	−.30
Dating frequency	−.16	−.39	−.18	−.18
Satisfaction with quality of friendships	−.32	−.59	−.39	−.55
Satisfaction with number of friends	−.29	−.45	−.35	−.36
Satisfaction with friendship clique	−.26	−.34	−.29	−.32
Perceived popularity	−.07	−.44	−.37	−.22
Overall satisfaction with social network	−.45	−.58	−.49	−.53

Note. Higher scores indicate greater satisfaction or frequency. Correlations exceeding .15 are significant ($p < .05$).

suggest that the determinants of social adjustment may differ in transitional and stable periods.

CONCLUSION

Wohlwill (1970) observed that the subject matter of developmental research is "behavior *change* taking place as the individual grows from birth to old age, or more particularly the properties and characteristics of these changes and the variables governing them" (p. 151). For years, developmental psychologists have been concerned with the question of continuity and discontinuity in development. Progress has been made, but most work has remained at a descriptive level. At this point, we must go beyond simply describing continuities and discontinuities and begin to specify the mechanisms underlying developmental change.

The study of transitions may prove to be particularly fruitful for understanding sources of discontinuity and continuity. In particular, transitional periods provide unique windows for viewing the organism as it undergoes change. Studies of transitions may lead to insights into the endogenous and exogenous factors that promote the maintenance or reorganization of patterns of functioning in individuals.

In the present chapter we have tried to define and describe the characteristics of transitions. Additionally, we have suggested a series of methodological procedures for studying them. We hope that our efforts to describe these phenomena more fully will stimulate further theoretical and empirical work on this critical topic.

Acknowledgments

The first author wishes to thank W. Dale Dannefer for helpful comments on earlier versions of this chapter.

REFERENCES

Baltes, P. B., & Goulet, L. R. Status and issues of a life-span developmental psychology. In L. R. Goulet & P. B. Baltes (Eds.), *Life-span developmental psychology*. New York and London: Academic Press, 1970.

Baltes, P. B., & Nesselroade, J. R. Multivariate longitudinal and cross-sectional sequences for analyzing ontogenetic and generational change: A methodological note. *Developmental Psychology*, 1970, 2, 163–168.

Baltes, P. B., Reese, H. W., & Nesselroade, J.R. *Life span developmental psychology: Introduction to research methods*. Monterey, California: Brooks/Cole, 1977.

Baltes, P. B., Reese, H. W., & Lipsitt, L. P. Life-span developmental psychology. *Annual Review of Psychology*, 1980, 31, 65–110.

Blasi, A. Concept of development in personality theory. In. J. Loevinger (Ed.), *Ego development*. San Francisco: Josey-Bass, 1976.

Block, J. *Lives through time*. Berkeley, Calif.: Bancroft, 1971.

Brim, O. G., Jr., & Kagan, J. (Eds.) *Constancy and change in human development*. Cambridge, Mass.: Harvard University Press, 1980.

Brim, O. G., Jr., & Ryff, C. D. On the properties of life events. In P. B. Bales & O. G. Brim, Jr., *Life-span development and behavior* (Vol 3). New York: Academic Press, 1980.

Buhrmester, D., Shaver, P., Furman, W., & Willems, T. *Social skills and relationship development during a life transition*. Paper presented at the American Psychological Association, Washington, D.C., 1982.

Cattell, R. B. The data box: Its ordering of total resources in terms of possible relational systems. In R. B. Cattell (Ed.), *Handbook of multivariate experimental psychology*. Chicago: Rand McNally, 1966.

Coan, R. W. Child personality and developmental psychology. In R. B. Cattell (Ed.), *Handbook of multivariate experimental psychology*. Chicago: Rand McNally, 1966.

Connell, J. P. *A model of relationships among children's self-related cognitions, affects and academic achievement*. Unpublished doctoral dissertation, University of Denver, 1981.

Connell,J. P., & Goldsmith, H. H. A structural modeling approach to the study of attachment and strange situation behavior. In R. Emde & R. Harmon (Eds.), *The development of attachment and affiliative systems*. New York: Plenum, 1982.

Connell, J. P., & Tero, P. Aspects of continuity and change in children's self-related cognition and affects within the academic domain. Unpublished manuscript, University of Rochester, 1982.

Danish, S. J., Smyer, M. A., & Nowak, C. A. Developmental intervention: Enhancing life-event processes. In P. B. Baltes & O. G. Brim, Jr. (Eds.), *Life-span development and behavior* (Vol. 3). New York: Academic Press, 1980.

Emde, R., Gaensbauer, T., & Harmon, R. Emotional expression in infancy: A bio-behavioral study. *Psychological Issues* (Vol. 10, No. 37). New York: International Universities Press, 1976.

Emmerich, W. Continuity and stability in early social development. *Child Development*, 1964, 35, 311–332.

Emmerich, W. Personality development and concepts of structure. *Child Development*, 1968, 39, 671–690.

Eurich, A. C. (Ed.). *Major transitions in the human life cycle*. Lexington, Mass.: Lexington, 1981.

Flavell, J. H. Stage-related properties of cognitive development. *Cognitive Psychology*, 1971, 2, 421–453.

Furman, W., Buhrmester, D., Shaver, P., & Willems, T. *Changes in relationships and social networks during the transition to college.* Paper presented at the International Conference on Personal Relations, Madison, Wisconsin, 1982.

Glaser, B. G., & Strauss, A. L. *Status passages.* Chicago: Aldine, 1971.

Goldberger, A. S. Structural equation models: An overview. In A. S. Goldberger & O. D. Duncan (Eds.), *Structural equation models in the social sciences.* New York: Seminar Press, 1973.

Gottlieb, G. The role of experience in the development of behavior and the nervous system. In G. Gottlieb (Ed.), *Neural & behavioral specificity.* New York: Academic Press, 1976.

Harter, S., & Connell, J. P. A model of the relationship among children's academic achievement and their self-perceptions of competence, control, and motivational orientation. In J. Nicholls (Ed.), *The development of achievement motivation.* Greenwich, Conn: JAI Press, in press.

Hultsch, D. F., & Plemons, J. K. Life events and life-span development. In P. B. Baltes & O. G. Brim, Jr. (Eds.), *Life-span development & behavior* (vol. 2). New York: Academic Press, 1979.

Jöreskog, K. G., & Sörbom, D. *LISREL-IV: Analyses of linear structural relationships by the method of maximum likelihood.* Chicago: National Educational Resources, 1978.

Kagan, J. The three faces of continuity in human development. In D. A. Goslin (Ed.), *Handbook of socialization theory and research.* Chicago: Rand McNally, 1969.

Kimmel, D. *Adults and aging.* New York: Wiley, 1974.

McCall, R. B. Challenges to a science of developmental psychology, *Child Development*, 1977, 48, 333–344.

McCall, R. B., Eichorn, D. H., & Hogarty, P. S. Transitions in early mental development. *Monographs of the Society for Research in Child Development*, 1977, 42 (3, serial No. 171).

Magnusson, D., & Endler, N. S. *Personality at the crossroads: Current issue in interactional psychology.* Hillsdale, N.J.: Lawrence Erlbaum, 1977.

Meredith, W. Rotation to achieve factorial invariance. *Psychometrika*, 1964, 29, 187–206.

Mischel, W. *Personality and assessment.* New York: Wiley, 1968.

Modell, J., Furstenberg, F., & Hershberg, T. Social change and transitions to adulthood in historical perspective. *Journal of Family History*, 1976, 1, 7–33.

Nesselroade, J. R. Temporal selection and factor invariance in the study of development and change. In P. B. Balter & O. G. Brim, Jr. (Eds.), *Life span development and behavior.* (Vol. 5). New York: Academic Press, in press.

Neugarten, B. L. Dynamics of transition of middle age to old age: Adaptation and the life cycle. *Journal of Geriatric Psychiatry*, 1970, *4*, 71–87.

Reese, H. W., & Smyer, M. A. *The dimensionalization of life events*. Paper presented at the West Virginia University Conference on Life-Span Developmental Psychology: Nonnormative Life Events, Morgantown, West Virginia, 1980.

Sackett, G. P., Sameroff, A., Cairns, R. B., & Suomi, S. J. Continuity in behavioral development: Theoretical and empirical issues. In K. Immelmann, G. Barlow, L. Petrinovich, & M. Main (Eds.), *Behavioral development: The Bielefeld Interdisciplinry Project*. Cambridge: Cambridge University Press, 1981.

Sameroff, A. J., & Chandler, M. J. Reproductive risk and the continuum of caretaking casualty. In F. D. Horowitz (Ed.), *Review of Child Development Research* (vol. 5). Chicago: University of Chicago Press, 1975.

Sells, S. B., Roff, M., Cox, S. H., & Mayer, M. *Peer acceptance–rejection and personality development*. Office of Education: final report of project OE 5–0417, Contract OE 2–10–051, January, 1967.

Shaver, P., Furman, W., Buhrmester, D., & Willems, T. *State and trait loneliness during the transition to college*. Paper presented at the meetings of the American Psychological Association, Los Angeles, 1981.

Stewart, A. J. The course of individual adaptation to life changes. *Journal of Personality and Social Psychology*, 1982, *42*, 1100–1113.

Strauss, S. Inducing cognitive development and learning: A review of short-term training experiments. I: The organismic developmental approach. *Cognition*, 1972, *1*, 329–357.

Van Gennep, A. *The rites of passage*. Chicago: University of Chicago Press, 1960.

Wohlwill, J. F. Methodology and research strategy in the study of developmental change. In L. R. Goulet & P. B. Baltes (Eds.), *Life-span developmental psychology: Research and theory*. New York: Academic Press, 1970.

Wohlwill, J. F. *The study of behavioral development*. New York: Academic Press, 1973.

CHAPTER 8

Self-produced Locomotion

AN ORGANIZER OF EMOTIONAL, COGNITIVE, AND SOCIAL
DEVELOPMENT IN INFANCY

Bennett I. Bertenthal, Joseph J. Campos, and Karen Caplovitz Barrett

In the last quarter of the first year the baby is no longer an observer of the passing scene. He is in it. Travel changes one's perspective. A chair, for example, is an object of one dimension when viewed by a six-month-old baby propped up on the sofa, or by an eight-month-old baby doing push-ups on a rug. It's even very likely that the child of this age confronted at various times with different perspectives of the same chair would see not one chair, but several chairs, corresponding to each perspective. It's when you start to get around under your own steam that you discover what a chair really is. . . .

We can multiply such studies in the nature of objects to include nearly everything accessible to him. It is a colossal undertaking, a feat of learning of such magnitude in such a brief time that we have no analogies in later life which compare in scale.

Selma Fraiberg, *The Magic Years* (pp. 52–54)

INTRODUCTION

A few years ago, Emde, Gaensbauer, and Harmon (1976) highlighted two periods of rapid developmental reorganization in infancy. These

Bennett I. Bertenthal • Department of Psychology, Gilmer Hall, University of Virginia, Charlottesville, Virginia 22904.   **Joseph J. Campos** • Department of Psychology, University of Denver, University Park, Denver, Colorado 80210.   **Karen Caplovitz Barrett** • Department of Psychology, University of Denver, University Park, Denver, Colorado 80210. The work on this chapter was supported by NIMH grant MH-23556, NICHD grant HD-16195, and a fellowship from the American Association of University Women.

periods were characterized by dramatic changes in perceptual, cognitive, and especially emotional functions. The period from 7 to 9 months of age is one of these times of rapid reorganization. It is marked by numerous changes in sensorimotor intelligence, including the beginnings of representation, changes in object permanence, new modes of understanding spatial relationships, more complex forms of imitation, and the beginnings of concept formation. This period also appears to be characterized by a burgeoning of fear: Infants at this age react aversively to separation, strangers, heights, looming stimuli, and various unfamiliar toys and objects (Scarr & Salapatek, 1970). The inverse of fear—security—also begins to be clearly evident. The child becomes capable of using the attachment figure as a "haven of safety" and as a "secure base for exploration" (Ainsworth, 1973; Bowlby, 1969). The important changes taking place in the attachment relationship herald major changes in other social contexts as well, including peer and sibling relationships and sociability to strangers (Campos, Barrett, Lamb, Goldsmith, & Stenberg, 1983).

These changes involve so many psychological functions, influence processes which are so different from one another, and take place in such a narrow time span that an underlying neurophysiological maturation has been postulated as a single unifying determinant. Additional support for the belief in maturational determination of the 7-month developmental shift comes from the consistent reports of low and nonsignificant correlations among the various newly emerging behavioral capacities (Fischer, 1980), as well as the improbability that classical learning mechanisms can produce such profound changes in such a short time (Kagan, Kearsley, & Zelazo, 1978). The most prevalent maturational interpretation of the 7-month developmental shift postulates a new level of organization of function by the frontal lobes of the brain. The frontal lobes control planning, foresight, delayed performance, anticipatory emotions such as fear, and concept formation—in short, the very processes which appear to be undergoing rapid changes. The frontal lobes also control other processes like handedness and speech production, changes which are now beginning to be identified in the second half-year of life (Ramsay, 1984).

Maturation is not the only mechanism of developmental transition, however, and certainly it does not operate in an experiential vacuum. Most developmental approaches have recently emphasized epigenetic pathways for the emergence of new cognitive, neurophysiological, and emotional abilities (Gottlieb, 1983). Epigenetic conceptualizations stress how the development of a nodal skill or process sets the stage for the emergence of other processes which depend, in part, upon inputs from the nodal skill for their own development. In the present chapter, we propose that the acquisition of self-produced locomotion—specifically, forward movement on all four limbs—is such an epigenetic event and that its emergence is fundamentally linked to many of the developmental changes taking place between 7 and 9 months of age. We will present data demonstrating the role of crawling as a determinant, and not merely a correlate, of sensorimotor and emotional developments. Our goal in this chapter is not to suggest that self-produced locomotion is the sole organizer of the 7-month shift. It is only one such process, but nonetheless it is an important one. Nor do we deny the importance of cerebral maturation in the 7-month shift, but we believe that maturation is more complexly interwoven with experience than hitherto assumed. Indeed, there is now evidence that self-produced locomotion may *facilitate* brain development and recovery of brain function following lesions (Dru, Walker, & Walker, 1975).

SELF-PRODUCED LOCOMOTION IN THEORETICAL PERSPECTIVE

There are few events in the early development of the infant that are greeted with as much enthusiasm by parents as the development of self-produced locomotion, the earliest forms of which are creeping and crawling. Such parental enthusiasm is understandable, since locomotion creates a new level of interaction between the baby and the environment. For example, the baby is able more directly to control interactions with other people by crawling to or away from them. Also, as attested to by the common practice of "child-proofing" a house, the baby is able to expand its arena of exploration to objects within several meters' distance. The development of locomotion provides further confirmation to parents that their baby is developing normally and becoming more similar to them. From the infant's standpoint, the emergence of crawling expands his or her perspective in a way that can only be described as a new world view.

The significance of this event has not escaped scrutiny by psychologists. For diverse reasons, which will be elaborated later, theorists as different as Piaget (1937/1954), Held and Hein (1963), Gibson (1979), Ainsworth (Ainsworth, Blehar, Waters, & Wall, 1978), Spitz (1965), and Mahler (Mahler, Pine, & Bergman, 1975) consider the development of locomotion to be an important behavioral accomplishment. However, theoretical interest has only rarely been converted into programmatic empirical research. According to Kopp (1979), there was some interest in the development of this skill during the 1930s when investigators such as Gesell (1929) were concerned with cataloguing the behavioral changes in the infant. Although there was a brief flirtation with relating locomotion and intellectual development, this interest quickly dissipated when it was learned that gross correlations between the two domains were notoriously unreliable. Until recently, there was little else relevant to the topic.

This situation is beginning to change. No doubt the growing interest in extensions of Piagetian theory into new directions (Haith & Campos, 1977) and in new conceptualizations of the relationship between specific skills and cognitive development (e.g., Fischer, 1980; Siegler, 1981) has contributed to the revival of empirical interest in locomotion. Also, research methodologies are becoming more sophisticated and more sensitive to the sometimes subtle effects of early development. The effects of these new advances are most obvious in relationship to a similar area of motoric development, visually guiding reaching (e.g., Von Hofsten, 1983). However, persistent problems have impeded the study of locomotion, including the difficulty of devising tasks that specifically reflect the influence of locomotion (for exceptions, see Benson & Uzgiris, 1981, and Corter, Zucker, & Galligan, 1980). As a result, few researchers have investigated the consequences of the emergence of crawling.

It is our contention that the most important (although not the only) contribution of self-produced locomotion to early development is that it represents a *setting event* for the development of other skills. That is, infants' new mobility greatly increases the probability of their encountering a host of skill-enhancing experiences. We do not suggest that the skills normally enhanced through self-produced locomotor experience cannot be developed through any other route. Rather, we emphasize that in the child's ordinary and expectable environment, self-produced movement mediates important developmental changes during the third quarter of the first year of life. In this chapter, we review our current research documenting the role of newly acquired locomotion in the development of wariness of heights, in spatial cognition, in concept formation, and in social communication. We then speculate on the implications of these and related findings for an understanding of the development of the self and the elaboration of the baby's affectivity.

DEVELOPMENT OF SPATIAL SENSORIMOTOR PERFORMANCE: FEAR OF HEIGHTS

Much anecdotal evidence, along with considerable theoretical speculation, suggests that self-produced locomotion plays a facilitative role in the development of an observer's understanding of spatial relationships in a new environment. At a subjective level, we can all relate to the problem presented by being driven to a new location only to discover after we have arrived that we must drive ourselves home. Our confidence in the return trip is much greater when we have had the opportunity to form a spatial map through the action of driving to the new location. Similarly, in discussing a number of recent findings, Bremner (1980) has postulated that the infant's ability to remember specific locations where a favored object has been hidden is facilitated by active movement. The developmental question emerging from such considerations is whether the acquisition of self-produced locomotion is, in fact, responsible for changes in spatial understanding.

It is somewhat paradoxical to find so little research in the literature on this issue in light of the classic experiments of Held and Hein that demonstrated such a dramatic effect of self-produced locomotion upon the behavior of young kittens and monkeys. In these experiments (e.g., Held & Hein, 1963), kitten littermates and their mothers were reared in completely darkened rooms from birth until selective exposures to visual stimulation began at 8–12 weeks. All visual exposure took place in a specially designed apparatus known as a kitten carousel (see Figure 1) that equated the visual experience received by each littermate. In this apparatus, one littermate was harnessed to an arm that rotated around the inside of a vertically striped cylinder while the other was placed in a gondola attached to a second arm. The movement of the gondola was yoked to the movement of the harnessed kitten; thus one kitten received visual feedback with self-produced locomotion while the other received feedback from passive movement.

At the end of each exposure period, both kittens were tested for visual placing and response to an approaching object. As soon as one member of each pair showed positive responses to both of these measures, kittens were tested for avoidance of an apparent drop-off on the deep side of a visual cliff. The results revealed that after receiving equal exposure to patterned visual stimulation, all actively reared kittens showed avoidance, whereas none of the paired littermates did. In fact, even after a much longer period of exposure (300–400 hours), passively reared kittens still did not demonstrate visual-motor coordination (Hein, 1972).

These results cannot be explained as generalized response inhibition given the results of a subsequent study (Hein, Held, & Gower, 1970), which employed monocular stimulation, using each kitten as its own control. One eye was exposed during active transport while the other was exposed during passive transport. The results showed that subsequent visual-motor coordination was specific to the eye exposed during active movement. Accordingly, these and related findings (Hein,

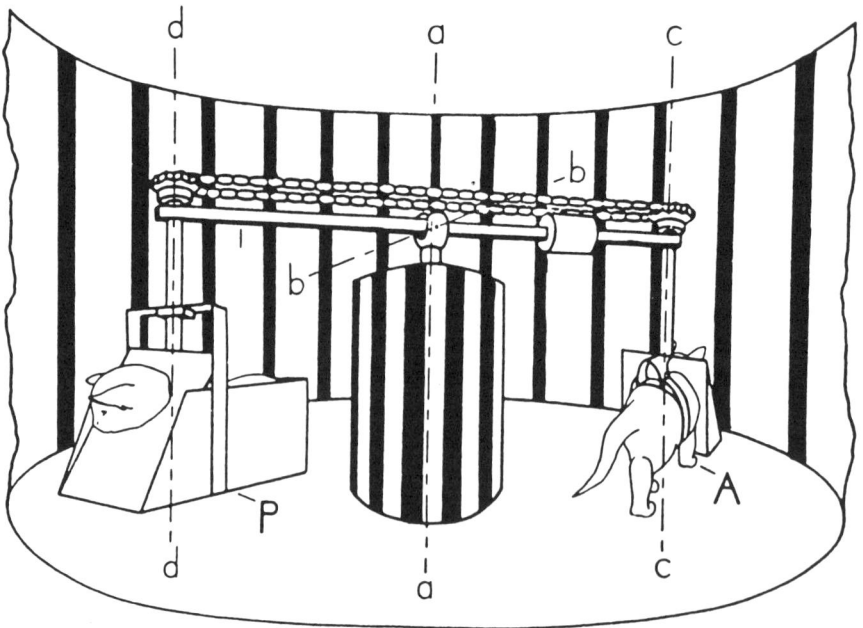

Figure 1. The kitten carousel apparatus used by Held and Hein (1963) to assess the effects of self-produced locomotion on the development of spatially appropriate behavior. (From "Movement-Produced Stimulation in the Development of Visually-Guided Behavior" by R. Held and A. Hein, *Journal of Comparative and Physiological Psychology*, 1963, *81*, 394–398. Copyright 1963 by the American Psychological Association. Reprinted by permission.)

1972) suggest that visually guided behavior is comprised of component systems that may be acquired separately during development.

Perhaps the major reason why these findings have not been extrapolated to the human infant relates to questions raised about the ecological validity of the Held and Hein studies. Walk (1978), for instance, has argued that the effect obtained in the kitten carousel studies may be less the result of self-produced movement *facilitating* perceptual-motor development than it is the result of passive movement *disrupting* the animal's visual coordination. Furthermore, the human infant, in contrast to the dark-reared kitten, has ample self-produced activity that can be correlated with visual input (reaching, for instance) well before crawling is acquired. It thus would appear that postulation of a major influence of self-produced locomotion upon psychological functioning in humans requires that prior motor acquisitions have very localized consequences. Most psychologists seem unprepared to entertain such an assumption. Thus, the extrapolation of findings from kittens and monkeys to human infants requires unjustified assumptions. More direct assessments using human infants are needed. The empirical data which we are about to review, derived from a large program of such assessments, demand a reconsideration of the effects of locomotion and a conceptualization of the ways in which this skill plays a role in development during the third quarter of the first year of life.

Visual Cliff Research

The behavior of human infants on the visual cliff represents one of the most extensively studied, but also one of the most widely misrepresented, phenomena in the field of human infancy. Since the publication of Walk and Gibson's (1961) classic monograph, most reviewers have perpetrated the conclusion that avoidance of height is innate in the human infant and merely requires the onset of locomotion for the pre-wired avoidance capacity to be manifest. Most reviewers have also assumed a one-to-one correspondence between depth perception and fear of heights in the human infant. These misrepresentations have endured to a large extent because it makes biological, adaptive sense for humans to possess an innate wariness of heights (Bowlby, 1973); falls, of course, can lead to serious injury and even death. It is now clear, however, that the onset of wariness of heights takes place some time *after* the acquisition of locomotion (Campos, Hiatt, Ramsay, Henderson, & Svejda, 1978; McGhee, 1979; Rader, Bausano, & Richards, 1980). All recent studies with the cliff, regardless of whether one assesses wariness by the locomotor crossing method (i.e., having the mother urge her locomotor infant to cross to her over the deep side of the cliff; Scarr & Salapatek, 1970) or by the direct placement method (i.e., whereby an experimenter places the infant directly atop the deep side of the cliff and measures nonlocomotor indices of fearfulness, such as heart rate acceleration; Schwartz, Campos, & Baisel, 1973) point to a developmental shift between 5 and 9 months of age toward increased wariness of heights. The most recent research from our laboratory has clearly implicated self-produced locomotion as one of the most significant factors in accounting for the developmental shift.

In one of the first studies relevant to this issue, we (Campos *et al.*, 1978) conducted a longitudinal study of the responses of 15 infants on the visual cliff. Infants were first tested 1–3 weeks after they began to crawl, and were tested once every 10–14 days until avoidance of the deep side was shown. Infants ranged in age from 6.3 to 8.3 months when first brought to the lab and were tested using the standard locomotor crossing method developed by Walk and Gibson (1961). As we mentioned before, this method involves the mother's calling the infant to cross to her over the deep or shallow sides of the cliff table. In violation of the common belief that infants will show avoidance of the deep side from the onset of locomotion, all 15 infants in this study crossed to the mother on the deep side of the cliff on at least one of the first two testing occasions. This behavior showed a dramatic change, however, sometime during the next few weeks of testing. All infants began to show avoidance of the deep side either by refusing to cross or by crossing the deep side using detour behavior. This latter behavior involved crawling on the centerboard of the cliff until reaching the far side, and then either holding onto the wall while crossing the glass to the mother, or otherwise minimizing the distance to cross to the mother over the deep side.

Additional support for the role of locomotor experience was provided by an examination of the difference in infants' latency to cross to the mother over the deep versus the shallow side of the cliff. For this analysis infants were first assigned equally to one of three groups defined by crawling experience; the mean age was virtually identical for each of the three groups. Infants with the briefest amount of crawling experience at the first testing (7 days) showed the least difference in latency to cross the two sides, whereas infants with more experience (14 or 21 days) showed a significantly greater difference in latency scores. Taken together, these results demonstrated that avoidance was not shown until crawling began and that the magnitude of the avoidance response was systematically related to crawling experience (at least during the first month of crawling). Moreover, the latency difference scores demonstrated that infants with more crawling experience differentiated the two sides, rather than merely showing increased inhibition of crawling on both sides of the visual cliff table.

Clearly, one interpretation for these results is that crawling experience is related to the development of fear of heights; however, the small sample size in this study renders tenuous any firm conclusions. In addition, a second interpretation must also be considered. It is possible that infants begin to react aversively to heights prior to crawling but that the initial acquisition of a new skill leads to a temporary disorganization of behavior. (A similar idea has been suggested by a number of cognitive-developmental theorists, such as Flavell, 1970, to explain a temporary regression in the child's cognitive performance when undergoing a transition to a new level of cognitive functioning.)

We therefore present the results from a study employing a second paradigm (Svejda & Schmid, 1979) that provide confirmation for the influence of locomotion experience on visual cliff wariness and also serve to disconfirm the latter proposed interpretation. This paradigm was designed specifically to test the role of locomotion in the developmental shift that occurs on the visual cliff. Infants were studied at only one age (7.3 months) so that locomotor experience would not be confounded with age. A total of 92 infants were tested (half locomoting and half not yet locomoting) using a variation of the direct placement procedure. This new procedure was called the descent paradigm because heart rate was assessed during the 3 seconds that the baby was being lowered down to the surface of the cliff as well as during the first second of contact with the surface. (Note that the baby's body is held parallel to the surface with eyes pointed downward while being lowered down to the cliff.) This particular paradigm is extremely well suited to measuring the infants' response to height because the assessment is conducted prior to the time when cardiac responses are contaminated by the infant's movements on the glass.

The results from this study, which are presented in Figure 2, strengthened our conviction that the development of crawling is linked to fear of heights. Prelocomotor infants showed nonsignificant changes in cardiac responsiveness on both the shallow and deep sides. On the other hand, locomotor infants showed significant heart rate acceleration during placement on the deep side, but little change on the shallow side. We attribute the shift to cardiac acceleration on the deep side to the development of wariness of heights. If our interpretation of the heart

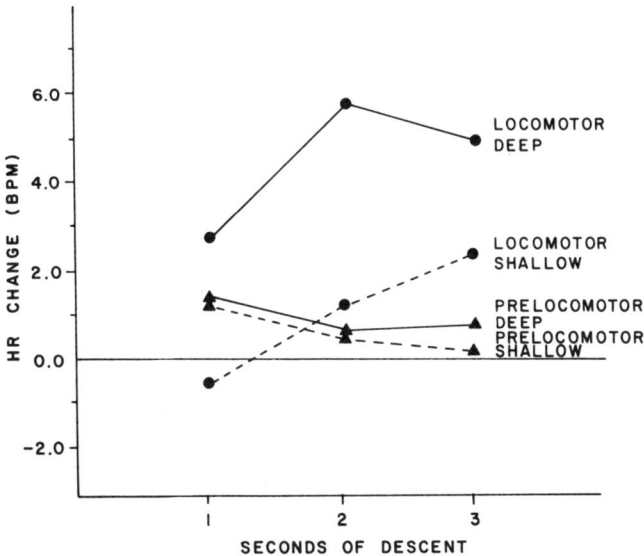

Figure 2. Heart rate changes during descent toward the deep and shallow sides of the visual cliff in 7.3-month-old infants as a function of whether the infants had had crawling experience or not. Data are plotted to show the second-by-second heart rate changes during descent toward each side of the cliff.

rate response is correct, it implies that the previously discussed alternative interpretation received no support from our data: There is no evidence that fear of heights is manifested by most subjects prior to locomotion.

In both of the preceding studies, however, crawling experience was confounded with age of onset of locomotion, a variable which Rader, Bausano, and Richards (1980) and Richards and Rader (1981) found to be the strongest predictor of avoidance of the deep side of the visual cliff. We thought it important to examine the potential influence of age of onset of locomotion more systematically during the early weeks of crawling experience. Therefore we assessed separately the effects of age, locomotor experience, and their interaction on the development of *avoidance* of crossing to the mother over the deep side of the visual cliff. In this study (Barrett & Campos, 1983), mothers were contacted when their infants were approximately 5½ months of age and were instructed to notify us as soon as each infant had begun to move about on his or her own. Infants were brought into the lab for testing after they had had either 11 days or 41 days of locomotor experience. We then categorized the infants according to their age at the onset of self-produced locomotion. Specifically, we divided the infants into those who began to crawl at 6.5 months (± 2 weeks), 7.5 months (± 2 weeks), or 8.5 months (± 2 weeks), giving us a total of six groups of subjects, with approximately 16 subjects in each group, half of each sex:

a. 6.5-month-olds with 11 days locomotor experience
b. 7.5-month-olds with 11 days locomotor experience
c. 8.5-month-olds with 11 days locomotor experience
d. 6.5-month-olds at locomotion onset, with 41 days crawling experience
e. 7.5-month-olds at locomotion onset, with 41 days crawling experience
f. 8.5-month-olds at locomotion onset, with 41 days crawling experience

By crossing age of acquisition of locomotion with duration of locomotor experience, we were able to test Rader *et al.*'s (1980; Richards & Rader, 1981) hypothesis that age and not locomotor experience accounts for visual cliff avoidance behavior. The results of this study revealed a clear effect of locomotor experience that was independent of the age when self-produced locomotion was first manifested. This effect of experience was evident regardless of whether we used nominal data (the number of infants who descended onto the deep side of the cliff) or interval data (the latency to descend from the center board of the visual

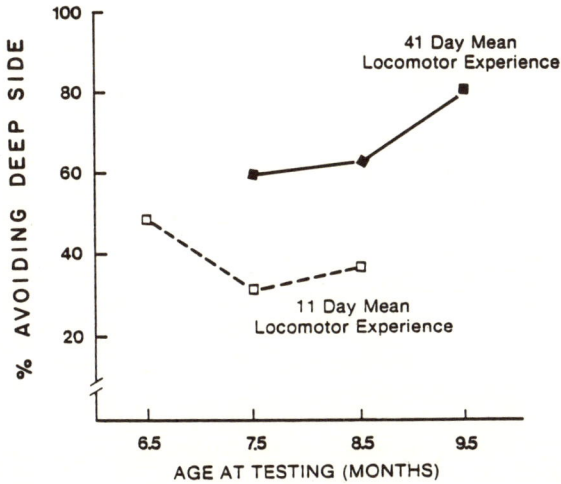

Figure 3. The percentage of subjects who avoided any descent onto the deep side of the cliff as a function of both locomotor experience and age of testing.

cliff onto the deep side minus the latency to descend onto the shallow side of the cliff).

Both the nominal and the interval data are presented in Figures 3 and 4. (We used only first-trial data since this could not be influenced by prior experience on the cliff. Moreover, since we found a significant trial by side effect for latency scores, it did not seem reasonable to average latency scores across the two trials.) As can be seen in Figure 3, regardless of whether the infant developed the ability to crawl at 6.5, 7.5, or 8.5 months of age, approximately 30%–50% of the infants avoided descending onto the deep side on the first trial when they had only 11 days of locomotor experience. However, the infants who had 41 days of locomotor experience when tested for the first time on the visual cliff showed avoidance in about 60%–80% of the cases, again regardless of the age of acquisition of self-produced locomotion.

When the data were quantified in terms of the differences in latency to leave the center board and descend onto the deep versus the latency to descend onto the shallow side of the cliff (Figure 4), a similar effect of locomotor experience was observed which was again independent of the effects of age. Infants locomoting only 11 days descended onto the shallow side about 30 to 50 seconds faster than they descended onto the deep side, whereas those with 41 days of locomotor experience descended onto the shallow side 65 to 90 seconds faster than they did onto the deep side. Although the Figures show an apparent age trend in

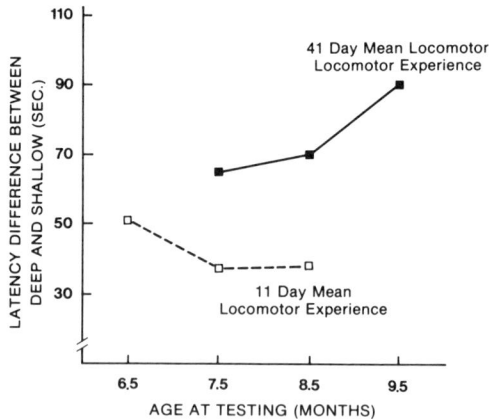

Figure 4. Latency differences to descent onto the deep or the shallow sides as a function of both age and locomotor experience.

these data, it should be noted that this trend did not reach significance in any of the analyses executed for either locomotor experience group. The outcome of both types of analyses, then, demonstrates that locomotor experience plays a role independent of age in the development of avoidance of heights.

The discrepancy between the findings of these studies and the findings of Rader (Rader *et al.*, 1980; Richards & Rader, 1981) may be better understood after considering one critical difference between them. In the Rader studies, most infants were brought into the laboratory after having had extensive locomotor experience. Few of the infants had had less than 30 days locomotor experience, and most had had considerably more. It appears, however, that the relation between visual cliff behavior and locomotor experience is strongest in the first five weeks after crawling onset and reaches asymptote not long thereafter because most of the new learning afforded by crawling is realized by then (Bertenthal & Campos, in press). Accordingly, it is not surprising that Rader and her associates reported only a modest effect for self-produced locomotor experience since most infants were tested at a time when the relation between crawling experience and cliff performance would have already reached asymptote. Our research group, on the other hand, which has studied the transition from no locomotion to locomotion and the effects of immediate locomotor experience, repeatedly finds consistent and large magnitude effects.

Despite the strength of these findings, there is a need to provide even more direct evidence of a functional relationship between the experience of self-produced locomotion and the development of fear of heights. It is possible that the development of locomotion and the emergence of wariness are both determined by a common third factor, which causes the emergence of wariness of heights to appear to follow the emergence of crawling. This problem is particularly evident in age-held constant designs, such as the Svejda and Schmid (1979) study described earlier, but also applies to any study that does not directly manipulate locomotor experience, as we were able to do in another study.

This manipulation was achieved by providing artificial self-produced locomotor experience to infants with wheeled walkers and testing them after they had obtained 40 hours of locomotor experience (Campus, Svejda, Bertenthal, Benson, & Schmid, 1981). Since some infants began locomoting without walker assistance prior to testing on the cliff, the final experimental design involved two groups of infants with walker experience: One group of 18 infants was tested prior to crawling (prelocomotor walkers) and another group of 16 infants was tested after less than 7 days of spontaneous crawling (locomotor walkers). The performance of infants in these two groups was compared to the performance of age-matched controls. One control group of 18 infants had had no crawling experience (control prelocomotors), and the other group of 16 infants had all been crawling for up to 7 days (control locomotors). Neither control group had received any artificial self-produced locomotor experience.

All babies were tested using the descent paradigm on the deep side of the cliff; no shallow trials were attempted since the cardiac response showed no significant change on that side in previous studies and it was important to minimize subject attrition resulting from extended testing. The results from this study are summarized in Figure 5. As can be observed, babies with walker experience showed a significantly greater acceleratory response than did the controls. Note also that control infants without crawling experience were the only group to show consistent (although nonsignificant) heart rate deceleration and that these scores were significantly lower than the scores of infants in the prelocomotor walker group. Furthermore, the data demonstrate a greater level of cardiac acceleration for the locomotor walkers than for locomotor controls, suggesting further that the walker experience was important in the development of this response. We conclude, therefore, that locomotion and fear are more than simple covariates of maturation; rather, it is more accurate to view the early development of crawling as functionally related to the onset of fear of heights.

In sum, combining the Rader studies and our own studies provides a reasonably complete developmental picture of the development of avoidance of heights: There is a rapid acquisition of avoidance related to

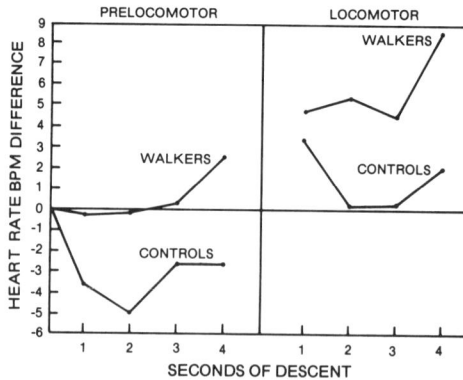

Figure 5. Second-by-second heart rate changes of subjects as they were lowered to the deep side of a visual cliff. Data are grouped according to whether subjects had had locomotor experience provided by walkers or not, and also whether they had had spontaneous locomotor experience or not. The ages of subjects in the four groups did not differ from one another.

self-produced locomotion, following which a period of little further increase in avoidance is evident. Such a spurt–plateau developmental function characterizes many developmental processes. It appears clearly to apply to development of avoidance of heights. We will speculate on the way in which self-produced locomotion might produce these effects shortly.

Spatial Orientation Research

One of the fascinating features of human infants' behavior at approximately 6 months of age is a tendency to code the location of objects in relation to their bodies, rather than more objectively in relation to stable environmental landmarks. A good example of this tendency is the well-known "AB̄ error" in object permanence testing, whereby a child who succeeds in finding an object at location A continues to search for it at A despite clearly observing that the object is now hidden at location B. The infant's tendency to behave in this way is considered particularly important by Piaget, who uses this phenomenon to build a strong case for the origins of knowledge in motoric activity (see Lamb & Campos, 1982, for a detailed discussion of this point).

There are now several studies documenting a developmental shift in the infant's coding of object location between 6 and 9 months of age. Most studies report egocentric coding (objects are localized relative to one's body) at 6 months of age and allocentric or objective coding (objects are localized relative to environmental landmarks) beginning around 9 months of age, especially if the environmental landmarks are very salient or very familiar (Acredolo, 1978; Acredolo & Evans, 1980; Bremner, 1978; Bremner & Bryant, 1977; Cornell & Heth, 1979). Bremner, following Piaget (1937/1954), suggested that this developmental shift might be related to the onset of locomotion since an egocentric frame of reference must be continually updated once the infant begins moving around in the environment. As a consequence, egocentric codes are recognized as relatively inefficient and are replaced by allocentric codes, which are independent of the infant's orientation in space.

Recently we put this hypothesis to an empirical test. Infants were tested in a paradigm developed by Acredolo and Evans (1980). The testing took place in a relatively small room lined with green material from floor to ceiling. A small round table was placed in the middle of the room and a movable highchair was placed between one wall and the table so that the infant faced the table when seated. The two walls parallel to the sides of the highchair each contained a centrally located window at a height corresponding to eye level for the baby. One of the windows was unremarkable in appearance except for the presence of a white cloth covering the opening. The other window, however, was surrounded by a series of flashing lights and orange stripes on the wall. These lights and stripes were designed to serve as a salient landmark for that window.

There were two phases to the procedure. In the first phase, the infant was trained to orient toward the window with the lights and stripes every time a buzzer (located under the table) was sounded. The training was accomplished by having an experimenter appear at the window approximately 3 seconds after the buzzer was sounded on each trial. The experimenter would then talk to the baby for about 10 seconds before disappearing behind the window curtain. As soon as the infant anticipated the appearance of the experimenter on four out of five trials, the second phase was initiated. This phase began with the infant being wheeled around to the opposite side of the room so that the infant's orientation to the room was reversed by 180°. The use of an egocentric code would lead infants to look at the opposite window during Phase 2. On the other hand, the positions of the lights and stripes are unaffected by the infant's reorientation; thus infants should still orient toward the same window in Phase 2 if they are using a landmark code. This phase consisted of five trials without feedback; thus the experimenter no longer appeared at the window following the buzzer that was sounded on each trial.

Three groups of infants were tested with this procedure: locomotor infants (X_{age} = 35 weeks), prelocomotor infants (X_{age} = 33.7 weeks), and prelocomotor infants with walker experience (X_{age} = 33.2 weeks).

Although data collection is not complete for this study, certain trends are evident. As summarized in Table 1, locomotor infants were scored as using a landmark code on 74% of the trials whereas an egocentric code was used on only 26% of the trials. On the other hand, prelocomotor infants used an egocentric code on 45% of the trials and an allocentric code on 56% of the trials. In this regard it is interesting to note that these mean percentages were not a function of half the infants using an egocentric code and the other half using a landmark code; rather, these means reflected inconsistent code use at the level of the individual subject. Thus it appears that the prelocomotor infant, at least on this task, does not systematically implement an egocentric code but rather interchangeably uses both codes (or fails to apply any systematic strategy).

Finally, the data for the prelocomotor infants with walker experience (range 4–10 weeks) provide the most convincing evidence for the role of locomotion in this task. The data for this group of infants show that a landmark code was used on 95% of the trials. This pattern of responding is markedly different from the responses shown by the prelocomotor infants without any walker experience.

It therefore seems reasonable to conclude (although somewhat tentatively because of the incomplete state of this study) that the emergence of self-produced locomotion does lead to more consistent use of a landmark code. Of course, this is not to imply that other factors are unimportant; stimulus salience, familiarity of the environment, and affective state all have been shown to influence spatial coding as well (Acredolo, 1982; Acredolo & Evans, 1980; Bremner, 1978). We interpret the preceding results to suggest that locomotion represents one of a number of different organismic and environmental variables that influence the infant's understanding of spatial orientation.

Delayed Locomotion: A Case Study

A major goal in assessing functional relations in development is to demonstrate that two skills are more than merely correlated in their

Table 1. Frequency of Responses Involving Each Spatial Code as a Function of Locomotor Experience

	Egocentric	Landmark
Locomotor ($n = 24$)	.26	.74
Prelocomotor ($n = 8$)	.45	.56
Prelocomotor with walker experience ($n = 6$)	.05	.95

emergence—that the development of one skill has a significant effect on the development of the other. Two principal methods exist for assessing such a relation, enrichment and deprivation procedures (Bertenthal, 1981b). Our testing of prelocomotor infants who had received walker experience represents the application of a specific enrichment technique. However, it is always more convincing when both enrichment and deprivation techniques are applied to the same problem, since the comparability of any experimentally manipulated experience with normal experience is always subject to question (Wohlwill, 1973). As might be expected, this goal is rarely met since ethical considerations all but eliminate opportunities for deprivation experiments. The exception arises when it is possible to test a group of naturally deprived children such as those who are incapable of locomoting because of a *peripheral* orthopedical handicap. (Tests of infants with central nervous system deficits resulting in locomotor handicaps, of course, confound psychological deficits produced by the lack of self-produced movement with those produced directly by the central nervous system lesion.) This approach was suggested to us by Marilyn Svejda, a pediatric nurse practitioner and a collaborator in our research project.

Last year, we had the opportunity to examine longitudinally an orthopedically handicapped child. This particular child was born with two congenitally dislocated hips and after an early operation was placed in a full body cast. We began testing this child on the visual cliff and spatial orientation task (using the same procedures described previously) at 6 months of age and continued testing once a month until the child was 10 months old. During the first two months of testing, the child remained in the cast and showed no evidence of crawling or creeping. Prior to the third visit the cast was removed and the child was placed in a Pavlik harness. (This orthopedic device is much lighter than a cast and allows for movement of the upper body.) Although the mother reported that her child showed some creeping while wearing this device, it appeared from the mother's description that this creeping was extremely effortful and, we suspect, quite limited—not likely to have provided the same increment in attention to the surround as would ordinary crawling. We never observed the creeping in the laboratory.

Two weeks prior to the fourth visit (i.e., at 8.5 months of age) the parents were instructed by their physician to remove the harness for four hours a day, and the mother reported that her child began showing much more proficient creeping when unencumbered by the harness. By the fifth visit, the harness had been removed and the child was crawling.

The results from the two assessments are summarized in Figure 6 and in Table 2. As can be observed, the difference between shallow and

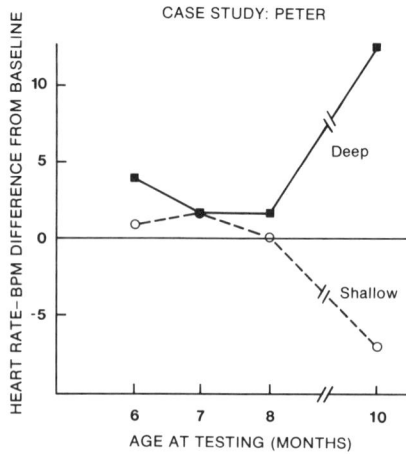

Figure 6. Mean heart rate changes of an orthopedically handicapped infant at different ages, demonstrating a cardiac developmental shift on the visual cliff as a function of the removal of a cast and the initiation of crawling. (At 8.5 months Pavlik Harness was removed 4 hours per day, permitting self-produced locomotion.)

deep cardiac responsiveness was unremarkable until the last visit (10 months) when cardiac acceleration showed a significant increase on the deep side coupled with a deceleration on the shallow side. Similarly, the child's spatial coding of the correct window in the spatial orientation experiment was either egocentric or mixed until the opportunity for consistent experience with self-produced locomotion occurred; henceforth, a landmark code was used.

What can be concluded from these results? Most important, these data are entirely consistent with the notion that self-produced locomo-

Table 2. Frequency of Response for Each Spatial Code as a Function of Age for Orthopedically Handicapped Infant

Age (months)	Egocentric	Landmark
6	1.00	.00
7	.60	.40
8	.60	.40
9	.20	.80
10	.00	1.00

tion does influence the development of spatial cognition. Of course, there are limits as to how much can be generalized from a case study, especially since the orthopedic device delayed locomotion for only a month and a half beyond the mean age at which crawling emerges. Still, when the concordance in performance between tasks and the systematic relation to the emergence of locomotion is considered in the context of all the other confirmatory data, these data merit consideration as additional support for the relation between self-produced locomotion and spatial cognition.

Interpretation

The preceding data clearly demonstrate a functional relation between the development of self-produced locomotion and phenomena involving spatial cognition, such as the development of wariness of heights and the shift from egocentric to allocentric localization of objects. But precisely how does self-produced locomotion bring about such effects? The interplay of several factors appears necessary, if for no other reason than that some of the developmental changes are primarily cognitive whereas others are primarily emotional. We will discuss three factors which we feel are likely to be involved in these developmental changes. These factors are (a) changes in selective attention to the environment resulting in improved extraction of information, (b) more precise calibration of distance emerging from the coordination of visual-motor cues such as convergence and monocular parallax with movement of the body in space, and (c) greater likelihood of emotional communication from the mother during periods of danger or uncertainty. We will elaborate on each of these factors in turn.

Crawling and Changing Attention to the Environment. The acquisition of self-produced movement is likely to facilitate the perception of a constant and stable visual environment. Infants' experience with change in spatial layout is comparatively restricted before the onset of crawling, both because infants spend most of their time in a stationary location and because when transported infants tend to be visually inattentive (Campos, Svejda, Campos, & Bertenthal, 1982). Infants are not inattentive however, when they are crawling about and orienting toward the object or event motivating the locomotion. Moreover, as infants move about, they observe continuous changes in the optic array specifying persistent and varying properties of the spatial layout (Gibson, 1979). As a consequence, infants may become increasingly sensitive to the variant and invariant forms of information in the environment (Bertenthal, 1981a). One such invariant is the location of an object in relation to a landmark; thus the acquisition of voluntary movement may result in the beginnings of consistent allocentric coding of spatial locations.

Self-produced movement may also facilitate the detection of some types of visual information specifying depth. Although infants certainly possess depth perception by the time they are 3 or 4 months of age (e.g., Fox, Aslin, Shea, & Dumais, 1980), it is nonetheless possible that once crawling starts there is further growth of the visual field in the third dimension. By this we mean both that infants attend more frequently to events occurring increasingly far away from their bodies and also that infants become sensitive to new and more subtle sources of information about changes in depth. For instance, although very young infants may be sensitive to the explosive expansion of a surface specifying imminent collision (e.g., Bower, 1974; Yonas, Pettersen, & Lockman, 1979), they may be relatively insensitive to the subtler changes in texture accretion and deletion, or magnification of texture, that specifies the gradual approach of a surface from a greater distance. It seems plausible that once crawling begins these other changes in depth-specifying information may start to become more salient to the child, especially because of their importance in visual guidance of locomotion (Gibson, 1979).

Crawling and Calibration of Distances.   Not all sensorimotor changes resulting from the development of crawling are necessarily the result of changes in attentiveness and the resulting improved detection of visual information. There is reason to believe that a correlation of active movement with binocular convergence or monocular parallax may be needed to calibrate the distance of an object from the observer (Campos *et al.*, 1982; Kaufman, 1979; Piaget, 1937/54). Some have argued, for instance, that information from the traditional primary cues for depth perception provides the perceiver only with information about the relative distances of objects from one another. Thus, in order to determine the precise distance of an object or the precise size of an unfamiliar object, the person must relate the degree of motion parallax or binocular convergence to previous instances of movement toward an object (see Campos *et al.*, 1982, for an elaboration of this point).

If this argument is valid, then prelocomotor infants, who have had experience calibrating distances only within arm's reach and slightly beyond, would show progressively poorer distance estimation capacities the farther away an object is. Once the development of crawling has begun, however, the coordination of movement with depth-specifying information calibrates the perceptual system to perceive greater distances with reasonable accuracy. One testable proposition that emerges from this line of reasoning is that prelocomotor and locomotor infants should differ in size constancy abilities (one of the few perceptual processes that require absolute distance information, but only when the target object is unfamiliar and at large distances from the infant).

Crawling and Emotional Communication about Danger. Neither an attentional hypothesis nor a hypothesis about calibration of distances can entirely account for why self-produced locomotion results in wariness of heights. Some other factor besides possible changes in depth perception or spatial cognition capacities must also play a role.

At one time, it was widely believed that perception of heights innately triggered fear (e.g., Freedman, 1974). Our data, however, raise the possibility of a much more complicated picture, since the emergence of wariness of heights tends to occur some time after emergent locomotion capacities have placed the infant at some risk of suffering serious falls. A plausible alternative interpretation to the innate fear hypothesis is that wariness of heights is developed through a process of associating edges or heights with falling accidents. This hypothesis also appears too simplistic, however. Studies by Walk (1978), Scarr and Salapatek (1970), and our own laboratory (unpublished data) all demonstrate a minimal role of falling accidents on cliff avoidance behavior. Accordingly, we have speculated that emotional communication in situations of danger, but in which the infant is not actually experiencing a fall, may help account for the development of wariness of heights. There is no doubt that encounters with dangerous heights become much more likely following the acquisition of locomotion and that vigilant caregivers will prevent infants from falling by hovering over them or by removing them from dangerous locations into which they crawl. It also seems likely that caregivers often emit an emotional expression sufficiently salient to mediate the infant's acquisition of avoidance of heights. Support for this interpretation comes from recent work demonstrating the power of emotional expressions (whether the mother's or another adult's) to regulate the infant's behavior (Sorce, Emde, Campos, & Klinnert, in press; Klinnert, Campos, Sorce, Emde, & Svejda, 1983), and the carryover effects of such emotional communication on subsequent behavior regulation in similar circumstances (Svejda & Campos, 1982).

Neither the selective attention hypothesis, the calibration of distance hypothesis, nor the emotional communication hypothesis as explanations for the phenomena described earlier has been tested empirically. However, there is now evidence that locomotion is indeed related to increased sensitivity to emotional communication, as well as to increased utilization of such information in regulation of behavior. We will now turn to evidence of this phenomenon.

SOCIAL COMMUNICATION

A number of theorists (Ainsworth et al., 1978; Bowlby, 1973; Fraiberg, 1959; Mahler et al., 1975; Spitz, 1965) have noted that the onset of crawling leads to functional changes in the attachment relationship between infant and mother. As infants begin to take advantage of their new locomotor abilities for exploring the environment, there is a need to balance the loss of proximal security with new forms of communication with the mother. One such form of communication is *social referencing* (Campos & Stenberg, 1981; Klinnert *et al.*, 1983), defined as the search for emotional information from the face, voice, or gesture of another (e.g., the mother) to help one (e.g., an infant) resolve uncertainties or disambiguate situations. Such a process demands maternal availability for sending signals and infant sensitivity for receiving them. Furthermore, the infant must possess certain decoding rules if the maternal signal is to be understood.

Our research on the relation between locomotion and social communication is much more modest than the previously described research on spatial cognition; yet, the results are no less impressive. Much of what we now know has come from an extensive study conducted by Garland (1982). This study was designed to test explicitly whether or not the roots of social referencing could be traced to the emergence of crawling.

The testing procedure involved five phases and was based in part upon a paradigm devised by Carr, Dabbs, and Carr (1975). During the first two phases, each lasting 6 minutes, the infant and the mother were placed together in a playroom. The child was seated on the floor with a number of toys located around her general vicinity. The mother was seated in a chair situated diagonally across the room from the child. In one phase the mother faced the child and in the other the chair was turned around so that the mother's back faced the child. During Phase 3, the mother left the room for 3 minutes and returned for a reunion (Phase 4). The fifth and last phase began as soon as the child was again willing to sit down and play with the toys while the mother returned to her chair. During the fifth phase a stranger entered the room and the mother was instructed (and previously trained) either to smile or to show distress to the presence of the stranger, but not to communicate in any other fashion. After a minute the stranger left, signaling the end of the study.

Subjects were 16 prelocomotor and 16 locomotor infants assigned evenly to a younger ($M = 6.4$ months) and older ($M = 8.5$ months) age group. Younger and older infants showed no significant difference in locomotor experience (4.5 and 5.1 weeks, respectively). Locomotor skill was rated on a 4-point scale (McGraw, 1943) such that the variability in early adeptness in crawling could be systematically examined rather than simply treated as error variance. The younger and older locomotor groups in this study showed no significant difference in mean locomotor efficiency scores (2.62 and 3.25, respectively).

There are two sets of findings that we wish to discuss. The first set serves to extend the findings from the previous section. In that section, we suggested that infants show increasing sensitivity to the invariant information in the environment. The present results suggest that locomotor infants also develop greater sensitivity to particular forms of varying information. A detailed coding of what infants looked at during the first two phases of this study revealed that prelocomotor infants spent as much time looking at the walls, floors, and pictures in the room as they did looking at their mother and the toys. On the other hand, locomotor infants spent significantly more time looking at their mother or at the toys. An important characteristic of both the mother and the toys is that they present varying as well as nonvarying forms of information and hence may be more interesting and important than the walls, floor, and pictures, which present only static information.

One possibility suggested by this finding is that locomotor infants should show greater sensitivity to the varying orientation of the mother, especially if they seek to communicate nonverbally with the mother. As can be observed in Figure 7, locomotor infants looked at their mothers significantly more often when mothers were facing the rear of the room, possibly because they were trying to reinitiate communication with their caregiver. There is good reason to suspect that this situation had a disorganizing effect on behavior because locomotor infants showed significantly less play with the toys when the mother was facing away from them.

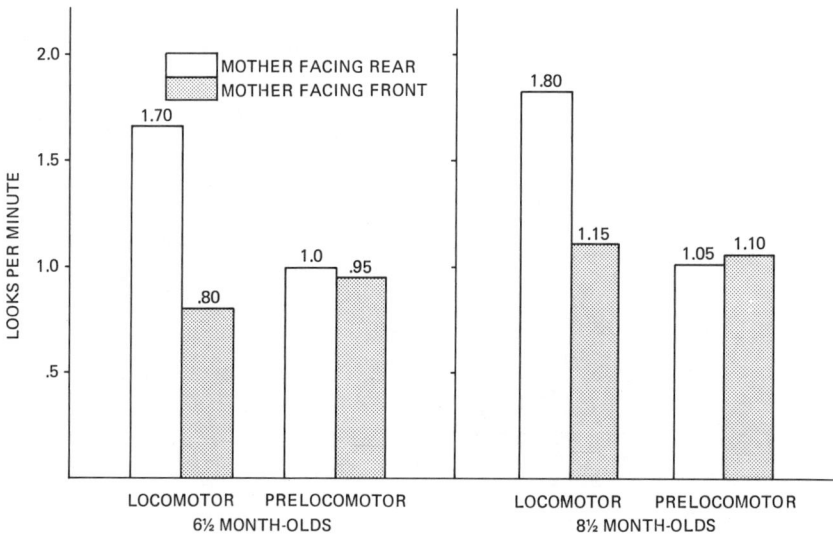

Figure 7. Frequency of looks to the mother as a function of locomotor status, age, and mother's orientation toward the infant in the room.

The influence of locomotor experience upon social communication might not be manifested only in the *perception of availability of signals,* however. The infant who is locomoting must rely on distal communication processes (such as emotional expressions of others) more frequently, since he or she will often be too far away from the caregiver to be accessible to proximal signals (such as being physically removed from a danger or stroked for reassurance). Might locomotion be the setting event for an increase in susceptibility to others' emotional signals in social referencing situations?

The second set of findings from Garland's study suggests this is the case. Data from the last phase of the study revealed that locomotor, but not prelocomotor, infants were likely to use the affective information signaled by the mother in their affective responses to the stranger. As can be observed in Table 3, five of eight locomotor infants smiled at the stranger if their mother was smiling and six of eight locomotor infants showed a sober response if their mother expressed fear. A chi square analysis revealed that this pattern of responding was significantly different than chance, $\chi^2(2) = 11.20$, $p < .03$, whereas the pattern of responding for prelocomotor infants was no different from chance responding, $\chi^2(2) = .4$, $p < .80$.

In sum, the research conducted thus far strongly supports a relation between locomotion and social communication. It is true that this relation must be viewed as somewhat tentative since locomotor experience has yet to be unconfounded with maturation in its relation to social communication. Yet two sources of evidence mitigate against the possibility of a maturational confound: (1) This possibility has been extensively investigated in regard to locomotion and fear of heights, but no evidence to support such a confound has been found; and (2) in a related study Gustafson (1981) investigated behavioral changes associated with locomotor experience and reported changes in visual behavior that are very similar to those described in the above study. Most important, the changes were observed not only with a group of locomotor infants, but also with a group of prelocomotor infants with walker experience.

Table 3. Effects of Maternal Expression on Infants' Reactions to Strangers

Mother	Locomotor infants			Prelocomotor infants		
	None	Smile	Sober	None	Smile	Sober
Happy	3	5	0	3	2	3
Fear	2	0	6	4	1	3

On the basis of the findings collected so far, it is our impression that locomotion may represent an extremely important process in the development of social communication. Since this particular skill plays such an integral role in early social-affective development, we suspect that the relation between locomotion and social communication will receive further scrutiny in the near future.

CONCEPT FORMATION

Another important skill that develops during the second half of the first year of life is concept formation. As defined in the literature on infancy, this skill involves the detection of an invariant property across a set of exemplars (generally, these exemplars vary in regard to other properties). One example of this skill involves extracting the property of gender from a set of same-sexed faces, a skill which, according to Cohen and Strauss (1979), is present by 7 months of age. Another example involves slightly older infants who are found to detect number equivalence across heterogeneous arrays of two to four objects (Starkey, Spelke, & Gelman, 1980; Strauss & Curtis, 1980).

There is one particular finding relevant to this literature that we considered especially intriguing. Ruff (1978) conducted a study in which infants were presented in a paired preference paradigm with a series of solid objects, all sharing the same form but differing on dimensions of color, size, and orientation. After a period of familiarization to different exemplars of the same form, infants were presented with two test objects. Both differed from the familiar objects in regard to color, size, and orientation, but one also differed in regard to form while the other maintained the original form. If infants could extract the invariant form from the set of familiar objects, then the novel form should be preferred on test trials since the other would be viewed as another exemplar of the more familiar form. Ruff reported that 9-month-olds did show preference for the novel form whereas 6-month-olds did not. Since these ages bracketed the period during which locomotion normally emerges, an obvious question for us was whether or not the onset of locomotion affects children's performance on this task.

In order to answer this question, 30 prelocomotor and 30 locomotor infants were tested with age held constant at 7.5 months (Campos, Bertenthal, & Benson, 1980). The locomotor infants had all been moving for approximately 6 weeks, according to maternal reports. Our procedure was identical to the one used by Ruff (1978, Experiment 4). During the familiarization phase, infants were presented with pairs of exemplars of the same form that differed from each other both within and between trials. There were six 30-second familiarization trials followed by two 10-second test trials. The test trials paired a novel form along with an exemplar of the familiarized form; both forms were presented in a novel size, color, and orientation, which was the same for the two forms.

As can be observed in Figure 8, locomotor infants looked significantly longer at the novel form ($M = 6.3$ seconds) than at the familiar form ($M = 4.4$ seconds) on the test trials. On the other hand, prelocomotor infants looked at the two forms for virtually the same amount of time. In essence, then, the locomotor infants showed a pattern of performance comparable to the performance of the 9-month-olds in the original study, and prelocomotor infants appeared to be more like the previously tested 6-month-olds in their pattern of performance.

These results thus suggest that early concept formation may also be influenced by the development of locomotion. Yet, in spite of these empirical findings, the mechanism underlying the relation between these two skills remains unclear. Perhaps the ability to explore the world in a new way and the concomitant increase in the acquisition of new information serve as catalysts for detecting common attributes so that the infant can more efficiently explore and understand the visual environment. It is also possible that locomotion is related to concept formation because the infant is given the opportunity to gain much greater experience with the same object in different perspectives and contexts. As a consequence, the infant learns to view a new perspective or context as another exemplar of the same invariant. Regardless of the mechanism, these findings are most significant in that they add to our con-

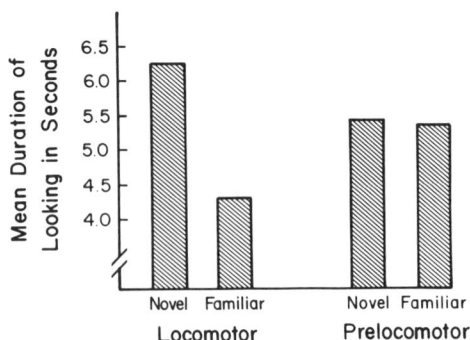

Figure 8. Duration of looking on test trials to the new exemplar of a familiar form as opposed to a new form.

tention that the effects of locomotion on early development are wide-spread.

CONCLUDING REMARKS ON EMPIRICAL STUDIES

In view of the preliminary nature of some of the previously reviewed research, we feel compelled to view this chapter more as a progress report than as a summary of a completed project. Any attempt to integrate our findings into a unitary framework would thus be premature. Still, given the consistent patterns of findings obtained to date, we wish to suggest a preliminary position regarding the apparent functional relationship between locomotion and the other phenomena investigated.

Self-produced Locomotion as a Mediator of Development

According to Flavell (1972), two behaviors or skills may be functionally related in their development according to any of five different processes. The first two processes, *addition* and *substitution,* refer to the development of alternative procedures for solving a task. The latter three processes are *modification,* in which a skill undergoes further differentiation, generalization, and stabilization; *inclusion,* in which one skill is coordinated or integrated with another; and *mediation,* in which one skill facilitates the development of another without becoming part of the skill.

It is our impression that the specific process best characterizing the functional relations revealed through our studies is the process of mediation. The development of locomotion provides the infant with an opportunity to modify earlier patterns of interaction with the caregiver by serving as a catalyst for the development of new forms of social communication. Also, the infant's understanding of the spatial layout shows dramatic changes after the development of locomotion. In particular, there is strong evidence that the infant's sensitivity to and delineation of variant and invariant visual information is facilitated by the opportunity to move continuously and observe systematic changes in the perspective structure of the environment. Generally speaking, in the research described in this chapter, locomotion was found to play at least a facilitative role in the development of other skills.

We wish to underscore, however, that locomotion may be neither necessary nor sufficient for the development of other skills. In previous sections, we pointed out how locomotion represents one of many factors facilitating development. Also, there is some evidence that the different skills investigated in our laboratory can develop even in prelocomotor infants. For example, some infants show wariness of heights early, others show precocious social referencing. Accordingly, we would predict that even infants without limbs would develop these same skills, although we would expect the absence of locomotion to slow down the development of these capacities. Nevertheless, given the ubiquity of locomotion, this skill is an extremely likely mediator of early development.

The suggestion that *mediation* is the major process connecting locomotion and other developments is by no means meant to imply that other processes are not relevant. Certainly, the process of inclusion must occur in the development of those behaviors involving visual-motor coordination (cf. Held & Hein, 1963), and it is hypothesized by some theorists (e.g., Piaget, 1937/1954) as a mechanism linking locomotion more directly to the development of a number of different skills (e.g., depth perception). Still, as of now, our own research provides direct support for the existence of only one developmental process—mediation. It is for this reason that we view the emergence of locomotion primarily as a setting event for the development of new perceptual-cognitive and social-affective behaviors.

Implications for the Development of the Self

Throughout this chapter we have confined our presentation to a discussion of specific skills that are related to the emergence of locomotion. It is our impression, however, that there exists other more general developments which are indirectly related to locomotion. Perhaps the most important of these is the elaboration of the self. Self-produced locomotion should dramatically influence the development of at least two aspects of this sensorimotor self. One aspect which has important implications for infants' *cognitive* development is the differentiation of self from environment. The other aspect which profoundly influences *emotional* development is the elaboration of infants' own sensorimotor *goals* and the appraisal of events as relevant to these personal goals. We will examine each of these aspects in turn.

Self-produced Locomotion and Differentiation of Self from World. According to many developmental theorists (e.g., Piaget, 1937/1954; Werner & Kaplan, 1963), self and world are initially undifferentiated in the child's understanding. There is good reason to posit that the emergence of self-produced locomotion contributes to the process of differentiation, since it provides new procedures for distinguishing between transformations produced by the observer and transformations produced independently from the observer. Both Held and Hein

(1963) and Gibson (1979) discuss mechanisms that explain how this differentiation may be achieved.

Held and Hein (1963) propose that the coordination of visual-motor responses provides the observer with the opportunity to correlate motor commands with the sensory input associated with those commands (what they call *reafferent stimulation*). As a consequence of this correlation, subsequent observations of movement in the visual world can be identified as produced either by movements of the self or by some external force. If the motion on the retina is produced by a movement of the observer, then the correlation between the command and reafferent stimulation is perfect, whereas externally produced visual information will be uncorrelated or incompletely correlated with the movements of the observer.

The Gibsonian mechanism for differentiation is fundamentally different from the preceding one. According to Gibson (1979), movement of the self need not be specified by proprioceptive feedback, nor by information stemming from motoric commands. He argues that there is direct information specifying movement of the self in the optic array. Whenever an observer moves toward a particular point in space, there occurs a radially expanding texture pattern emanating from that point, with a particular acceleratory function congruent with the speed of locomotion (Warren, 1976). Although an object approaching an observer also undergoes expansion, it is possible to differentiate between self-movement and object movement. Consider a situation in which a mother is standing in the corner of a room and the baby is situated a fair distance away. If the mother begins moving toward the baby, the projected image of the mother undergoes optical expansion; yet at the same time the background texture, provided by the floor and walls, remains constant. Alternatively, if the baby begins moving toward the mother (whether actively or passively), not only will the mother undergo optical expansion, but so will all other portions of the projected structure, for instance, the background texture on the walls. Accordingly, the flow and nonflow of perspective structure contains information about the self since the flow specifies that a point of observation is moving and nonflow specifies the opposite. Gibson refers to the detection of this information as visual proprioception. Although such visual proprioception does not distinguish *active* (self-produced) locomotion from *passive* locomotion, it appears probable that the active observer will be more likely to detect this information.

In theory, either one or both of these mechanisms for specifying self-produced locomotion could become operative with the development of locomotion. The development of sensitivity to the differences between self-produced changes and environmentally produced changes represents a significant advance in the process of differentiating self from the world. In turn, this new level of differentiation should contribute to the child's recognition of his or her own *instrumentality*.

According to many theorists, such a sense of agency and instrumentality is a crucial component of the cognitive self. Piaget (1937/1954) and others (e.g., Kagan *et al.*, 1978; Lewis & Brooks-Gunn, 1979) have noted that during the third quarter-year of life infants become increasingly aware that their actions produce effects upon the world. We hold that locomotion provides an extremely important avenue for the development of such an appreciation of control over the environment. Locomotion allows the infant to achieve instrumentally many goals because the baby is no longer a "prisoner" of a particular position. Thus, for example, the infant is for the first time able to satisfy a desire for food by crawling to the kitchen pantry and taking out a bag of cookies. In a similar vein, the locomotor infant is able to exercise a greater control over the frequency and duration of interactions with others since he or she is able to crawl over to other people as well as away from them. As children gain increasing experience with locomotion, their understanding of themselves as causal agents should develop accordingly, and that understanding will in turn facilitate further differentiation of self from the world.

Self-produced Locomotion and the Affective Self. Locomotion, however, does not merely provide an avenue for growth of infants' *comprehension* that they are different from the environment, nor even that they can influence other objects; it also may lead to a blossoming of affectivity. We have become increasingly cognizant that goal-directed behavior and self-regulation are intimately linked to affectivity (Campos & Barrett, 1983; Campos *et al.*, 1983). This notion is implicit in the concept of emotional appraisal, as proposed by Arnold (1960) and Lazarus (1982) and explicit in the concept of affective appreciation proposed by Campos and Barrett (1983). Appreciation is the process by which events are related to the self—determined to be congruent or incongruent with a person's attainment of significant goals. Neither appreciation nor goal in this conceptualization need be conscious; many goals may be preadapted (Izard, 1977) and many appreciations are extremely rudimentary. However, as both self and type of goal change in the course of development, so does the potential level of appreciation. We believe that self-produced locomotion may catalyze the development of this appreciation process by enabling infants to encounter more and more situations in which the outcome of their actions is made relevant or meaningful in relation to their own motoric goals.

In particular, by placing infants in a wider range of mastery experiences, self-produced locomotion provides them with a wide range of intense emotions. Many theorists (e.g., Harter, 1983; White, 1959) have highlighted the pleasurable feelings of efficacy ensuing from successful mastery attempts. We propose that these feelings are crucial motivators of exploration, rehearsal of new skills, and consolidation of memory (Bower, 1981). Self-produced locomotion must lead to a dramatic increase in the incidence of such pleasure, because so many more efficacious actions can be executed. Crawling itself may lead to intense pleasure just because it involves the successful execution of a new skill. Moreover, crawling brings infants into direct controlling contact with novel animate and inanimate objects. Such encounters not only are pleasant in themselves but set the stage for the use of these objects in efficacious ways later.

Sometimes, however, such encounters are not pleasurable. Some objects and animals move abruptly and/or loudly and inexplicably. On some occasions, the infant's actions lead to angry scolding or to gasps of fear from the caregiver. Each of these experiences may inculcate fear in the infant. On the other hand, some of the infant's attempts to control objects are unsuccessful, either because of the infant's own limitations or because of proscriptions and frustrations imposed by the caregiver. Such experiences may create anger. It is thus clear that self-produced locomotion can lead to a dramatic increase in the incidence, the quality, and the complexity of emotional experience and expression.

Locomotion also provides new means of coping with affectively charged events. For example, locomoting infants may control their distance from novel persons or objects and may retreat to their mothers as a haven of safety (Bowlby, 1969; Mahler et al., 1975). Moreover, such coping may change the infant's emotional reaction to the event. Reviews of the stranger distress literature (e.g., Horner, 1980) have noted that infants who are free to control their distance from a stranger are rarely distressed. Similarly, locomotor infants may back away from the edge of a visual cliff, and when doing so they frequently smile rather than show fear. The motoric avoidance of the drop-off produces, in Lazarus's (1968) terms, a secondary appraisal. As a result of such coping, behavior motivated by one emotion (fear) leads to expressions that may reflect a hedonically different emotion (one related to happiness).

EPILOGUE

The theme of this chapter has been that self-produced locomotion is one of the nodes playing a determining role in the 7- to 9-month developmental shift documented by Emde et al. (1976). We have tried to demonstrate how its consequences are both very specific (e.g., influencing visual cliff performance) and very broad (creating a new level of self-functioning and awareness). This chapter would be incomplete, however, if we did not allude to the reverse side of the organismic consequence of the acquisition of self-produced movement. We wonder what the psychological consequences are for the baby who, for neurological or orthopedic reasons, is impeded from locomotion. Disorders like cerebral palsy and meningomyelocele are prevalent and have devastating consequences for the acquisition of locomotion. When does the child's emerging sense of self begin to show the effects of this impediment? How does the child come to develop an adequate sense of spatial cognition? Or does he or she ever do so? If the child with such an impediment has different sets of goals than does a mobile child, how does this influence the motorically impeded child's emotionality? At what age do these effects become evident?

It is surprising to us that despite an impressive literature on motoric impediments (cf. Campos *et al.*, 1982) so few studies have measured the cognitive, emotional, or self-regulatory processes to which we have pointed in this chapter as being directly related to the acquisition of self-produced movement. Our single venture into the study of orthopedically handicapped infants was an attempt to clarify a theoretical point. However, it seems time to embark on a much more thorough investigation of the clinically-significant consequences of locomotor impediments.

Acknowledgments

We gratefully acknowledge the many contributions of Mrs. Charlotte Henderson to all phases of this work, as well as that of Dr. Marilyn Svedja, Ms. Donna Schmid, Ms. Nancy Benson, Ms. Anne Thomas, and Ms. Patricia East.

REFERENCES

Acredolo, L. P. The development of spatial orientation in infancy. *Developmental Psychology*, 1978, *14*, 224–234.

Acredolo, L. P. The familiarity factor in spatial research. *New Directions for Child Development*, 1982, *15*, 19–30.

Acredolo, L. P., & Evans, D. Developmental changes in the effects of landmarks on infant spatial behavior. *Developmental Psychology*, 1980, *16*, 312–318.

Ainsworth, M. The development of infant–mother attachment. In B. Caldwell & H. Ricciuti (Eds.), *Reviews of child development research* (vol. 3). Chicago: University of Chicago Press, 1973.

Ainsworth, M., Blehar, M., Waters, E., & Wall, S. *Patterns of attachment*. Hillsdale, N.J.: Lawrence Erlbaum, 1978.

Arnold, M. *Emotion and personality* (vols. 1 & 2). New York: Columbia University Press, 1960.

Barrett, K., & Campos, J. *Wariness of heights: An outcome of locomotor experience or age?* Paper presented at the meetings of the Society for Research in Child Development, Detroit, April 1983.

Benson, J. B., & Uzgiris, I. C. *Self-produced movement in spatial understanding.* Paper presented at the meetings of the Society for Research in Child Development, Boston, Massachusetts, April 1981.

Bertenthal, B. I. *Differentiation of self and the development of visual proprioception.* Paper presented at the meetings of the International Society for the Study of Behavioral Development, Toronto, Canada, August 1981. (a)

Bertenthal, B. I. The significance of developmental sequences for investigating the what and how of development. *New Directions for Child Development,* 1981, *12,* 43–54. (b)

Bertenthal, B. I., & Campos, J. J. A reexamination of fear and its determinants on the visual cliff. *Psychophysiology,* in press.

Bower, G. Mood and memory. *American Psychologist,* 1981, *36,* 129–148.

Bower, T. G. T. *Development in infancy.* San Francisco: W. H. Freeman & Co., 1974.

Bowlby, J. *Attachment and loss* (vol. 1). New York: Basic Books, 1969.

Bowlby, J. *Attachment and loss: Separation* (vol. 2). New York: Basic Books, 1973.

Bremner, J. G. Egocentric versus allocentric spatial coding in nine-month-old infants: Factors influencing the choice of code. *Developmental Psychology,* 1978, *14,* 346–355.

Bremner, J. G. The infant's understanding of space. In M. V. Cox (Ed.), *Are young children egocentric?* London: Conena (1980).

Bremner, J. G., & Bryant, P. E. Place versus response as the basis of spatial errors made by young infants. *Journal of Experimental Child Psychology,* 1977, *23,* 162–171.

Campos, J. J., & Barrett, K. C. Towards a developmental theory of emotion. In C. Izard, J. Kagan, & R. Zajone (Eds.), *Cognition, emotion and behavior.* New York: Cambridge University Press, 1983.

Campos, J. G., Stenberg, C. Perception, appraisal, and emotion: The onset of social referencing. In M. Lamb & L. Sherrod (Eds.), *Infant social cognition.* Hillsdale, N.J.: Lawrence Erlbaum, 1981.

Campos, J., Hiatt, S., Ramsay, D., Henderson, C., & Svejda, M. The emergence of fear on the visual cliff. In M. Lewis & L. Roseblum (Eds.), *The development of affect.* New York: Plenum Press, 1978.

Campos, J., Bertenthal, B., & Benson, N. *Self-produced locomotion and the extraction of form invariance.* Paper presented at the meetings of the International Conference on Infant Studies, New Haven, Connecticut, April, 1980.

Campos, J., Svejda, M., Bertenthal, B., Benson, N., & Schmid, D. *Self-produced locomotion and wariness of heights: New evidence from training studies.* Paper presented at the meetings of the Society for Research in Child Development, Boston, Massachusetts, April 1981.

Campos, J., Svejda, M., Campos, R., & Bertenthal, B. The emergence of self-produced locomotion: Its importance for psychological development in infancy. In D. Bricker (Ed.), *Intervention with at-risk and handicapped infants.* Baltimore: University Park Press, 1982.

Campos, J., Barrett, K., Lamb, M., Goldsmith, H., & Stenberg, C. *Socioemotional development.* In M. Haith & J. Campos (Eds.), *Carmichael's handbook on child psychology: Infancy and developmental psychobiology.* New York: Wiley, 1983.

Carr, S., Dabbs, J., & Carr, T. Mother–infant attachment: The importance of the mother's visual field. *Child Development,* 1975, *46,* 331–338.

Cohen, L. B., & Strauss, M. S. Concept acquisition in the human infant. *Child Development*, 1979, *50*, 419–424.

Cornell, E. H., & Heth, C. D. Response versus place learning by human infants. *Journal of Experimental Psychology: Human Learning and Memory*, 1979, *5*, 188–196.

Corter, C. M., Zucker, K. J., & Galligan, R. F. Patterns in the infant's search for mother during brief separation. *Developmental Psychology*, 1980, *16*, 62–69.

Dru, D., Walker, J. P., & Walker, J. B. Self-produced locomotion restores visual capacity after striate lesions. *Science*, 1975, *187*, 265–266.

Emde, R., Gaensbauer, T., & Harmon, R. Emotional expression in infancy: A bio-behavioral study. *Psychological Issues* (Vol. 10, No. 37). New York: International Universities Press, 1976.

Fischer, K. A theory of cognitive development: The control and construction of hierarchies of skills. *Psychological Review*, 1980, *87*, 477–531.

Flavell, J. H. Concept development. In P. H. Mussen (Ed.), *Carmichael's manual of child psychology*. New York: Wiley, 1970.

Flavell, J. H. An analysis of cognitive-developmental sequences. *Genetic Psychology Monographs*, 1972, *86*, 279–350.

Fox, R., Aslin, R. N., Shea, S. L., & Dumais, S. T. Stereopsis in human infants. *Science*, 1980, *207*, 323–324.

Fraiberg, S. *The magic years*. New York: Scribner, 1959.

Freedman, D. *Human infancy: An evolutionary perspective*. Hillsdale, N.J.: Erlbaum, 1974.

Garland, J. B. *Social referencing and self-produced locomotion*. Paper presented at the meetings of the International Conference on Infant Studies, Austin, Texas, March 1982.

Gesell, A. *Infancy and human growth*. New York: McMillan, 1929.

Gibson, J. J. *The ecological approach to visual perception*. Boston, Mass.: Houghton Mifflin, 1979.

Gottlieb, G. The psychobiological approach to developmental issues. In M. Haith & J. Campos (Eds.), *Carmichael's handbook on child psychology: Infancy and developmental psychobiology*. New York: Wiley, 1983.

Gustafson, G. E. *Effects of locomotion on infants' social and exploratory behavior: An experimental study*. Paper presented at the meetings of the Society for Research in Child Development, Boston, Massachusetts, April 1981.

Haith, M., & Campos, J. Human infancy. *Annual review of psychology*, 1977, *28*, 251–293.

Harter, S. Developmental perspectives on the self-system. In M. Hetherington (Ed.), *Carmichael's handbook on child psychology* (vol. 3). New York: Wiley, 1983.

Hein, A. Acquiring components of visually guided behavior. In A. Pick (Ed.), *Minnesota symposia on child psychology* (vol. 6). Minneapolis: University of Minnesota Press, 1972.

Hein, A., Held, R., & Gower, E. Development and segmentation of visually controlled movement by selective exposure during rearing. *Journal of Comparative and Physiological Psychology*, 1970, *73*, 181–187.

Held, R., & Hein, A. Movement-produced stimulation in the development of visually-guided behavior. *Journal of Comparative and Physiological Psychology*, 1963, *81*, 394–398.

Horner, T. Two methods of studying stranger reactivity in infants: A review. *Journal of Child Psychology and Psychiatry*, 1980, *21*, 203–219.

Izard, C. E. *Human emotions*. New York: Plenum Press, 1977.

Kagan, J., Kearsley, R., & Zelazo, P. *Infancy: Its place in human development*. Cambridge, Mass.: Harvard University Press, 1978.

Kaufman, L. *Perception: The world transformed*. New York: Oxford University Press, 1979.

Klinnert, M., Campos, J., Sorce, J., Emde, R., & Svedja, M. Emotions as behavior regulators: Social referencing in infancy. In R. Plutchik & H. Kellerman (Eds.), *Emotions in early development* (vol. 2 of *The Emotions*). New York: Academic Press, 1983.

Kopp, C. B. Perspectives on infant motor system development. In M. H. Bornstein & W. Kessen (Eds.), *Psychological development from infancy: Image to intention*, Hillsdale, N.J.: Lawrence Erlbaum, 1979.

Lamb, M., & Campos, J. *Development in infancy*. New York: Random House, 1982.

Lazarus, R. Thoughts on the relations between emotion and cognition. *American Psychologist*, 1982, *37*, 1019–1024.

Lewis, M., & Brooks-Gunn, J. *Social cognition and the acquisition of self*. New York: Plenum Press, 1979.

Mahler, M., Pine, F., Bergman, A. *The psychological birth of the human infant*. New York: Basic Books, 1975.

McGhee, P. *Humor: Its origin and development*. San Francisco: W. H. Freeman, 1979.

McGraw, M. *The neuromuscular maturation of the human infant*. New York: Columbia University Press, 1943.

Piaget, J. The construction of reality in the child. New York: Basic Books, 1954. (Originally published, 1937.)

Rader, N., Bausano, M., & Richards, J. On the nature of the visual-cliff avoidance response in human infants. *Child Development*, 1980, *51*, 61–68.

Ramsay, D. Onset of duplicated syllable babbling and unimanual handedness in infancy: Evidence for developmental change in hemispheric specialization. *Developmental Psychology*, 1984, *20*, 64–71.

Richards, J., & Rader, N. Crawling-onset age predicts visual cliff avoidance in infants. *Journal of Experimental Psychology*: Human Perception and Performance, 1981, *7*, 382–387.

Ruff, H. Infant recognition of the invariant forms of objects. *Child Development*, 1978, *49*, 293–306.

Scarr, S., & Salapatek, P. Patterns of fear development during infancy. *Merrill-Palmer Quarterly*, 1970, *16*, 53–90.

Siegler, R. S. Developmental sequences within and between concepts. *Monographs of the Society for Research in Child Development*, 1981, *46* (2, Serial No. 189).

Sorce, J., Emde, R., Campos, J., & Klinnert, M. Maternal emotional signaling: Its effects on the visual cliff behavior of one-year-olds. *Developmental Psychology*, in press.

Spitz, R. *The first year of life*. New York: International Universities Press, 1965.

Starkey, P., Spelke, E., & Gelman, R. *Number competence in infants: Sensitivity to numeric invariance and numeric change*. Paper presented at the meetings of the International Conference on Infant Studies, New Haven, Connecticut, April 1980.

Strauss, M. S., & Curtis, L. E. *Infant perception of numerosity*. Paper presented at the meetings of the International Conference on Infant Studies, New Haven, Connecticut, April 1980.

Svejda, M., & Campos, J. *The mother's voice as a regulator of the infant's behavior*. Paper presented at the International Conference on Infant Studies, Austin, Texas, March 1982.

Svejda, M., & Schmid, D. *The role of self-produced locomotion on the onset of fear of heights on the visual cliff*. Paper presented at the meetings of the Society for Research in Child Development, San Francisco, March 1979.

von Hofsten, C. *Developmental changes in the organization of prereaching movements*. Paper presented at the meetings of the Society for Research in Child Development, Detroit, Michigan, April 1983.

Walk, R. Depth perception and experience. In R. Walk & H. Pick (Eds.), *Perception and experience*. New York: Plenum Press, 1978.

Walk, R., & Gibson, E. A comparative and analytical study of visual depth perception. *Psychological Monographs*, 1961, *75* (15, whole no. 519).

Warren, R. The perception of egomotion. *Journal of Experimental Psychology: Human Perception and Performance*, 1976, 2, 448–456.

Werner, H., & Kaplan, B. *Symbol formation: An organismic-developmental approach to language and the expression of thought.* New York: Wiley, 1964.

White, R. Motivation reconsidered: The concept of competence. *Psychological Review*, 1959, 66, 297–333.

Wohlwill, J. H. *The study of behavioral development.* New York: Academic Press, 1973.

Yonas, A., Pettersen, L., & Lockman, J. Young infants' sensitivity to optical information for collision. *Canadian Journal of Psychology*, 1979, 33, 268–276.

CHAPTER 9

Effects of Hospitalization on Early Child Development

David A. Mrazek

Improved understanding of the psychological effects of pediatric hospitalizations requires a developmental perspective. Empirical studies suggest that children between 6 months and 4 years of age are at greater risk than older children for both short-term (Schaffer & Callender, 1959; Prugh, Straub, Sands, Kirschbaum, & Lenihan, 1953) and long-term negative consequences of hospitalization (Douglas, 1975; Quinton & Rutter, 1976). By the time most children reach 6 years of age they appear to be able to manage the experience of being admitted to the hospital with less difficulty, although some variability exists during the transitional period of 4 to 6 years. In fact, there is some evidence that children older than 6 years of age can learn to cope more effectively with the stresses related to physical illness and its treatment as a result of a sensitive and therapeutic hospital experience. This change in coping ability, which may reflect a shift in the child's ability to process the experience of hospitalization, will be a primary focus of this chapter. In some sense, this shift reflects a discontinuity in development that can be conceptualized as the ending of an early period of heightened vulnerability.

Over the past generation the belief that hospitalization early in life has a negative psychological effect on children has become an established clinical axiom. The origins of this concept can be traced to Spitz (1945), who focused primarily on the effects of institutionalization rather

David A. Mrazek • National Jewish Hospital and Research Center, Denver, Colorado 80206. This research was supported in part by the Developmental Psychobiology Research Group Endowment Fund.

than short-term medical hospitalizations. However, considering only medical admissions, Prugh *et al.* (1953) demonstrated that children between 2 and 4 years of age showed more severe immediate reactions to hospitalizations than older children. He described generalized high levels of anxiety, angry or frightened protests to separations from parents, acute panics, and prolonged periods of crying. Such responses were vividly portrayed in the classic film, "A Two-year-old Goes to the Hospital" (Robertson, 1952). Schaffer and Callender (1959) demonstrated that a period of increased vulnerability to the effects of hospitalization could be demonstrated in children after 6 months of age. Their study compared two groups of infants. The first group was hospitalized prior to 6 months of age and the second during the second 6 months of life. The mean length of hospitalization was only 15.4 days. The older group resembled the preschool children in the Prugh study and reacted to hospitalization with overt distress, whereas the younger sample responded to the hospital environment with minimal protest. More recently, specific characteristics of hospitalizations have been described with reference to their having a differential impact on the child's subsequent development (Quinton & Rutter, 1976). More work is needed in this area and a number of specific variables require clarification. The following factors will be considered in this review: (1) the child's age at the time of hospitalization; (2) length and frequency of hospitalization; (3) degree of distress due to illness, diagnostic investigations, and medical treatment; (4) hospital characteristics; (5) parental involvement; (6) tempermental child characteristics; and (7) family adjustment prior to hospitalization.

In a review of the literature on hospitalization effects, two critical concepts are helpful. The first is the somewhat paradoxical position that hospitalization need not be seen as a negative experience, but rather may have positive effects on the child's physical condition and psychological status. The second is that the potential negative effects of hospitalization are age-dependent and a consequence of a shift in the specific developmental capacities of the child. Data from an ongoing study of the emotional development of asthmatic preschool children will be used to illustrate an empirical approach to clarifying specific effects of two aspects of the hospital experience. Other variables that play a role in modifying the negative effects of hospitalization will then be reviewed from a developmental perspective.

THE LITERATURE ON CHILDHOOD HOSPITALIZATION

The thesis that severe psychopathology could result from early institutional experiences was demonstrated graphically a generation ago by Spitz (1945). Earlier clinical reports of the fate of infants and preschool children in large infant asylums revealed an extremely high rate of infant mortality in the early twentieth century (Chapin, 1915). Even a generation later, the infant mortality rate at Bellevue Hospital was still nearly 10% (Bakwin, 1942). These descriptive accounts have had a substantial emotional impact on pediatricians and hospital staff. Subsequently, long-term institutionalization of infants has been perceived as dangerous, and even shorter hospital stays have been viewed as potentially harmful for young children. However, it is necessary to consider the relevance of these historical reports to current pediatrics. On a strictly physiological level, the high infant mortality statistics reported in these early studies occurred in a preantibiotic era. Furthermore, the environmental conditions of long-term medical hospitalizations during that period have little in common with the current hospital experiences of infants and young children.

The observation that children younger than 4 years of age were most likely to demonstrate emotional and behavioral problems during hospitalization was confirmed by Illingworth and Holt (1955), who along with Prugh *et al.* (1953) first discussed characteristics of the hospital experience as mediating variables. These characteristics included visiting hours, preparation for medical procedures, and therapeutic play opportunities. As parental separation was seen as one of the primary stressors of hospitalization, investigators began to focus on modification of negative effects by increasing children's contact with their parents and providing surrogate parental figures when parent contact was not possible (Bowlby, 1973; Bowlby, Robertson, & Rosenbluth, 1952; Fagin, 1966; Robertson & Robertson, 1971). Success in decreasing the distress associated with parental separation led to a series of studies focusing on improving pediatric nursing care as a means of mediating emotional distress. A strategy of providing a single nurse who could anticipate the child's anxiety and actively intervene with the child was shown to be more effective in reducing self-reported anxiety and increasing cooperation than was routine supportive care (Visintainer & Wolfer, 1975). The specific preparation of school-age children and their families for hospitalization was also shown to decrease maternal anxiety and lead to a decrease in behavior problems after hospitalization. It was less clear whether these interventions were as helpful for preschool children (Ferguson, 1979; Wolfer & Visintainer, 1979). Although these studies highlighted intervening variables that might affect outcome, they shared a pathologic orientation, with their primary focus defined as minimizing expected negative effects.

A controversy began when Douglas (1975) showed that a single hospitalization occurring in the first four years of life was not associated with increased emotional problems, but rather that multiple early hospitalizations were needed to demonstrate even a short-term effect. Additionally, Douglas (1973) demonstrated that a specific delayed effect, enuresis in adolescence, was associated with multiple hospitalizations during the preschool years. The finding that multiple early admissions were associated with both emotional and conduct disorder later in childhood was subsequently demonstrated by Quinton and Rutter (1976). Furthermore, Rutter (1979) addressed the specific issue of whether chronic family problems rather than multiple hospitalizations might be responsible for the development of these later problems. He found that children with multiple early hospital admissions, even in the absence of chronic family difficulties, still showed greater emotional disturbance later in childhood. However, children from disadvantaged homes were more likely to have had multiple early admissions and were also more likely to have suffered from long-term adverse effects. In other words, the negative effect of multiple hospitalizations was *potentiated* by chronic family difficulties. A recent retrospective study showed that adult patients suffering from depression and chronic pain were more likely to have experienced hospitalizations early in childhood (Pilowsky, Bassett, Begg, & Thomas, 1982), further suggesting potential long-lasting or delayed consequences.

The entire question of to what degree early stressful experience contributes to later psychological adaptation was taken up by Clarke and Clarke (1976), who challenged Douglas's findings. Their hypothesis that early experience may have only a modest implication for later adjustment has been supported by recent studies that have not demonstrated a relationship between early hospital experience and later psychological problems. A recent study found that there was no increased rate of psychiatric disturbance following a single orthopedic hospitalization for children beyond the preschool period (Brown, Chadwick, Shaffer, Rutter, & Traub, 1981). Perhaps of greater interest, Ferguson (1979) found that children who experienced their first hospitalization at 6 or 7 years were actually able to use this experience adaptively and reduce later fears of hospitalization in quite a different way than could 3- and 4-year-old children who experienced subsequent hospitalizations with increased anxiety. Such findings provide a direct challenge to hospital care providers to reorient their views of hospitalization and consider the mechanisms by which such positive growth can be stimulated, rather than to be content with merely minimizing negative reactions.

PSYCHOLOGICAL EFFECTS OF HOSPITALIZATION ON YOUNG, SERIOUSLY ILL, ASTHMATIC CHILDREN

To investigate further the effects of hospitalization, data from an ongoing study of the emotional development of preschool asthmatic children were used, focusing on the variables of age at time of hospitalization and frequency of hospitalization. All the children described in this study were seriously asthmatic, and all but two had been hospitalized repeatedly during the first four years of life. The primary variables being studied were the children's current behavioral adjustment and emotional interactions with their parents. The potential impact of frequent hospitalizations was studied using two strategies. First, any variability in the children's behavior that was associated with the developmental stage of the children when previous hospitalizations took place was examined. Second, associations between the total number of previous hospitalizations and current behavior were explored.

Asthma is a particularly distressing illness and was expected to have an effect on the child's emotional development independent of hospitalization. The specific characteristics of asthma that combine to make it a particularly stressful illness include: (1) onset in the first year of life, (2) a sudden frightening onset of breathlessness, (3) helplessness of both child and parent during an attack, (4) variable reversibility of attacks even with medical intervention, (5) persistent recurrence of attacks, and (6) threat of complete apnea and death. Although these aspects of the illness appeared to be sufficiently stressful to place a child at risk, it was expected that repeated and often precipitous medical hospitalizations might further disrupt the parent–child relationship.

Thirty-three severely asthmatic preschool children between 36 and 72 months of age who had been sequentially admitted to National Jewish Hospital were evaluated using two methodologies. The rationale for selecting this age group was based on the objective of documenting immediate and short-term consequences of the illness in a young population of patients who had recently been diagnosed as having asthma. Parental interviews were conducted that incorporated the questions included in the Behavior Screening Questionnaire (B.S.Q.) developed by Richman and Graham (1971). This instrument was recently validated using a United States population by Earls, Jacobs, Goldfein, Silbert, Beardslee, and Rivinus (1982). Videotaped observations of mother–child interactions were also coded using a sequential analytic strategy (Mrazek, Dowdney, Rutter, & Quinton, 1982). A comparison sample of 22 healthy children was also studied.

An overall score reflecting the degree of behavioral disturbance was derived from the interview by compiling scores indicating the degree of difficulty in 12 areas. The observational data were reduced to examine three aspects of mother–child interaction: overt distress on the part of the child, oppositional behavior, and maternal positive affect and reinforcement.

The asthmatic children showed more frequent and severe behavioral problems by parent report. These were particularly related to depressed mood, fearfulness, and difficulties in sleeping (Mrazek, Anderson, & Strunk, 1984). Additionally, they were more frequently distressed and more oppositional. Little difference in maternal positive affect expression or reinforcement was found. This report will include the results of subsequent data analyses that begin to distinguish the effects of the asthmatic illness from those of the subsequent hospital treatment.

VARIABLES AFFECTING RESPONSE TO HOSPITALIZATION

Variables Illustrated with Empirical Results from the Study of Asthmatic Children

Age at Time of Hospitalization. Most studies of the hospitalization of children have focused on the negative effects of the experience when it occurs before 4 years of age. Although some controversy exists as to whether hospitalization in the first six months of life has any persistent negative result, short-term effects of admission to the hospital between the ages of 6 months and 4 years include: (1) withdrawal or depressive responses, (2) increased irritability and decreased frustration tolerance, (3) immature or regressive behaviors, and (4) increased opposition and behavioral problems (Ferguson, 1979; Prugh et al., 1953; Prugh & Eckhardt, 1980). Furthermore, these studies have suggested that the hospitalization of 2- and 3-year-old children results in more pronounced distress than hospitalizing older children and may contribute to greater risk for the development of later problems.

The question of whether hospitalization occurring in the first six months of life results in less overt disturbance could only be addressed indirectly using the data from this study of asthmatic children. The severely asthmatic sample as a whole demonstrated more overt distress and frequent oppositions during the observational assessments. However, those children who were hospitalized during this first six-month period were more compliant and less likely to demonstrate any overt distress than children with later multiple hospital stays ($p < .005$ by chi square). Although this was in some ways an adaptive response, there was some clinical concern that these children were less spontaneous and that their passivity was a result of their continuous medical adversity.

The most striking short-term hospital effects have been noted in the literature when children were hospitalized between 24 and 48 months of age. The question of whether multiple hospitalizations during this period would result in more persistent difficult behavior was addressed more directly. The sample of severely asthmatic preschool children was divided into two groups on the basis of whether they had been hospitalized on more than two occasions between 24 and 48 months. An analysis of variance comparing these two asthmatic samples and the nonasthmatic comparison sample showed that significant group differences ($F = 8.4$, $p < .001$) existed when the frequency of behavioral problems as measured by the Behavior Screening Questionnaire was the dependent variable. To determine whether these three groups differed from each other on this variable, linear contrasts were performed for each pair of groups. Significant differences were shown between the comparison group and the asthmatic group without multiple hospitalizations between 24 and 48 months of age ($p = .05$), as well as between the comparison group and the asthmatic group with multiple hospitalizations during the same age period ($p < .001$). Additionally, a linear contrast between the two asthmatic groups demonstrated that the group with multiple hospitalizations had more behavioral problems than did the others, a comparison also reaching statistical significance ($p < .05$). In summary, the experience of having severe asthma, with or without having had multiple hospitalizations during the predicted period of increased vulnerability, was associated with more reported behavioral problems; furthermore, multiple hospitalizations during this vulnerable period appeared to compound the effect of the illness alone.

A second analysis of variance was done to determine whether persistent noncompliance during the observational assessment differed between the three groups. The dependent variable, persistent noncompliance, was derived by examining all confrontations that were coded over the course of the observation and calculating the percentage of these conflicts in which the parents were unsuccessful in gaining the child's compliance. Group differences were demonstrated by the preliminary analysis of variance ($F = 4.03$, $p < .05$). Linear contrasts demonstrated significant differences between the group with more than two hospitalizations between 24 and 48 months and both the asthmatic group with fewer hospitalizations during this period ($p < .025$) and the comparison sample ($p < .01$). No difference was found on this measure between the comparison group and those children who had not experienced more than two hospitalizations between 24 and 48 months. In summary, this analysis indicated that only those severely asthmatic children with multiple hospitalizations during the predicted period of increased vulnerability demonstrated the behavioral characteristic of persistent noncompliance.

These two findings taken together, the severely asthmatic children who had experienced multiple hospitalizations during their preschool years showed both more behavioral difficulties and greater non-compliance than did healthy children or severe asthmatics who had not been hospitalized on multiple occasions during this apparently vulnerable period of development. In contrast, while the mothers of the less frequently hospitalized asthmatic children still reported that they had a higher frequency of behavioral difficulties than did the comparison sample, their children were not more difficult to control during the standardized observational assessment. One could tentatively conclude that severe asthma alone may increase a preschool child's risk for the development of emotional difficulties, but that the additional experience of repeated hospitalizations during the third and fourth years of life may further disrupt the parent–child relationship, leading to a greater likelihood of the child's becoming persistently noncompliant.

Length and Frequency of Hospitalization. An initial step toward understanding the effects of hospital admission is empirically to define the range of experiences that constitute a hospitalization. There is little in common between an overnight confinement because of confusing abdominal symptoms and a prolonged I.C.U. admission. Unfortunately, many studies do not make a distinction between these disparate experiences.

A single short hospitalization in childhood appears rarely to result in either serious or lasting difficulties. However, the occurrence of a second hospitalization appears to alter the risk for subsequent emotional difficulties (Douglas, 1975; Quinton & Rutter, 1976). The idea that the first hospitalization sensitizes a young child to later experiences follows from this observation.

The data on the described sample of asthmatic children could not address this issue at the level of whether two or three hospitalizations had a developmental effect, as these children had experienced a greater number of hospitalizations (mean number of hospitalizations = 8.3). More than 90% of these admissions to the hospital were for control of respiratory symptoms. Therefore it was possible to examine the differential effect of frequent hospitalizations of a fairly homogeneous type.

The correlations between the number of hospitalizations each child had experienced and both overall behavioral problem scores and degree of oppositional behavior were calculated. Increased frequency of hospitalization was associated with more behavioral problems as identified by the B.S.Q. ($r = .42$, $p < .025$). Furthermore, analysis of the interactions revealed that mothers of children with more frequent hospitalizations were less likely to be reinforcing or provide approval to their child ($r = .36$, $p < .05$). To understand these difficulties further, we used a chi square analysis to examine the 12 specific problem areas covered in the B.S.Q. Children with more hospitalizations demonstrated greater dependency ($p < .02$), more peer problems ($p < .05$), and more confrontation regarding going to bed ($p < .05$) than the asthmatic children with less frequent hospitalizations.

Additional Variables Reported in the Literature

Degree of Distress Resulting from Medical Illness, Diagnostic Procedures, and Treatment. Most studies of hospital effects have looked at heterogeneous groups of children with a wide variety of diagnoses. Studies that have looked at single short-term acute admissions for tonsillectomies have tended to demonstrate that those children experienced few serious problems. A study of older children who had a single orthopedic procedure showed they had few difficulties and that some children demonstrated increased competence in coping with medical procedures following the hospitalization (Brown et al., 1981).

The fact that the effects of the illness itself is usually ignored is a serious methodological shortcoming. Minde and his co-workers (Minde, Perrotta, & Orter, 1982; Minde, Whitelaw, Brown, & Fitzhardinge, 1984) have recently developed a "morbidity scale" to monitor the severity of illness and psychophysiological status of infants within a newborn intensive care unit. This approach documents daily stress and includes the notation of intrusive or painful procedures and other treatment effects. A similar strategy of assessing the medical status and the effects of treatment procedures with older children would allow for the control of this variable in future analyses. An alternative strategy would be to select samples of children with relatively homogeneous pediatric problems and interventions and then compare samples that vary regarding the severity of illness on a similar set of outcome measures.

Parental and Family Involvement during the Hospitalization. Since the initial studies prior to 1960, there has been a dramatic increase in the emphasis on parental involvement during the course of a pediatric hospitalization. This can be seen in the abandonment of visiting hours on most pediatric units and the encouragement of visits by extended family members, including grandparents and siblings. Nowhere is this more evident than in the treatment of asthma. For more than 30 years, many treatment centers discouraged parent involvement, and prolonged parental separations were generally felt to be therapeutic (Purcell, Brady, Chai, Muser, Molk, Gordon, & Means, 1969). Currently, family therapy is increasingly being suggested as an appropriate intervention for asthmatic children (Lask & Kirk, 1979; Lask & Matthew, 1979; Minuchin, Rosman, & Baker, 1978), and intensive family involvement over the course of the hospitalization is seen as critical for the maintenance of therapeutic gains. Emphasis on parent availability throughout pediatric hospital stays has resulted in (1) decreased anxiety on the part of the parents and the child (Ferguson, 1979), (2) elimination of prolonged parental separations from young children, (3) greater opportunity for the medical staff to assist parents in becoming actively involved in the child's treatment, and (4) facilitation of the child's transition from the hospital environment to home by maximizing parental awareness of medical and psychological aspects of the child's care.

Child's Temperament. Temperamental characteristics have usually been defined as constitutional characteristics and thought to be associated with some of the variation in the range of children's later adaptability (Thomas & Chess, 1977). The influence of temperament on children's reactions to hospitalization has not been directly studied, but it would appear to be an important variable. Unfortunately, virtually all estimates of children's temperament are based on parental report. Recent studies have shown that children with certain chronic illnesses do demonstrate specific temperamental characteristics (Kim, Ferrara, & Chess, 1980; Kim, Ferrara, Mattsson, & Chess, 1981). Epileptic children were found frequently to fit the category of "difficult," showing high levels of activity, intensity of reactions and distractibility, positive mood, and low rhythmicity and persistence. In contrast, asthmatic children appear to fit the description of "slow to warm up," showing low levels of rhythmicity, adaptability, intensity of reaction, persistence, and negative mood. Children with eczema demonstrate a temperamental profile quite similar to that of the healthy comparison group. This promising classification of potential constitutional differences may provide added understanding of children who are at risk for later negative effects. However, prospective studies that document temperament using more objective criteria prior to the initial onset of illness are needed to address the question of whether these behavioral patterns are the effects of the specific illness or actually predisposing constitutional characteristics.

Quality of Family Interaction Prior to Hospitalization. As illustrated by recent developmental studies, any response to stress must be considered within the context of the prestress state of the individual or system. Hinde and McGinnis (1977) demonstrated that responses of primate infants to separation were dependent on the preseparation character of the mother–child relationship. In a somewhat parallel manner, Dunn and Kendrick (1980) showed that children's responses to the birth of a new sibling could be predicted by the quality of the previous mother–child relationship. The quality of mothering has been postulated as having a differential effect on how well the child can manage separation while in the hospital (Fagin, 1966), but studies focusing on the parent–child interaction have only begun to be completed (Marton, Minde, & Ogilvie, 1981; Minde, Perrotta, & Marton, 1984). For a better understanding of how parents can mediate the hospital experience for the child, it is necessary to assess the prehospitalization family system with regard to the quality of the interpersonal relationships.

Hospital Characteristics. The characteristics of the hospital to which a child is admitted are a critical variable that has not been systematically addressed in the literature. The following characteristics are commonly recognized as important: (1) policies regarding parents and family support, (2) quality and sensitivity of caretaking staff, (3) consistency of staff, (4) preparation of children for procedures, and (5) appropriate and consistent stimulation and activity. The experience of an admission to a hospital that excludes family, has no pediatrically trained, child-oriented staff, permits random rotation of caretakers, and does not provide appropriate preparation or stimulation would appear at face value to have an altogether different effect than a hospital stay in a more enlightened modern pediatric center. Although the quality of most hospitals has improved, no studies directly compare different pediatric hospitals.

The modern pediatric facility has many advantages as a therapeutic milieu that are not available in outpatient pediatric management. Families can be seen as units, helped to learn about the child's problems, and assisted in implementing therapeutic approaches. Particularly for disorganized families with substandard housing and multiple social problems, there is a real advantage in thoroughly evaluating the child's medical condition in a safe, humane, and comfortable hospital setting. This advantage is compounded if the parents have felt overwhelmed by the task of trying to manage a difficult or dangerously ill child. As programs become more family-centered, the popular perception that they are potentially threatening may rapidly change. Although comparison of different pediatric inpatient settings would require a collaborative research effort among many centers, such studies are needed to document the relative effectiveness of family-centered programs and traditional medical facilities with more limited resources.

DISCUSSION

The premise of this review is that traditional hospitalizations that occur during the preschool years represent an emotional stress for the child. However, after the age of 4 or 5 years the experience appears to have a diminished pathologic potential. This transition is probably facilitated by the child's increased developmental capacities in a number of areas. From a cognitive perspective, the child masters the Piagetian stage of preoperational thinking and begins to be able to make sense of cause-and-effect relationships using concrete operations. Thus the understanding that painful treatments will have longlasting benefit can become a reality for the child. From a social-emotional perspective, the child is at a stage at which he is able to turn to a number of peers and adults for security and reassurance, whereas at earlier stages of development only the primary attachment figures were able to provide this support. Once the child has a well-established sense of object constancy and an appreciation of himself as an independent entity, it is much easier to help him cope with the unavoidable separations from parents and loved ones that are necessitated by a hospital stay. From a psychodynamic perspective, the child's misinterpretations of fantasies and wishes regarding his parents can be dealt with in a more objective and effective manner after the age of four or five. Specifically, it can be made clear to the child that neither the hospitalization nor his illness is in any way a punishment for perceived unacceptable thoughts or impulses.

The first months of life were at one time thought to be a period when infants were less sensitive to maternal separation and consequently somewhat immune to interpersonal stresses created by the hospital experience. Clearly, the sensory-motor period of cognitive development is one in which only relatively simple associations or schematas are possible. Additionally, the documentation of separation and stranger distress occurring only after 6 months of age (Emde, Gaensbauer, & Harmon, 1976) has led some clinicians to believe that very early hospitalizations may well be less disruptive to the child's development. Clinical observations such as those by Minde, Perrotta, and Marton (1984) and Harmon and Culp (1981) regarding the relatively normal development of many premature infants further support the notion that the very young child can endure considerable stress without long-term negative impact. Observation of preschool severely asthmatic children has been somewhat less reassuring. Infants with early-onset asthma who then experienced persistent illness demonstrated a more subdued and passive style of interaction when evaluated during the preschool years and compared to either children with a later onset of asthma or healthy children. For these children the early hospital experience was not isolated and circumscribed, but rather the beginning of a life-long chronic illness requiring repeated hospital treatment.

If hospitalizations are to be considered stressful stimuli for young children, then it is sensible to attempt to define the severity of the stress. The quantification of the effects of the multiple variables demonstrated to influence children's responses to hospitalization is a first step toward the clarification of the intensity of this stress. In this regard, multiple hospitalizations would be seen as more damaging than brief or isolated admissions. Painful or frightening illnesses would have more negative effect than benign or circumscribed problems. Pain and discomfort would compound the disruptive impact on development. In a similar manner, some constitutionally stronger children would be able to cope with higher degrees of stress than less fortunate peers. Finally, emotional support provided for a child by families and caregivers would decrease stress. Positive effects of these environmental influences depend not only on the nature of the child's relationship with adults and peers prior to hospitalization, but also upon the ongoing nature of their interactions throughout the course of the treatment.

This conceptualization leads directly to the development of an appreciation of the interactive effects of stress (Rutter, 1979). More positively, to the degree that specific variables can be seen as playing an important role in the adaptation of a particular child and family, their identification can lead to an early intervention strategy with multiple components. In most cases, our understanding of the interplay between these variables is still insufficiently precise to be able to adopt highly specific intervention strategies. Review of the current level of knowledge in the area of early prediction highlights the need for careful, systematic research (Rutter, 1982). Previous failures to provide effective treatment will, one hopes, not lead to discouragement, but rather will provide the building blocks with which to develop a more sophisticated approach to the medical care of young children.

CONCLUSIONS

A developmental shift appears to occur between 4 and 5 years of age which results in a decrease in the young child's vulnerability to the negative consequences of hospitalization regardless of other considerations. The child's increased competence in cognitive, emotional, and social development apparently allows the child to place the hospital and illness experience in a perspective that allows more adequate coping. However, understanding the effects of a specific hospitalization on a given child is clearly dependent on a wide variety of factors that clinical experience and empirical studies have identified as being associated with increased risk. What is now needed is a more comprehensive understanding of the relative effects of these variables, which requires a multivariate research strategy. The benefits of such an endeavor would include (1) the development of an at-risk index, (2) early identification of vulnerable patients, and (3) development of early interventions to minimize negative effects for high-risk patients.

REFERENCES

Bakwin, H. Loneliness in infants. *American Journal of Diseases of Children*, 1942, *63*, 30–40.

Bowlby, J. *Attachment and loss* (Vol. 2). New York: Basic Books, 1973.

Bowlby, J., Robertson, J., & Rosenbluth, D. A two-year-old goes to hospital. *The Psychoanalytic Study of the Child*, 1952, *7*, 82–94.

Brown, G., Chadwick, O., Shaffer, D., Rutter, M., & Traub, M. A prospective study of children with head injuries. III. Psychiatric sequelae. *Psychological Medicine*, 1981, *11*, 63–78.

Chapin, H. D. Are institutions for infants necessary? *The Journal of the American Medical Association*, 1915, *64*, 1–3.

Clarke, A. M., & Clarke, A. D. B. *Early experience: Myth and evidence*. London: Open Books, 1976.

Douglas, J. W. B. Early disturbing events and later enuresis. In I. Kolvin, R. C. MacKeith, & S. R. Meadow (Eds.), *Bladder control and enuresis* (Clinics in Developmental Medicine No. 48/49). London: Heinemann, 1973.

Douglas, J. W. B. Early hospital admissions and later disturbances of behaviour and learning. *Developmental Medicine and Child Neurology*, 1975, *17*, 456–480.

Dunn, J., & Kendrick, C. The arrival of a sibling. *Journal of Child Psychology and Psychiatry*, 1980, *21*, 119–132.

Earls, F., Jacobs, G., Goldfein, D., Silbert, A., Beardslee, W., &Rivinus, T. Concurrent validation of a behavior problems scale to use with three-year-olds. *Journal of the American Academy of Child Psychiatry*, 1982, *21*, 47–57.

Emde, R. N., Gaensbauer, T. J., & Harmon, R. J. Emotional expression in infancy: A biobehavioral study. *Psychological Issues*, 1976, *10(1)*, 37.

Fagin, C. M. R. N. *The effects of maternal attendance during hospitalization on the post-hospital behavior of young children*. Philadelphia: F. A. Davis, 1966.

Ferguson, B. F. Preparing young children for hospitalization. *Pediatrics*, 1979, *64*, 656–664.

Harmon, R. J., & Culp, A. M. The effects of premature birth on family functioning and infant development. In I. Berlin (Ed.), *Children and our future*. Albuquerque: University of New Mexico Press, 1981.

Hinde, R. A., & McGinnis, L. Some factors influencing the effect of temporary mother–infant separation: Some experiments with rhesus monkeys. *Psychological Medicine*, 1977, *7*, 197–212.

Illingworth, R. S., & Holt, K. S. Children in hospital: Some observations on their reactions with special reference to daily visiting. *Lancet*, 1955, *2*, 1257–1262.

Kim, S. P., Ferrara, A., & Chess, S. Temperament of asthmatic children. *Journal of Pediatrics*, 1980, *97*, 483–486.

Kim, S. P., Ferrara, A., Mattsson, A., & Chess, S. *Comparative temperament profiles of children with chronic diseases*. Presented at the 28th annual meeting of the American Academy of Child Psychiatry, 1981.

Lask, B., & Kirk, M. Childhood asthma: Family therapy as an adjunct to routine management. *Journal of Family Therapy*, 1979, *1*, 33–49.

Lask, B., & Matthew, D. Childhood asthma: A controlled trial of family psychotherapy. *Archives of Disease in Childhood*, 1979, *54.*, 116–119.

Marton, P., Minde, K., & Ogilvie, J. Mother–infant interactions in the premature nursery: A sequential analysis. In S. L. Friedman & M. Sigman (Eds.), *Preterm birth and psychological development*. New York: Academic Press, 1981.

Minde, K., Perrotta, M., & Orter, C. The effect of neonatal complications in same-sexed premature twins on their mother's preference. *Journal of the American Academy of Child Psychiatry*, 1982, *21*, 446–452.

Minde, K., Perrotta, M., & Marton, P. Maternal caretaking and play with full-term and premature infants. *Journal of Child Psychology and Psychiatry*, 1984 (in press).

Minde, K., Whitelaw, A., Brown, J., & Fitzhardinge, P. The effects of neonatal complications in premature infants on early parent–infant interactions. *Developmental Medicine and Child Neurology*, 1984, (in press).

Minuchin, S., Rosman, R., & Baker, L. *Psychosomatic families: Anorexia nervosa in context.* Cambridge: Harvard University Press, 1978.

Mrazek, D. A., Anderson, I. S., & Strunk, R. C. Disturbed emotional development in severely asthmatic preschool children. In J. E. Stevenson (Ed.), *Recent research in developmental psychopathology (Journal of Child Psychology and Psychiatry* Book Supplement No. 4). Oxford: Pergamon Press, 1984.

Mrazek, D. A., Dowdney, L., Rutter, M. L., & Quinton, D. L. Mother and preschool child interaction: A sequential approach. *Journal of American Academy of Child Psychiatry*, 1982, *2*, 453–464.

Pilowsky, I., Bassett, D. L., Begg, M. W., & Thomas, P. G. Childhood hospitalization and chronic intractable pain in adults. *International Journal of Psychiatry in Medicine*, 1982, *12*, 75–84.

Prugh, D. G., & Eckhardt, L. O. Children's reactions to illness, hospitalization, and surgery. In H. I. Kaplan, A. M. Freedman, & B. J. Sadock (Eds.), *Comprehensive textbook of psychiatry III* (vol. 3). Baltimore: Williams & Wilkins, 1980.

Prugh, D. G., Straub, E. M., Sands, H. H., Kirschbaum, R. M., & Lenihan, E. A. A study of the emotional reactions of children and families to hospitalization and illness. *American Journal of Orthopsychiatry*, 1953, *23*, 70–106.

Purcell, K. Brady, K., Chai, H., Muser, J., Molk, L., Gordon, M., & Means, J. The effect on asthma in children of experimental separation from the family. *Psychosomatic Medicine*, 1969, *31*, 144–164.

Quinton, D., & Rutter, M. Early hospital admissions and later disturbances of behavior: An attempted replication of Douglas' findings. *Developmental Medicine and Child Neurology*, 1976, *18*, 447–459.

Richman, N., & Graham, P. A behavioural screening questionnaire for use with 3-year-old children. *Journal of Child Psychology and Psychiatry* 1971, *12*, 5–33.

Robertson, J. Film: *A Two-year-old Goes to the Hospital.* London: Tavistock Child Development Research Unit; New York: New York University Film Library, 1952.

Robertson, J., & Robertson, J. Young children in brief separation. *Psychoanalytic Study of the Child*, 1971, *26*, 264–315.

Rutter, M. Protective factors in children's responses to stress and disadvantage. In M. W. Kent & J. E. Rolf (Eds.), *Primary prevention of psychopathology: Vol. 3. Social competence in children.* Hanover, N.J.: University Press of New England, 1979, pp. 49–74.

Rutter, M. Prevention of children's psychosocial disorders. *Pediatrics*, 1982, *70*, 883–894.

Schaffer, H. R., & Callender, W. M. Psychologic effects of hospitalization in infancy. *Pediatrics*, 1959, *24*, 528–539.

Spitz, R. A. Hospitalism: An inquiry into the genesis of psychiatric conditions in early childhood. *The Psychoanalytic Study of the Child*, 1945, *1*, 53–74. New York: International University Press.

Thomas, A., & Chess, S. *Temperament and development.* New York: Brunner/Mazel, 1977.

Visintainer, M. A., & Wolfer, J. A. Psychological preparation for surgical pediatric patients: The effect of children's and parents' stress responses and adjustment. *Pediatrics*, 1975, *56*, 187–202.

Wolfer, J. A., & Visintainer, M. A. Prehospital psychological preparation for tonsillectomy patients: Effects on children's and parents' adjustment. *Pediatrics*, 1979, *64*, 646–655.

PART III

STUDIES OF CONTINUITIES

CHAPTER 10

The Development of Representation from 10 to 28 Months

DIFFERENTIAL STABILITY OF LANGUAGE AND SYMBOLIC PLAY

Inge Bretherton and Elizabeth Bates

INTRODUCTION

This paper is based on the premise that biological systems are self-regulating and self-organizing and that their functioning is governed by laws which are in principle understandable.

The analysis of developing biological systems is, however, hampered by their great complexity. At any one point in time the total system is composed of a hierarchy of subsystems which are linked to one another through a variety of positive and negative feedback loops. New properties of the whole system can arise out of the particular manner in which developing subsystems come to exert reciprocal influences on one another.

Excerpts from this chapter were originally presented at the Biennial Retreat of the Developmental Psychobiology Research Group, Estes Park, Colorado, June 1982 and at the Annual Meeting of the American Psychological Association, Anaheim, California, August 1983.

Inge Bretherton • Department of Human Development and Family Studies, Colorado State University, Fort Collins, Colorado 80523.   Elizabeth Bates • Department of Psychology, University of California-San Diego, LaJolla, California 92093. The project discussed in this chapter was funded by grant BNS76-1724 from the National Science Foundation and by a grant from the Spencer Foundation.

Genetic and environmental inputs may affect the rate of development and the functioning of different subsystems in different ways
(Tanner, 1970). In addition, subsystems need not necessarily develop at
comparable rates across individuals or at a constant rate in the same
individual. Differential acceleration and deceleration of developmental
rates would lead to instability of individual differences over time, despite the fact that the instability was in principle capable of explanation.
For example, Wilson (1978) documented almost identical fluctuations in
the intelligence quotients of monozygotic twins over the first six years of
life. Yet the temporal pattern of these developmental fluctuations differed substantially from one MZ twin pair to the next. This suggests that
genetic factors may be responsible for variations in the rate of development, an interpretation supported by the concurrent finding that the
developmental profiles of dizygotic twins and of siblings were quite
divergent. Thus, orderly development is not incompatible with instability of individual differences.

On the other hand, although we can think of a number of reasons
why individual differences at one age may not necessarily predict individual differences at a later age, the fact that instability *can* occur does
not imply that it *must* occur. Since explanations of stability will be simpler than explanations of instability, the best approach to the question of
continuity versus discontinuity in development may be to look for continuity first.

Types of Stability

Studies concerned with stability of individual differences over time
have concentrated on two general areas:

1. The assessment of behavioral style (temperament, personality
   traits, dyadic interaction)
2. The assessment of ability (verbal, spatial, mathematical, artistic)

The material to be presented in this paper is based on repeated
assessments of infants' developing representational ability as demonstrated in language and symbolic play. However, much of our subsequent discussion also applies to studies of behavioral style.

Stability of individual differences has traditionally been investigated
in two ways (McCall, Eichorn, & Hogarty, 1977): by assessing the same
variable repeatedly over time (homotypic progression) or by assessing
the relationship of variable A at time 1 to variable B at time 2 (heterotypic
progression). If the term *homotypic* is interpreted in a very strict sense
(the same measure is applied at time 1 and time 2) the documentation of
homotypic progressions becomes exceedingly difficult. Repeated mea-

surement of height in inches is one of the few examples of a true homotypic progression. Even if one is willing to use the term *homotypic* more loosely, problems of definition still arise. For example, if the development of intentional communication is the domain under study, an investigator might regard a progression as homotypic if preverbal gestures (such as pointing, showing, and giving) are assessed at 9 months of age and first words at 12 months of age. If a researcher wishes to emphasize the shift in communicative mode, however, the same progression might be viewed as heterotypic

In fact, very few progressions which document the relationship of variable A at time 1 to variable B at time 2 are heterotypic in a very strict sense (which would imply that A and B have nothing in common). When an investigator attempts to demonstrate such a relationship it is usually because the two variables and their underlying constructs are seen as standing in some kind of relationship to one another (perhaps A is regarded as a prerequisite or necessary component of B). For example, the understanding of means–end relations which manifests itself concurrently in simple tool use and in intentional gestural communication (social tool use) around 9 months of age is seen by some investigators as a prerequisite to verbal communication (Bates, Benigni, Bretherton, Camaioni, & Volterra, 1979).

McCall *et al.* (1977) distinguished several forms of homotypic progression, which are depicted in Figure 1. Case 1 (order) illustrates progressions in which specific milestones develop in a particular invariant sequence but the rate of development differs from individual to individual so that precise predictions as to when an individual will reach a particular level are not possible. To take a hypothetical example: multisyllabic babbling might always precede the acquisition of conventional words, but if Case 1 obtained we would find that assessing the complexity of multisyllabic babbling at 7 months did not improve prediction of vocabulary size at 15 months. Case 2 (weak stability) holds where as-

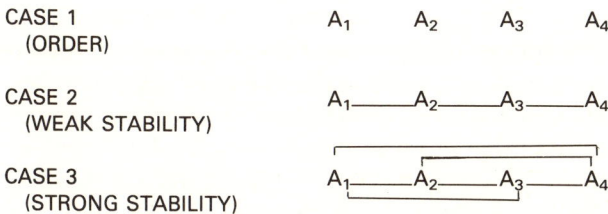

Figure 1. Three types of homotypic progression. Lines between assessments of A (e.g., A_1 and A_2) indicate hypothesized correlations. (Adapted from McCall *et al.*, 1977.)

sessment of ability A at time 1 is correlated with A at time 2, but where assessments more distant from one another (A at time 1 and time 3 or 4) are not correlated. Returning to our hypothetical example, we would find that babbling at 7 months predicted vocabulary size at 15 months, but not mean length of utterance (MLU) at 30 months. In other words, whatever factor led to similar rates of development between 7 and 15 months would no longer operate between 15 and 30 months. Case 3 (strong stability) allows for very little flexibility in development: assessments of ability A at times 1, 2, 3, and 4 are all correlated with each other. Thus babbling at 7 months would predict not just vocabulary size at 15 months and MLU at 30 months but also verbal SAT scores at 18 years (please remember that this is a hypothetical example). Obviously Cases 2 and 3 are ideal cases. In real data one would expect to find many intermediate examples. Furthermore, approximations to Case 2 are more likely when the time span between assessments is large, and conversely, approximations to Case 3 are likely when the timespan between consecutive assessments as well as between the first and last assessment are relatively small. Heterotypic progressions can be categorized in a similar way (order, weak stability, strong stability), except that successive measurements now represent different variables (see McCall *et al.*, 1977).

As noted before, heterotypic progressions are generally studied because of some well-based hunch that variable A and variable B have some underlying factor in common, or that A is a necessary precursor to the development of B. In cases in which a relationship between two heterotypic variables is expected to continue over a period of time the two homotypic progressions (A and B) and their intercorrelations can be investigated as a hybrid homotypic-heterotypic progression. Three hypothetical correlational patterns are of special interest to us here, for reasons which will become apparent shortly. To describe them we have borrowed the term *homology* from the ethological literature, where it is used to characterize the evolution of somewhat dissimilar surface structures from a common ancestral form: A duck's and a penguin's wings are homologous; a porpoise's flippers and a cod's fins are not (Eibl-Eibesfeldt, 1970). In the context of this paper we use the term *homology* when there is reason to believe that the same underlying factors may be responsible for generating correlations between two homotypic progressions. At least three types of homology are possible:

1. *Remote homology.* Correlations of A and B are due to general maturational factors. If remote homology obtained, one would expect correlations between A and B as well as correlations of A and B with a host of other maturationally driven variables.
2. *Specific homology.* Both A and B are outward manifestations of a

specific, much more narrowly defined underlying factor C. Specific homology requires fairly strong intercorrelations of A and B as well as relative independence of these two variables from other variables not influenced by C.

3. *Local homology.* A and B are related to each other at some points in time but not others, because C, an underlying capacity, has several distinct subcomponents. During periods when A and B draw on a common subset of C's components, both will be highly correlated. During periods when A and B draw on few common components of C, they will be weakly correlated. This is the model proposed in Fischer's (1980) theory of skill development.

The findings which we will present in this chapter are based on the concurrent assessment of two homotypic progressions (language and symbolic play). Intercorrelations between these two progressions are expected on theoretical grounds. The first model, remote homology, will not be examined here, since other data from the same project show that it does not hold (for example, combinatorial play is not correlated or negatively correlated with symbolic play and language). Our chief aim is to compare and contrast the specific and local homology models.

Piaget (1962) interpreted the coemergence of verbal and enactive representation at the end of the first year as two manifestations of but one underlying symbolic or semiotic function. Hence we believe that he would have supported the specific homology position. Striking parallels in the development of language and symbolic play have been noted not only by Piaget but by others who were inspired by his work (e.g., Bates, Bretherton, Shore, & McNew, 1983; Fein, 1979; Fenson & Ramsay, 1980; McCune-Nicolich, 1981; Volterra, Bates, Benigni, Bretherton, & Camaioni, 1979; Wolf, 1982). The most important of these are:

1. Language is a tool for the representation of absent persons, objects, and actions. Symbolic play is a tool for the representation of events (including persons, objects, and actions) outside their everyday context.
2. In language arbitrary sounds come to stand for objects, events, and persons. In symbolic play arbitrary objects are made to stand for other objects (a child may substitute a placeholder, such as a block or a counterconventional object such as a spoon, to represent a telephone).
3. In language and play the child begins by producing single units (vocal and enactive names) which are later combined into meaningful sequences (sentences or event schemata).

Unlike Piaget, Fischer (1980) would predict correlations between language and symbolic play only during those periods of development when both forms of representation make use of the same component skills. To determine when correlations between the two domains should be expected, proponents of the local homology model would therefore emphasize differences between enactive and verbal representation, rather than concentrating primarily on similarities. We have noted several such differences (although we do not regard our list as exhaustive):

1. In language temporal and spatial relations between agents and objects come to be expressed as separate morphemes indicating time, direction, and location (grammaticization). In symbolic play these relations are implicit in the enactment itself.
2. In symbolic play realistic toys provide a large measure of perceptual support as the child plans to enact an event outside its normal context. Some perceptual support is provided even by substitute objects used as placeholders for other objects. In language about absent persons, objects, or events such planning may be carried out with little or no perceptual support.
3. Although the gestures used in symbolic play are sometimes mimed without objects (drinking, pouring), it is rare for young speaking and hearing children to use enactive gestures to communicate about objects outside the play context.
4. Language is much more highly conventionalized than symbolic play although partners may develop idiosyncratic play conventions in the course of repeated interactions.
5. Language is an obligatory form of representation for hearing and speaking children, whereas symbolic play is an optional form. The skills associated with the latter are therefore not so consistently and frequently practiced.

There are two ways in which parallels in the development of representation through verbal and enactive means can be studied. The first solution (absolute comparison) is to identify the concurrent onset of specific equivalent behaviors in both domains (e.g., McCune-Nicolich, 1981; Volterra et al., 1979; Wolf, 1982b). The second solution (relative comparison) is to evaluate the level of development in both domains in a variety of ways and to correlate the resultant scores. When the first strategy can be theoretically supported it is obviously preferable. However, well-based theoretical grounds for some of the hypothesized correspondences between language and symbolic play are not always available. It has been suggested, for example, that two-word speech should coincide with the onset of two-scheme sequences in symbolic play. Yet it is not clear, a priori, why the verbal unit should be words rather than propositions. The choice of unit (word = scheme) derives from empirical observation, that is, from noticing approximately simultaneous onset of two-word combinations in language and two-scheme combinations in pretending, rather than from deeply compelling structural analyses. Indeed, Wolf (1982b) suggests that semantic rather than contentless structural correspondences between language and play should be investigated (e.g., talking about another person's action = pretending at another person's behavior). Fenson (1982) has also advocated semantic, rather than purely structural, comparisons.

The second strategy is less precise, but easier to implement. Building on language and symbolic play measures which represent the "state of the art" in both domains, a researcher can ask the question whether children who are relatively advanced in one mode of representation (or specific aspects of it) are also advanced in the other. The aims of Strategy 2 are thus more modest, but we see Strategy 2 as a good stepping-stone toward a more directed, less exploratory use of Strategy 1. *Post hoc* examination of the data on the basis of obtained correlational patterns may make it possible to sharpen theoretically based formulations which can then be further tested. In the study to be presented in this paper we chose Strategy 2.

THE CONCURRENT DEVELOPMENT OF LANGUAGE AND SYMBOLIC PLAY

Method

The findings to be reported here were obtained in the course of a longitudinal investigation of symbol development conducted by the authors. Although various, more circumscribed aspects of the data have been reported elsewhere, the longitudinal intercorrelations of language and symbolic play across the whole age span of the project (from 10 to 28 months) have not previously been examined.

Sample.   The initial longitudinal sample consisted of 16 boys and 16 girls from middle-class families who were identified from newspaper birth announcements. They were contacted by letter and follow-up telephone call, resulting in a 70% acceptance rate. The initial 32 children were studied at 10 and 13 months. Since 5 children subsequently moved away, 3 new children were invited to participate in the project at 20 months to bring the sample up to 15 boys and 15 girls. This sample of 30 children was also studied at 28 months.

Hence correlations from 10 to 13 months include 32 children; correlations from 10 and 13 months to 20 and 28 months include only 27 children, and correlations from 20 to 28 months are based on 30 children.

Procedures and Analyses.   Because detailed accounts of data collection and reduction procedures are available in a number of publications, they will be reviewed only in as much detail as is necessary to follow the main argument of the paper. At all four ages (10, 13, 20, and 28 months) children were observed in the home and in the laboratory in a variety of structured and unstructured contexts. The mother was present throughout. Almost all the observational data reported here are based on videotaped information. Intercoder reliabilities, based on point-to-point comparisons of the transcripts, exceeded 80% for all measures except language at 20 months. Because indistinct speech was so common at that age, we fell just short of attaining 80% intercoder agreement. The 20-month language data were therefore derived from those utterances on which 2 coders completely agreed. In addition to direct observations, we obtained information from checklists filled out by the mother, from maternal interviews, and from standardized tests. Pearson product-moment correlations (two-tailed) were used throughout.

The findings will be reported in three parts. The two homotypic developmental progressions (language and symbolic play) will be separately presented in the first two sections. Each of these sections will contain a discussion of within-domain correlations. In the third section we will document intercorrelations of the two homotypic progressions treated as a hybrid homotypic/heterotypic progression and consider implications of the findings for future studies of language and symbolic play. The paper will conclude by discussing the more general relevance of the results for longitudinal studies in which stability of individual differences is the major focus.

Longitudinal Stabilities in Language Development

Description of Variables.   The earliest language measures used in this study (10 and 13 months) were almost exclusively based on information obtained from the mother. We chose this approach because previous experience had shown that the collection of a representative speech sample at these ages would have been an extremely time-consuming and prohibitively costly enterprise with a sample of 30 children. Fortunately earlier studies (Bates *et al.*, 1979) had indicated that maternal interviews can yield very reliable data, provided mothers are asked to provide quite specific information pertaining to the infants' present behavior.

The reader who wishes to obtain further information on any of the procedures and variables listed below is referred to specific publications which provide more detailed descriptions. After each variable we indicate whether it was obtained from maternal report (R) or direct observation (O).

10 Months

Comprehension and Production Vocabulary (R). These two variables were derived from a vocabulary checklist based on findings from a previous study (Bates *et al.*, 1979) and filled out by the mother during the home visit (see Carlson-Luden, 1979). *Comprehension vocabulary* refers to the number of words the infant was reported to understand without gestural support. *Production vocabulary* refers to the number of words and expressions which the infant used spontaneously (excluding mere imitation).

13 Months

Comprehension and Production Vocabulary (R). These measures were based on an extensive list of words culled from several studies of child language (Bates *et al.*, 1979; Bloom, 1973; Nelson, 1973). This list was used in conjunction with a detailed, probing interview during which the mother was asked (1) whether the child comprehended or produced a specific word and (2) in what context comprehension or production was observed (i.e., did the mother gesture as she spoke, could comprehension of a command be due to the mother's tone of voice, did the infant tend to utter a particular word only in a specific context, location, or at a specific time of day?). Further details are available in Snyder, Bates, and Bretherton (1981).

Comprehension Box: Word Comprehension and Vocal Schemes (O). These measures are based on a structured task administered in the laboratory. Sets of three familiar objects were lined up in a plexiglass box with three compartments. Upon verbal request for one of the objects, a transparent lid covering the front of the box was lifted to allow the infant to make a choice. There were 16 trials, 8 with realistic and 8 with abstract object-arrays. *Word comprehension* refers to the number of correct choices an infant made, *vocal schemes* to the number of labels and appropriate sound effects produced by the infant (i.e., "cah" or "vroom-vroom" in conjunction with a toy car). Details of this task are reported in Bretherton, Bates, McNew, Shore, Williamson, & Beegly-Smith, (1981).

20 Months

Language Clusters (R). Four language clusters were derived from a statistical analysis of data obtained during a detailed maternal interview which was designed to capture significant developmental phenomena reported in the recent child language literature. Two of the language clusters were based on multiword speech, two on more general semantic abilities which could be demonstrated in one-word and/or multiword speech. The multiword *referential cluster* was defined by four summary variables all of which pertained to the ability to make specific reference to objects (for example, in telegraphic utterances such as "daddy car"). The multiword *grammatical morpheme cluster* was composed of four summary variables concerned with the expression of relations (that is use of grammatical morphemes such as prepositions, pronouns, articles, verb inflections). The two general clusters were a *dialogue cluster* and a *semantic-cognitive cluster*. The dialogue cluster included measures concerned with answering and asking questions, taking turns in simple conversations, rate of imitation, and labelling. The semantic-cognitive cluster was composed of two summary variables: reference to absent persons, objects, or events and the use of implicit case roles (e.g., possession) in one-word or multiword utterances. Detailed descriptions of these clusters can be found in Bretherton, McNew, Snyder, and Bates (1983).

Language Clusters (O). Three clusters were obtained from statistical analysis of the infants' spontaneous speech during 90 minutes of structured tasks and free play videorecorded in the laboratory and at home. Coding was based on the categories devised for the language interview described above. The referential cluster included items which were almost identical with those composing the interview-derived referential and semantic-cognitive clusters. The grammatical morpheme and the dialogue clusters resembled the corresponding interview clusters closely. Further details are reported in Bretherton *et al.* (in press).

Mean Length of Utterance (O). MLU was calculated from the 90-minute videotape transcripts following Brown's (1973) procedures. Grammatical morphemes such as *-ing, -ed,* and *-s* as well as lexical items are included in the calculation of MLU.

Production Vocabulary (O). The number of different words which an infant produced during the observations was tallied from the videotape transcripts.

Production Vocabulary (R). The mothers were asked to indicate on a checklist which of the approximately 400 words their child produced. The list was compiled from our own pilot data and from publications by Bloom (1973), Bloom, Lightbown, and Hood (1975), and Nelson (1973).

28 Months

Peabody Picture Vocabulary Test (O). The PPVT, a standardized test, was administered during the home visit. Raw scores were tabulated. Although the PPVT was designed as a measure of general intelligence, we included it here to assess language comprehension.

Mean Length of Utterance (O). MLU was based on the child's speech during four videotaped interactive episodes: (1) a 30-minute session of spontaneous and elicited symbolic play in the laboratory, (2) a 5-minute free-play session with the mother in the laboratory, (3) a 5-minute snack with the mother at home, and (4) a 5-minute bookreading session with the mother at home. These sessions are described in Bretherton, O'Connell, Shore, and Bates (1984) and Beeghly-Smith (1981).

Propositional Analyses According to Kintsch (O). The propositional analyses were carried out by Snyder and O'Connell (in preparation) using the same four interactive situations from which MLU was computed. The measures derived from the propositional analyses were, however, much more fine-grained assessments of semantic-syntactic language complexity than MLU. The coding procedure (based on Kintsch, 1974) was adapted for text analysis by Turner and Greene (1977) and here applied to child language for the first time. The *propositions* measure represents a count of all utterances containing both a predicate (state, action, condition) and an argument (agent, instrument, recipient). The *modifier* category is based on lexical items which modify the predicate in some way (quantifiers, qualifiers, negatives). The *connectives* measure includes conjunctions, disjunctions, causal connectives, and expressions of location, time, and manner. For a more detailed description of these variables and a second type of propositional analysis developed by Johnston (1981) on the basis of Slobin (1967), the reader is referred to Snyder and O'Connell (in preparation). A very readable explanation of propositional analysis is presented in Slobin (1979).

Production Vocabulary (O). The data used for the calculation of MLU and the propositional analyses were also subjected to a vocabulary count.

Internal State Words (R). A list of internal state words used by young children was compiled from Bretherton, McNew, and Beeghly-Smith (1981). This list formed the basis of an extensive maternal interview (administered in the home) regarding the use of internal-state language in appropriate contexts. The number of internal-state words reported for each child is included here (see Bretherton & Beeghly, 1982, for further details).

Implications of the Language Correlations. The correlations of the 28-month measures (PPVT, MLU, propositional analyses and vocabulary counts) with the 10, 13, and 20 months language variables are displayed in Table 1.

Although the very high proportion of significant correlations across age is the most striking finding, there is also some evidence for two somewhat differentiated strands of language development.

The four multiword clusters at 20 months (that is, the two referential and the two grammatical morpheme clusters obtained from interview and observation) were not correlated with comprehension (the PPVT) at 28 months (see Table 1). In addition, MLU at 20 months was unrelated to 28-month comprehension. Comprehension at 28 months

Table 1. Correlations of Language Variables at 10, 13, and 20 Months with Language Variables at 28 Months

| | PPVT[b] (O) | MLU[b] (O) | 28 months Kintsch Propositional Analysis (O)[a] | | | Vocabulary (O) | Internal state words (R) |
			Propositions	Modifiers	Connectives		
10 months (N = 27)							
Production vocabulary (R)[a]		.37[c]		.45[d]	.34[c]		.43[d]
Comprehension vocabulary (R)	.48[e]	.52[e]	.35[c]	.62[f]	.33[c]		.42[d]
13 months (N = 27)							
Production vocabulary (R)		.42[d]		.40[d]		.33[c]	.42[d]
Comprehension vocabulary (R)	.57[e]	.44[d]		.32[c]	.36[c]		.62[f]
Vocal schemes (O)							
Word comprehension (O)	.36[c]	.34[c]		.34[c]		.35[c]	.46[d]
20 months (N = 30)							
Referential cluster (O)		.66[f]	.43[d]	.45[d]	.56[f]	.64[f]	.43[d]
Referential cluster (R)		.69[f]	.46[e]	.57[f]	.54[e]	.66[f]	.74[f]
Grammatical morpheme cluster (O)		.63[f]	.45[e]	.42[d]	.41[d]	.52[e]	
Grammatical morpheme cluster (R)		.56[f]	.36[c]		.42[d]	.61[f]	
Dialogue cluster (O)	.39[d]	.67[f]	.39[c]	.36[c]	.46[e]	.67[f]	.67[f]
Semantic cognitive cluster (R)	.59[f]	.54[f]		.53[e]	.41[d]	.51[e]	.79[f]
Dialogue cluster (R)	.37[c]	.56[f]	.35[c]	.42[d]	.39[c]	.54[e]	.57[f]
MLU (O)	.48[e]	.48[e]	.31[c]		.35[c]	.47[c]	
Vocabulary (R)	.48[e]	.81[f]	.58[f]	.71[f]	.72[f]	.63[f]	.73[f]
Vocabulary (O)	.39[d]	.74[f]	.49[d]	.45[d]	.58[f]	.73[f]	.47[e]

[a] R = maternal report, O = observations.
[b] PPVT = Peabody Picture Vocabulary Test, MLU = Mean Length of Utterance.
[c] p < .10 (2-tailed).
[d] p < .05 (2-tailed).
[e] p < .01 (2-tailed).
[f] p < .001 (2-tailed).

was, however, correlated with the 20-month general semantic-cognitive cluster. This cluster assessed not language complexity, but the variety of meanings a child could express in one-word or multiword speech. Production at 28 months (MLU, reported and observed vocabulary), on the other hand, entered into highly significant correlations with the four multiword clusters at 20 months. Moreover, production vocabulary and MLU at 20 months were more highly correlated with 28-month production measures than with the PPVT which assessed comprehension. Additional evidence (presented in Bates, Bretherton, & Snyder, in preparation) indicates a parallel correlational pattern of 13-month reported comprehension and production with the 20-month measures.

Although it is common knowledge that language comprehension precedes production, the size of the comprehension–production gap in this study was quite variable. There were some children whose comprehension ability far outstripped their production ability. These same children were nevertheless able to convey complex meanings (e.g., to talk about absent persons or events) through one-word utterances at 20 months. By contrast, some 20-month-olds who used many two-word utterances (and therefore obtained high scores for multiword production) may have acquired them as whole phrases. On the basis of additional evidence which cannot be presented here, Bates *et al.* (in preparation) and Bretherton *et al.* (1983) interpret these correlational patterns as evidence for two language acquisition strategies, one based on analytic-reconstructive processes and the other based on a tendency to process phrases holistically.

A second interesting finding concerns the internal consistency of the language complexity measures. MLU is an extensively validated global assessment of semantic-syntactic complexity, whereas the propositional measures (based on Kintsch, 1974) evaluate separate aspects of this development. That the very fine-grained propositional measures at 28 months showed any significant relationships with language measures obtained at 10 and 13 months was quite surprising to us. Even more striking, however, were the correlations of the 20-month language variables with virtually all propositional measures.

In sum, the correlations displayed in Table 1 represent two very stable homotypic progressions based on language production and comprehension (see Figure 1, Case 3), which are also significantly intercorrelated at 10, 20, and 28 months and marginally correlated at 13 months (see Bates *et al.*, in preparation, for the within-age language correlations).

Although it is of more peripheral interest here, we should like to draw the reader's attention to the usefulness of maternal interviews in the study of child language. At 13 months the two interview measures

Table 2. Correlations of 28-Month Symbolic Play Variables with Symbolic Play at 10, 13, and 20 Months

	28-months scenarios (O)[a]								28-months free play	
	Total sequences	Total scheme variety	SP[b]	RL	PH	CC	Descriptive utterances	Pretend utterances	Alone	With mother
10 months (N = 27)										
Symbolic scheme diversity (R)[a]										
13 months (N = 27)										
1. Symbolic scheme diversity (R)			.52[e]				.37[c]			
2. Imitation task (Killen–Uziris, O)										
Spontaneous schemes										
Appropriate schemes										
Counterconventional schemes	.56[f]	.35[c]			.32[c]	.34[c]			.33[c]	
3. Comprehension box										
Total symbolic schemes			.41[d]							
Total vocal schemes	.36[c]	.47[d]	.49[e]			.43[d]				

4. Scenarios (Wolf, O)						
Spontaneous scheme diversity	.34c					
Realistic scheme diversity		.50e			-.38d	-.40d
Tiny scheme diversity		.35c	.51c			-.32c
Abstract scheme diversity						.42d
Placeholder scheme diversity			.34c	.38c		
Total scheme diversity (elicited)		.38c				
20 months (N = 30)						
1. Scenarios (O)						
Total sequences	.49e	.40d				
Total scheme diversity	.52e	.33c	.41d			
Total descriptive utterances	.40d					
Total pretend utterances	.74f	.40d	.34c	.37c		
Also SP, RL, PH, CC[b]						
2. Free play alone and with mother (O)						
Scheme diversity, alone				.42d		
Scheme diversity, with mother				.42d		

[a] R = maternal report, O = direct observation.
[b] SP = spontaneous, RL = real, PH = placeholder, CC = counterconventional scenario condition.
[c] $p < .10$ (2-tailed).
[d] $p < .05$ (2-tailed).
[e] $p < .01$ (2-tailed).
[f] $p < .001$ (2-tailed).

(comprehension and production vocabulary) were much better predictors of later language development than the observational comprehension and production measures obtained from the comprehension box task (see Table 1). Predictions of language development from 20 to 28 months were equally good for the interview- and observation-based variables. These are useful methodological findings since the analysis of the interview measures at 20 months was far less time-consuming than the transcription of the corresponding videotaped material. Even at 28 months a language interview can still be helpful. As Table 1 shows, a useful estimate of general vocabulary can be obtained by requesting information about a more limited domain (internal-state words). For researchers interested in a quick assessment of language ability, the longitudinal intercorrelations of the vocabulary measures derived from checklists and interviews are especially encouraging.

Finally, it is noteworthy that significant longitudinal correlations for language were obtained during a period when infant cognitive tests (e.g., the Bayley Developmental Test) have not been able to demonstrate stability of individual differences (McCall et al., 1977). The instability of DQ (developmental quotient) across the period spanning the transition from sensorimotor to symbolic functioning has sometimes been attributed to a discontinuity in the items. Early items rely heavily on fine-motor control whereas later items are much more cognitively based. In other words, it has been proposed that DQ may not constitute a valid homotypic progression. The data presented in Tables 1 and 2 give indirect support to this explanation. When a homotypic variable (language) is repeatedly assessed during the same period, respectable stability can be demonstrated.

On the other hand, the data in Table 1 do not corroborate the "scoop model" of early development which was recently proposed by McCall (1981). McCall's model is based on Waddington's (1957) metaphor of the epigenetic landscape. In a nutshell, McCall suggests that during the sensorimotor period, when development is heavily canalized, strong self-righting tendencies come into play whenever environmental influences drive the developing individual off its genetically determined course. This, claims McCall, generates *instability* of individual differences during the sensorimotor period, that is, up to about 18 months. With the onset of symbolic functioning, on the other hand, environmental influences are capable of inducing permanent shifts in the path of development, leading to stability of individual differences after 18 months. McCall's scoop model is not supported by the data presented in this chapter. We observed stability of individual differences in language development within the sensorimotor period as well as across the transition from sensorimotor to symbolic functioning.

Longitudinal Stabilities in the Development of Symbolic Play

Description of Variables.   Most of the symbolic play assessments used in our project were derived from direct observation and contained an experimental component. Observation was chosen as the preferred method of data collection because, even at one year of age, pretend play is easier to elicit than language. Infants were given several opportunities to engage in spontaneous symbolic play, but in addition a number of structured tasks were modeled for them. Some of these tasks were especially developed for this project, others were borrowed from published work. The decision to use modeling as an eliciting technique was based on the assumption that it would enhance performance, but not much beyond the child's spontaneous capacity. Piaget (1962) suggested that imitation and spontaneous production of behavior are closely linked. This claim is supported by Slobin and Welsh (1973) who found that linguistic imitation is highly constrained by what the child can produce spontaneously.

In other studies of symbolic play, modeling has indeed been shown to facilitate the performance of schemes (e.g., Fenson, 1982; Largo & Howard, 1979). We used modeling in our project not only to enhance play, but also to manipulate the realism versus abstractness of toys. Play with substitute objects (a block = a car) tends to be more difficult than play with realistic objects. By using a substitute object (whether spontaneously or in imitation), infants demonstrate their ability to *decontextualize*, that is, to represent a symbolic scheme with little or no perceptual support from the object itself (Piaget, 1962). It has been suggested that this behavior is analogous and related to the capacity of letting arbitrary sounds stand for objects and relations in language.

A symbolic schemes checklist and a symbolic play interview were administered to the mother at 10 and 13 months, but not—as in the case of language—to provide the primary data. We wished merely to verify whether such an instrument could be validated against observational data. Each of the variables described below is marked R or O, where R indicates maternal report and O indicates observation.

10 Months
Symbolic Scheme Diversity (R).   Symbolic schemes described in Volterra *et al.* (1979) were used to construct a checklist which the mother filled out during the home visit (see Carlson-Luden, 1979).

13 Months
Symbolic Scheme Diversity (R).   A standardized interview was constructed on the basis of findings by Volterra *et al.* (1979). Mothers were asked, during the laboratory visit, to describe symbolic schemes they had see their child perform spontaneously at home.

Killen–Uzgiris Imitation Task (O). During this task, devised by Killen and Uzgiris (1981), the child was given four familiar objects (a car, hairbrush, cup, and toy cat), one at a time. Spontaneous schemes with each object were observed first. For each object a conventional scheme (driving the toy car) and a counterconventional scheme (drinking from the toy car) were modeled silently in random order. Three measures were obtained: *spontaneous scheme performance, conventional imitation, and counterconventional imitation* (see Bates, Bretherton, Snyder, Shore, & Volterra, 1980, for further details).

Wolf-scenarios (O). These three scenarios consist of very simple sequences (eating, drinking, putting a doll to bed) developed by Wolf and Gardner (1979). A set of three toys per scenario was given to the infant for spontaneous play (e.g., a plate, spoon, and napkin for the eating scenario). Subsequently a brief sequence (scraping the plate, putting the spoon to the mouth, smacking the lips, wiping the mouth) was modeled four times, first with realistic objects, then with a substitute for one of the props (e.g., the spoon in the eating scenario). The four substitutes were (1) a tiny version of the object, (2) an abstract, pared-down version of the object, and (3) a placeholder such as a block, which had no special object-associated schemes. Coding took account of the quality of schemes and thus differed slightly from data reported in Bates *et al.* (1980). Scheme diversity scores (summed over the three scenarios) were computed for each of the five conditions: *spontaneous, realistic, tiny, abstract, and placeholder.* A measure of *total elicited scheme diversity* was also calculated by summing across the four elicited conditions (see Bates *et al.*, 1980, for further details about the scenario presentations).

Comprehension Box (O). A plexiglass box, already described in the language section of this paper, was used as a multiple-choice test for object-word comprehension. In the course of this test infants produced many appropriate symbolic schemes whether or not they chose the requested object. A *symbolic scheme diversity* measure and a *vocal schemes* measure were derived from this task. The latter consisted primarily of sound effects (e.g., "vroom" while rolling a toy car) but also included use of conventional labels (see Bretherton *et al.*, 1981, for further details).

20 and 28 Months

Scenarios (O). Three scenarios, more complex than those presented at 13 months, were developed (eating breakfast; bathing a doll and giving it a ride in the stroller; having a mother bear put a baby bear to bed). The child was first encouraged to play with the scenario-relevant toys (e.g., pitcher, cup, spoon, bowl, and napkin for breakfast). A standardized scenario was then modeled three times, with each modeling sequence followed by a period during which the child could play with the toys spontaneously. The first modeling took place with the realistic toy set. During the second and third modeling a placeholder (block) and counterconventional object (e.g., a comb instead of a spoon for breakfast) were substituted for one central toy in the scenario. Qualitative and quantitative measures were obtained from the transcripts of the videotaped sessions (*scheme frequency, scheme diversity, sequenced schemes* occurring in two-unit sequences or more, *longest sequence*). Although these measures were not independent, each assessed somewhat different aspects of symbolic play. Scores were computed in two ways: (1) by summing across scenarios and conditions (*total scenario*) and (2) by conducting separate tallies for each of the four conditions (*spontaneous, realistic, placeholder,* and *counterconventional*).

Play related language during the scenarios was also assessed: *descriptive utterances* were statements labeling actual props or actions ("dolly," "put in," "roll it"). *Pretend utterances* referred to pretend substances, objects, states, actions, and locations ("*drink juice*," "*he tired*," "*go to work*"). These measures were tallied for the scenarios as a whole (*total descriptive and pretend utterances*), as well as for the four separate conditions (*spontaneous, realistic, placeholder, counterconventional*). The aim was to examine the complementarity of language and play at the later ages, a topic which has been neglected until recently because of researchers' preoccupation with parallels in the development of representation in play and language (but see Fenson, 1984).

Since modeling of object substitution gave rise to several forms of protest (children searched for the realistic object, refused to use the substitutes, or pointed out their true purpose), separate measures of *behavioral and verbal protest* were also obtained. Contrary to expectations, verbal protest actually increased from 20 to 28 months in spite of the fact that symbolic schemes during the substitution conditions also rose (see Bretherton, O'Connell, Shore, & Bates, 1984, for details).

Symbolic Play, Alone and with Mother (O).   A set of toys suitable for combinatorial play (stacking, nesting, shape-sorting) and for symbolic play (human figures, furniture, a plexiglass doll's house) was presented to the child in two separate 5-minute sessions. During the home visit the child was observed in solitary play. Two or three days later mother and child together played with the same toys in the laboratory. *Symbolic scheme diversity, alone and with mother,* were tallied from these videotaped sessions (see O'Connell & Bretherton, 1984, for details).

Implication of the Symbolic Play Correlations.   Even cursory comparison of Table 2 with its language counterpart (Table 1) shows that there was considerably less longitudinal stability in symbolic play. In some ways the situation is even worse than it appears at first glance, since the two 28-month summary measures (total scheme diversity and total sequences) were derived from the same raw data and were therefore not independent. Total scheme diversity was a linear combination of the scheme diversity scores for the four contextual conditions (spontaneous, realistic, placeholder, and counterconventional). Both total scenario and separate condition measures were included in this analysis in order to test for internal consistency of play in the same way as we had already done for language (where we compared the global MLU measure with the more finegrained propositional analyses).

Total scenario sequences at 28 months were significantly correlated with three 13-month variables (each from a different task and none a summary variable) as well as with total sequences and total scheme diversity at 20 months (see Table 2). Curiously, the best predictor of 28-month total sequence was pretend language at 20 months, even though 28-month pretend language was not correlated with 20-month play.

The second 28-month summary measure displayed in Table 2, total scheme diversity, showed fewer significant correlations with earlier play variables. Inspection of the four separate scheme diversity measures by condition (see also Table 2) shows that consistent correlations with earlier measures occurred only for *spontaneous* scheme diversity. None of the three modeled 28-months scenario conditions (realistic, placeholder, counterconventional) yielded more than sporadic correlations with 20- and 13-month symbolic play assessments.

In addition, there was no consistent pattern of intercorrelation among the object substitution measures at 13, 20, and 28 months. Indeed, verbal protest to object substitution was positively, not negatively, correlated with 13-month play, r (26) = .52, $p < .01$. It appears that a 28-month-old's verbal protest to modeled use of placeholder or counterconventional objects cannot be simply interpreted as indicating low-level symbolic ability, a point to which we will return later.

Another intriguing finding was that vocal schemes from the comprehension box task at 13 months were correlated with some later symbolic play measures (see Table 2). Remember that vocal schemes at 13 months did not predict later language (see Table 1). That this measure was correlated with later symbolic play suggests that sound effects like "vroom-vroom" made in conjunction with rolling a car may be more akin to symbolic schemes than to verbal names (see Bretherton, Bates, McNew, Shore, Williamson, & Beeghly-Smith, 1981, for further discussion of this point).

The few negative correlations displayed in Table 2 remain a puzzle. The free-play assessments (scheme diversity alone and with mother) at 28 months accounted for two of these. Indeed, the free-play measures showed no consistent cross-age correlations at all.

In sum, Table 2 contains several intriguing local patterns but we cannot discern a coherent overall pattern in these correlations. Should we, therefore, conclude that performance of symbolic play is only marginally stable across age or that we are dealing with a weak homotopic progression? The answer to this question is no. We must also consider the possibility that our measures may be invalid or unreliable.

Let us address the question of validity first. Although the longitudinal stability of the scenario measures at 20 months left much to be desired, we did find significant age changes (in an upward direction) for all measures except scheme frequency and descriptive language (see Bretherton et al., 1984). We also obtained the predicted experimental effects in response to contextual manipulations at both ages. Scheme diversity increased after realistic modeling but then decreased again after modeling with substitute objects. These findings corroborate those of other investigators (e.g. Fein, 1975; Kagan, 1981; Largo & Howard, 1979; Lowe, 1975). They also attest to the validity of our symbolic play measures. We were further encouraged by the within-age stability or convergent validity of the different play assessments. These are only summarized here, but detailed tables are available from the authors. At 13 months 6 of 10 possible correlations among the Wolf-scenario conditions were significant at $p < .05$. For the other symbolic play measures obtained at 13 months (reported scheme diversity, measures from the Killen–Uzgiris imitation task, comprehension box measures, total elicited Wolf-scenario scheme diversity), 9 of 21 possible coefficients were significant at $p < .05$. At 20 months, 3 out of 6 possible correlations among the four scenario conditions were significant, although at 28 months only the two object substitution conditions were correlated with each other at $p < .05$. With regard to the total scenario and free-play measures at the later ages, 4 of 10 correlations were significant ($p < .05$) at 20 months and 6 of 10 at 28 months. It is interesting that the coherence among the separate measures did not increase with age as we found for the language variables at 20 and 28 months (see Bates et al., in preparation).

Measures can be valid but unreliable. In this study, unreliability could occur in a number of ways. First, coding of the symbolic schemes from videotapes might have posed serious problems. Yet, we encountered little difficulty in attaining at least 80% intercoder agreement on all the symbolic play measures used in our project. We are also not aware that other investigators of symbolic play have found categorization of the observed behaviors especially troublesome. Second, unreliability can be due to very infrequent occurrence (Waters, 1978) or considerable fluctuations of the behavior of interest if it is assessed over too brief a time span. Even though symbolic schemes were not infrequent at any age, the second alternative—substantial fluctuations in the quantity and quality of symbolic play—remains a possibility. If this hypothesis is correct, longer and/or repeated observation periods might yield better stability of individual differences. In support of this hypothesis, Kagan (1981) reports that when 3 consecutive monthly assessments of symbolic play obtained during the third year were summed and then correlated with scores for the next three consecutive assessments, very high longitudinal correlations emerged which had not been evident in the month-to-month correlations. Wolf (1982a) and McCune-Nicolich (1982) have also noted short-term fluctuations in the quality of symbolic play.

We believe that the problem of stability must be solved before we can tackle the second problem posed by our symbolic play findings: no correlations of the spontaneous scenario measures with the two object substitution conditions (placeholder and counterconventional) at 28 months. We believe that scores obtained during the object substitution conditions of our scenarios at 28 months may not be pure measures of representational ability but may also be assessing something like willingness to "toy" with reality. Some of our most verbal 28-month-olds passionately refused to carry out the modeled substitution (see Bretherton *et al.*, 1984, for further discussion of this point).

In sum, symbolic play from 10 to 28 months represents a much less coherent homotypic progression than language, with some significant correlations from 13 to 28 and from 20 to 28 months, but virtually none from 13 to 20 months except for four relationships with play-related language (not displayed here). We suspect, however, that longer observation periods might yield a more stable homotypic progression. There is also tentative evidence that spontaneous symbolic play and symbolic play after modeling with placeholders and counterconventional objects may assess somewhat different abilities.

Conjoint Stability of Language and Symbolic Play: Intercorrelations between Two Homotypic Progressions

The correlations between language and symbolic play measures are displayed in Tables 3, 4, and 5. For the sake of simplicity, two representative language measures at each age were chosen to demonstrate concurrent and longitudinal consistency of individual differences across the two domains. We felt that the known stability of the language assessments justified this procedure. Moreover, the correlational patterns for the other language variables—which could not be displayed here for lack of space—were quite similar.

We will first discuss concurrent and longitudinal correlations of symbolic scheme enactment with the general language measures. This will be followed by a consideration of concurrent and longitudinal correlations among the play-related and general language measures.

Table 3. Correlations Between Two 13-Month Language Variables and Symbolic Play at 10, 13, 20, and 28 Months

Symbolic play measures	13 months comprehension vocabulary (R)[a]	13 months production vocabulary (R)
10 months (N = 32)		
Symbolic scheme diversity (R)		
13 months (N = 32)		
1. Killen-Uzgiris task (O)[a]		
Spontaneous schemes	.30[c]	.36[d]
Appropriate imitation		.36[d]
Counterconventional imitation		.51[e]
2. Wolf-scenarios (O)		
Abstract scheme diversity	.32[c]	
Placeholder scheme diversity	.36[d]	
3. Comprehension Box (O)		
Symbolic scheme variety		.35[c]
Vocal scheme variety		.48[e]
4. Symbolic scheme diversity (R)		.39[d]
20 months (N = 27)		
1. Scenarios		
Pretend utterances, CC[b]		.35[c]
Behavioral protest to object substitution	−.43[d]	
2. Free play alone and with mother		
28 months (N = 27)		
1a. Scenario schemes (O)		
Scheme frequency, RL[b]	.48[d]	
Scheme diversity, RL	.44[d]	
Sequenced schemes, RL	.58[e]	
Scheme diversity, PH[b]		.33[c]
Longest sequence, CC		.45[d]
1b. Scenario utterances (O)		
Descriptive utterances, RL	.52[d]	
Pretend utterances, total	.48[d]	.39[d]
Pretend utterance, RL	.38[c]	.40[d]
Pretend utterance, PH	.41[d]	
Pretend utterances, CC	.43[d]	
2. Free play alone and with mother		
Scheme diversity (with mother)	.36[c]	

[a] R = reported, O = observed.
[b] RL = realistic, PH = placeholder, CC = counterconventional scenario condition.
[c] $p < .10$ (2-tailed).
[d] $p < .05$ (2-tailed).
[e] $p < .01$ (2-tailed).

Table 4. Correlations of Two 20-Month Language Variables and Symbolic Play at 20 Months ($N = 30$)

20-month symbolic play	20-month multiword referential cluster (R)[a]	20-month vocabulary (O)[a]
1a. Scenario schemes (O)[a]		
Scheme frequency, CC[b]	.42[d]	
Scheme frequency, all	.33[c]	
Scheme diversity, CC	.33[c]	
Longest sequence, RL[b]	.46[e]	
Longest sequence, CC	.42[d]	.44[d]
Sequenced schemes, RL	.31[c]	
Sequenced schemes, CC	.46[e]	
Sequenced schemes, all	.39[d]	
1b. Scenario utterances (O)		
Descriptive utterances, SP[b]	.45[d]	.58[f]
Descriptive utterances, RL	.43[d]	.34[c]
Descriptive utterances, PH[b]		.47[e]
Descriptive utterances, CC	.32[c]	
Descriptive utterances, total	.49[e]	.57[f]
Pretend utterances, RL	.44[d]	.39[d]
Verbal protest to object substitution	.35[c]	.47[e]
2. Free play alone and with mother		

[a] R = reported, O = observed.
[b] SP = spontaneous, RL = realistic, PH = placeholder, CC = counterconventional scenario condition.
[c] $p < .10$ (2-tailed).
[d] $p < .05$ (2-tailed).
[e] $p < .01$ (2-tailed).
[f] $p < .001$ (2-tailed).

Concurrent Correlations of General Language and Symbolic Schemes.   The within-age correlations for 13 months are presented in the first section of Table 3, those for 20 months in Table 4, and those for 28 months in the first section of Table 5. At all three ages we found a number of interesting within-age correlations.

Cohesiveness was greatest at 13 months when almost all of the various symbolic play measures were correlated with either vocabulary production or comprehension (see Table 3). At this age, infants' 13-month symbolic scheme repertoire and vocabulary repertoire tended to go together.

At 20 months we obtained correlations of the multiword referential cluster derived from the interview with several sequence and object substitution measures (see Table 4). That sequence measures should be correlated with multiword speech is sensible and supports the argument that a general combinatorial ability may emerge concurrently in lan-

Table 5. Correlations of Two 28-Month Language Variables with Symbolic Play Variables at 28, 20, and 13 Months

	28 months PPVT (O)	28 months MLU (O)
28 months (N = 30)		
1a. Scenario schemes (O)[a]		
Scheme frequency, RL[b]	.38[d]	
Scheme diversity, RL	.38[d]	
Longest sequence, RL		.40[d]
Sequenced schemes, RL	.44[d]	.34[c]
Scheme frequency, CC[b]	−.36[c]	
Scheme diversity, CC	−.31[c]	
1b. Scenario utterances (O)[a]		
Descriptive utterances, RL	.63[f]	.34[c]
Total descriptive utterances	.35[c]	
Pretend utterances, RL	.40[d]	.32[c]
Pretend utterances, PH[b]	.42[d]	
Total pretend utterances	.39[d]	
Verbal protest to object substitution	.42[d]	
2. Free play alone and with mother (O)		
Scheme diversity (with mother)	.38[d]	
Scheme diversity (alone)		.43[d]
20 months (N = 30)		
1. Scenario schemes (O)		
Longest sequence, RL		.35[c]
Descriptive utterances, SP[b]		.46[e]
Descriptive utterances, RL		.37[d]
Descriptive utterances, total		.48[e]
Pretend utterances, RL		.45[d]
Verbal protest to object substitution		.45[d]
Behavioral protest to object substitution	−.36[c]	
2. Free play alone and with mother		
13 months (N = 27)		
1. Symbolic scheme diversity (R)[a]	.52[e]	.40[d]
2. Killen–Uzgiris imitation (O)		
counterconventional schemes		.42[d]

[a]O = observed, R = derived from maternal report.
[b]SP = spontaneous, RL = realistic, PH = placeholder, CC = counterconventional scenario condition.
[c]$p < .10$ (2-tailed).
[d]$p < .05$ (2-tailed).
[e]$p < .01$ (2-tailed).
[f]$p < .001$ (2-tailed).

guage and symbolic play (see also Shore, O'Connell, & Bates, in press, who analyzed the same data using a somewhat different sequence measure with similar results). Twenty-month vocabulary, on the other hand, tended to show no relationships with symbolic scheme diversity. Scheme and vocabulary repertoire were no longer linked to one another, as they had been at 13 months.

At 28 months the comprehension measure (PPVT) was correlated with play following modeling with realistic toys, but not with any of the other scenario conditions (see Table 5). This makes us suspect that language comprehension may have influenced scenario enactment even though the scenarios could, in our judgment, be understood without language. Correlations with the counterconventional condition were actually negative, more evidence that acceptance of an imposed substitute object at 28 months is not necessarily related to general intelligence but may, at least in part, assess something like cognitive style. Twenty-eight-month language production (MLU) was also correlated with the realistic scenario condition, but not as consistently as the PPVT.

The two 28-month free-play measures were differentially related to 28-month language: play with mother was correlated with the PPVT, a comprehension measure, whereas play alone was correlated with MLU, a production measure (see Table 5). There is always a danger of overinterpreting such findings, but the divergent correlations do make sense here. Independent evidence suggests that comprehension skills played a role in scheme performance during play with mother. Scheme frequency and diversity were higher during this condition than during play alone, with the difference accounted for by children's positive responses to the mother's verbal commands (see O'Connell & Bretherton, 1984, for further details).

Longitudinal Correlations of General Language and Symbolic Play Schemes. Cross-age correlations from 13 to 28 months and from 20 to 28 months were very infrequent. There was only one marginal correlation of 13-month language with 20-month symbolic schemes (see Table 3). Similarly, we found only one marginal correlation of 28-month language with symbolic play schemes at 20 months (see Table 5).

By contrast, the 13-month vocabulary comprehension measure was related to three of the different symbolic scheme measures obtained during the 28-month realistic scenario condition (see Table 3). We found this pattern interesting since it was paralleled by the correlations of 28-month comprehension (the PPVT) with realistic scenario schemes at the same age (see Table 5). Note that in the intercomparisons of symbolic play (see Table 2) it was 28-month spontaneous—not realistic—scenario performance which was associated with earlier symbolic play (at 13 and

20 months). Thus, different components of symbolic play at 28 months were related to play and to language at 13 months.

Interestingly, both the PPVT and MLU at 28 months were correlated with the interview-derived measure of symbolic scheme diversity at 13 months (see Table 5). This finding made us regret that we had not devised symbolic play interviews for the later ages. The only other 13-month symbolic play variable which predicted later language was counterconventional scheme performance during the Killen–Uzgiris imitation task (see Table 5). This very same measure was also highly correlated with language production at 13 months (see Table 3), suggesting that imposed (elicited) object substitution may have a very different meaning at 13 than at 28 months. At the later age counterconventional object substitution was negatively related to language (see Table 5).

In addition, free play with mother at 28 months was marginally correlated with vocabulary comprehension at 13 months, whereas free play alone showed no significant relationships with either 13-month comprehension or production (see Table 3). Free play at 20 months was unrelated to either of the 28-month language assessments (see Table 5).

Play-related Language and General Language. We found many relationships of general language at 13, 20, and 28 months to play-related language at 20 and 28 months (see Tables 3, 4, and 5). Comprehension and production vocabulary at 13 months entered into a considerable number of correlations with play-related language at 28 months (see Table 3). The two 20-month language measures (see Table 4) were significantly correlated with 20-month play-related language during all scenario conditions. The 28-month language variables (PPVT and MLU) were correlated with some of the play-language assessments at 28 and 20 months (see Table 5). Overall, the correlations of general language with play-related language at 20 and 28 months were somewhat more frequent than those with symbolic schemes. It is, of course, hardly surprising that general measures of language comprehension and production should be associated with that aspect of symbolic play which requires language skills. We should like to emphasize, however, that at both 20 and 28 months there were considerably fewer pretend and descriptive play-utterances than play schemes (see Bretherton *et al.*, 1984), a fact which would lead one to expect lower stability for these variables.

If we had no prior knowledge of the homotypic progressions for language and symbolic play presented in the first two sections of this paper (see Tables 1 and 2), we would feel justified in claiming that the intercorrelations of language and symbolic play reported in this third section supported the local homologies position (see Fischer, 1980, and

Chapter 5, this volume), and indeed they may. Having established the striking contrast in the correlational pattern of language and symbolic play, however, we now feel much less confident in pursuing the plan which we formulated when embarking on this longitudinal analysis, that is, to inspect the raw data for very specific correspondences between the two domains. Instead, our findings lead us to suggest that a shift in strategy may be in order. Before engaging in further studies of language–symbolic play correspondences we must first discover whether longitudinal assessments of symbolic play would attain improved stability if data from several relatively long observation periods were pooled. In other words, we need replications of Kagan's (1981) finding. Without such information, the continued search for parallel developments in language and symbolic play is not likely to be meaningful.

It could, of course, be objected that many of the language measures employed in this longitudinal analysis were summary measures which combined assessments across several contexts, whereas the various symbolic play assessments were obtained during relatively short observation periods. This would not be a fair objection. For example, separate MLU scores for each of the four interactive sessions at 28 months would have yielded very similar correlational patterns to those obtained with the combined MLU measure used here (see description of variables in the language section). The MLU scores calculated for each of the three 5-minute interactions with the mother at 28 months had intercorrelations above .70, despite the fact that children produced far fewer than the 100 utterances recommended for computation of MLU (see Beeghly-Smith, 1981). The reason why we combined MLU and other language measures was indeed because very short-term assessments were stable across contexts. Finally, a recent study by Fenson (1982) completely corroborates the findings reported here. Extremely high stability of individual differences from 26 to 30 months was obtained for language and much lower stability for symbolic play, even though Fenson made an attempt to create semantically corresponding play and language measures and assessed both during the same session.

GENERAL IMPLICATIONS

Although the findings presented in this paper bear most strongly on the continuing debate of language–symbolic play links, they have a number of implications for longitudinal studies in general.

First, the contrasting correlational patterns for language and symbolic play underscore the need for an in-depth examination of the sampling properties of measures used in longitudinal studies of individual differences. In a longitudinal study of skill or ability we would generally expect to see a smooth upward developmental trend for group data. Yet, even though the general direction is upward, the individual developmental curves may be uneven and thus automatically lead to instability of individual differences. For some variables a larger sample of behavior may smooth individual developmental curves and make the demonstration of stability possible. Although this strategy may not always succeed, it must be tried before we can claim that there are no stable individual differences for a specific developing ability or skill. The same line of reasoning can be applied to longitudinal instabilities in personality measures (traits, temperament), although the developmental fluctuations of such measures would not be superimposed on a general upward trend.

Second, we suggest that systematic cost–benefit analyses be conducted on measures used in longitudinal studies. Whereas some variables may have to be assessed across relatively long time spans, others can be reliably measured with relatively little effort. For example, data from our project show that one is likely to obtain a very good estimate of total vocabulary at 20 months by requesting much more limited information from the mother (for example, the number of body parts which the child can name). Although many investigators are intuitively aware of such points, hard data on which to base decisions are hard to come by.

Unfortunately our recommendations have a serious drawback. Most investigators are not particularly eager to conduct purely methodological studies. One solution may therefore be to reanalyze existing data sets, in other words, to engage in the kind of *post hoc* data-snooping which we have undertaken here (when first designing this project we had planned to investigate specific within-age, but not longitudinal language–symbolic play correlations). For example, *post hoc* reanalyses of the robust longitudinal data on infant–mother attachment gathered by Ainsworth during many hours of detailed observation (see Ainsworth, Blehar, Waters, & Wall, 1978) might tell us whether substantial reductions in observation time would still have yielded the same stable correlational patterns and, if so, at what point the correlational patterns might begin to break down.

Finally, we must gain more insight into why sampling properties of variables differ so substantially from one behavioral domain to another, even when the domains appear to be closely related, as in the case presented here. So far, theoretical discussions have tended to focus on *periods of instability*, not on *instability of behavior within specific developmental domains*. For example, McCall's (1981) scoop hypothesis (described above) postulates that individual differences during the sensorimotor period are inherently unstable because the effect of environmental influences is cancelled out by a powerful tendency for self-righting. Neither our symbolic play data nor our language data support the scoop hypothesis. Individual differences in language development showed stability during the sensorimotor period and well into the representational period. By contrast, individual differences in the development of symbolic play showed low stability across both periods. A second, somewhat different hypothesis was also advanced by McCall (McCall et al., 1977). They proposed that there may be brief transitional periods during early development when breakdowns in the stability of individual differences tend to occur. These breakdowns are interpreted as evidence for a discontinuity or reorganization in the fundamental nature of human performance (e.g., from sensorimotor to symbolic functioning). Similar notions were advanced by Emde, Gaensbauer, and Harmon (1976). Two of the data collection periods (at 13 and 20 months) in the study reported here coincided with stage boundaries proposed by McCall et al. The only hint of a transitional instability in our findings is provided by the infrequent correlations of 13-month to 20-month symbolic play, and of 13- and 28-month language to 28-month symbolic play. No notable instability in the language to language correlations were observed at 13 and 20 months.

It is the contrast in the longitudinal stability of language and symbolic play which is the most striking aspect of the data reported here. None of the hypotheses proposed so far can account for stability in language and instability in symbolic play during the same period of development. Hence, theorizing about developmental transitions may have tended to be too general. It now seems clear that to characterize a particular homotypic progression as having order, weak stability, or strong stability (see Figure 1), or to define a hybrid homotypic/heterotypic progression as based on remote, specific, or local homology (see page 232–233), is a tenuous enterprise without more precise information about the sampling properties of the variables representing the domain or domains under study. As we said in the beginning, orderly development does not require stability of individual differences over time. Genetic and/or environmental factors can exert different influences on an individual's rate of development, and differential rates of development thus generated can in turn lead to instability of individual differences in longitudinal studies. There is a danger, however, that failure to assess a specific ability or trait thoroughly enough to yield stability of individual differences (where this exists) could lead to false conclusions about hypothesized periods of transition or reorganization (developmental shifts). We therefore believe that a closer examination of the differential reliability of measures commonly used in longitudinal studies of early development is a vital first step toward asking and answering precise questions regarding discontinuities in development.

Acknowledgments

We gratefully acknowledge the devoted and creative help of the following colleagues and students: Marjorie Beeghly, Vicki Carlson, Karlana Carpen, Ann-Claire France, Antony Gerard, Andrew Garrison, Kim Kirschenfeldt, Sandra McNew, Barbara O'Connell, Cynthia Rodacy, Cecilia Shore, Lynn Snyder, Elizabeth Teas, and Carol Williamson. We would also like to express our warmest thanks to the mothers and children without whom this study would have been impossible.

REFERENCES

Ainsworth, M. D. S., Blehar, M. C., Waters, E., & Wall, S. *Patterns of attachment: A psychological study of the strange situation.* Hillsdale, N.J.: Lawrence Erlbaum, 1978.

Bates, E., Benigni, L., Bretherton, I., Camaioni, L., & Volterra, V. *The emergence of symbols: Cognition and communication in infancy.* New York: Academic Press, 1979.

Bates, E., Bretherton, I., Snyder, L., Shore, C., & Volterra, V. Gestural and vocal symbols at 13 months. *Merrill-Palmer Quarterly,* 1980, *26,* 407–423.

Bates, E., Bretherton, I., Shore, C., & McNew, S. Names, gestures and objects: The role of context in the emergence of symbols. In K. Nelson (Ed.), *Children's language* (Vol. 4). Hillsdale, N.J.: Lawrence Erlbaum, 1983.

Bates, E., Bretherton, I., & Snyder, L. *Language development from 10 to 28 months.* Monograph in preparation.

Beeghly-Smith, M. *Developmental and cross-contextual stability of mother–child interactive behavior.* Unpublished doctoral dissertation, University of Colorado, 1981.

Bloom, L. *One word at a time: The use of single word utterances before syntax.* The Hague: Mouton, 1973.

Bloom, L., Lighbown, P., & Hood, L. Structure and variation in child language. *Monographs of the Society for Research in Child Development,* 1975, *40,* Serial No. 160(2).

Bretherton, I., & Beeghly, M. Talking about internal states: The acquisition of an explicit theory of mind. *Developmental Psychology,* 1982, *18,* 906–921.

Bretherton, I., Bates, E., McNew, S., Shore, C., Williamson, C., & Beeghly-Smith, M. Comprehension and production of symbols in infancy. *Developmental Psychology,* 1981, *17,* 728–736.

Bretherton, I., McNew, S., & Beeghly-Smith, M. Early person knowledge as expressed in verbal and gestural communication: When do infants acquire a 'theory of mind'? In M. E. Lamb & L. R. Sherrod (Eds.), *Infant social cognition: Empirical and theoretical considerations.* Hillsdale, N.J.: Lawrence Erlbaum, 1981.

Bretherton, I., McNew, S., Snyder, L., & Bates, E. Individual differences at 20 months: Analytic and holistic strategies in language acquisition. *Journal of Child Language,* 1983, *10,* 293–320.

Bretherton, I., O'Connell, B., Shore, C., & Bates, E. The effect of contextual variation on symbolic play: Development from 20 to 28 months. In I. Bretherton (Ed.), *Symbolic play: The development of social understanding.* New York: Academic Press, 1984.

Brown, R. *A first language: The early stages.* Cambridge, Mass.: Harvard University Press, 1973.

Carlson-Luden, V. *Causal understanding in the 10-month-old.* Unpublished doctoral dissertation, University of Colorado, 1979.

Eibl-Eibesfeldt, I. *Ethology: The biology of behavior.* New York: Holt, Rinehart, & Winston, 1970.

Emde, R. N., Gaensbauer, T. J., & Harmon, R. J. *Emotional expression in infancy: A biobehavioral study.* New York: International Universities Press, 1976.

Fein, G. G. A transformational analysis of pretending. *Developmental Psychology,* 1975, *11,* 291–296.

Fein, G. G. Play and the acquisition of symbols. In L. Katz (Ed.), *Current topics in early childhood education.* Norwood, N.J.: Ablex, 1979.

Fenson, L. Personal communication, August 1982.

Fenson, L. Contributions of action and speech to pretend play in infants and toddlers. In I. Bretherton (Ed.), *Symbolic play: The development of social understanding.* New York: Academic Press, 1984.

Fenson, L., & Ramsay, D. Decentration and integration of the child's play in the second year. *Child Development,* 1980, *52,* 171–178.

Fischer, K. W. A theory of cognitive development: The control and construction of hierarchies of skills. *Psychological Review,* 1980, *87,* No. 6(c).

Johnston, J. *Analyzing the content of children's language.* Paper presented to the California Speech and Hearing Association, 1981.

Kagan, J. *The second year: The emergence of self-awareness.* Cambridge, Mass.: Harvard University Press, 1981.

Killen, M., & Uzgiris, I. Imitation of actions with objects. *Journal of Genetic Psychology,* 1981, *138,* 219–229.

Kintsch, W. *The representation of meaning in memory.* Hillsdale, N.J.: Lawrence Erlbaum, 1974.

Largo, R., & Howard, J. Developmental progression in play behavior of children between nine and thirty months. *Developmental Medicine and Child Neurology,* 1979, *21,* 492–503.

Lowe, M. Trends in the development of representational play in infants from one to three years: An observational study. *Journal of Child Psychology Psychiatry,* 1975, *16,* 33–47.

McCall, R. B. Nature-nurture and the two realms of development: A proposed integration with respect to mental development. *Child Development,* 1981, *52,* 1–12.

McCall, R. B., Eichorn, D. H., & Hogarty, P. S. Transitions in early mental development. *Monographs of the Society for Research in Child Development,* 1977, *42*(3), Serial No. 171.

McCune-Nicolich, L. Towards symbolic functioning: Structure of early pretend games and potential parallels with language. *Child Development,* 1981, *52,* 386–388.

McCune-Nicolich, L. Personal communication, 1982.

Nelson, K. Structure and strategy in learning to talk. *Monographs of the Society for Research in Child Development,* 1973, *38,* Serial No. 149(1–2).

O'Connell, B., & Bretherton, I. Symbolic and cominatorial play, alone and with mother: A Vygotskyan approach. In I. Bretherton (Ed.), *Symbolic play: The development of social understanding.* New York: Academic Press, 1984.

Piaget, J. *Play, dreams and imitation in childhood.* New York: Norton, 1962.

Shore, C., O'Connell, B., & Bates, E. *First sentences in language and symbolic play.* Developmental Psychology, in press.

Slobin, D. I. (Ed.). *A field manual for cross-cultural study of the acquisition of communicative competence.* Berkeley: Language-Behavior Research Laboratory, University of California, 1967.

Slobin, D. *Psycholinguistics* (2nd ed.). Glenview, Ill.: Scott, Foresman, 1979.

Slobin, D. I., & Welsh, C. A. Elicited imitation as a research tool in developmental psycho-linguistics. In C. A. Ferguson & D. I. Slobin (Eds.), *Studies of child language development.* New York: Holt, Rinehart & Winston, 1973.

Snyder, L., & O'Connell, B. *A propositional analysis of speech at 28 months.* Manuscript in preparation.

Snyder, L., Bates, E., & Bretherton, I. Content and context in early lexical development. *Journal of Child Language,* 1981, *8,* 565–582.

Tanner, J. M. Physical growth. In P. H. Mussen (Ed.), *Carmichael's manual of child psychology* (3rd ed.). New York: Wiley, 1970.

Turner, A., & Greene, E. *The construction and use of a propositional text base.* Institute for the Study of Intellectual Behavior, Technical Report #63, University of Colorado, Boulder, Colorado, April 1977.

Volterra, V., Bates, E., Benigni, L., Bretherton, I., & Camaioni, L. First words in language and action: A qualitative look: In E. Bates *et al., The emergence of symbols: Cognition and communication in infancy.* New York: Academic Press, 1979, pp. 141–222.

Waddington, C. H. *The strategy of the genes.* London: Allen & Sons, 1957.

Waters, E. The reliability and stability of individual differences in infant–mother attachment. *Child Development,* 1978, *49,* 483–494.

Wilson, R. S. Synchronies in mental development: An epigenetic perspective. *Science,* 1978, *202,* 939–948.

Wolf, D. Personal communication, May 1982. (a)

Wolf, D. Understanding others: A longitudinal case study of the concept of independent agency. In G. Forman (Ed.), *Action and thought: From sensorimotor schemes to symbol use.* New York: Academic Press, 1982. (b)

Wolf, D., & Gardner, H. Style and sequence in early symbolic play. In N. Smith & M. Franklin (Eds.), *Symbolic functioning in children.* Hillsdale, N.J.: Lawrence Erlbaum, 1979.

CHAPTER 11

Developmental Transformations in Mastery Motivation

MEASUREMENT AND VALIDATION

George A. Morgan and Robert J. Harmon

INTRODUCTION

Mastery motivation and similar concepts such as effectance motivation, intrinsic motivation, and competence motivation have roots in an appealing belief that there is an intrinsic motive to control the environment, to master skills, and to be effective. It is usually assumed that individual differences in this motivation will show some continuity across infancy and early childhood and will influence later competence.

This chapter will review briefly and selectively the theoretical background and research that led Leon Yarrow, the authors, and several

George A. Morgan • Department of Human Development and Family Studies, Colorado State University, Fort Collins, Colorado 80523.   **Robert J. Harmon** • Department of Psychiatry, University of Colorado School of Medicine, Denver, Colorado 80262. Aspects of the authors' research on mastery motivation have been supported by the Grant Foundation Endowment Fund of the Developmental Psychobiology Research Group, Department of Psychiatry, University of Colorado School of Medicine, of which both authors are members. In addition, Dr. Morgan's research has also received support from a Colorado State University Faculty Research Grant and Biomedical Research Support Grants. Dr. Harmon's research has received support from BRSG-RR-0537 awarded by the Division of Research Resources, National Institutes of Health. Dr. Harmon is currently supported by Research Scientist Development Award 5 K01 MH00281 and Research Grant 5 R01 MH34005 from the National Institute of Mental Health. Both authors are members of the MacArthur Foundation network of investigators studying the transition from infancy to early childhood. Network funds and discussions helped in the final writing of this paper.

other colleagues to carry out a series of studies designed to develop measures of what we have called *mastery motivation*. We will outline these studies, several of which are still in progress, focusing on those that deal with continuities and discontinuities in the development of mastery motivation during infancy and early childhood.

We believe that this research is important for at least three reasons:

1. Motivation is an important determinant of success in school and in many aspects of life, but little is known about its behavioral precursors in infancy and toddlerhood. Until recently, there have been few attempts to develop measures of such motivation in very early childhood.

2. Much research effort has been devoted to trying to predict school-age competence and intelligence from infant experiences and developmental level, but these efforts have been generally unsuccessful (see McCall, 1979). This is in part because some children show marked changes in competence from infancy to school age relative to their age mates. Our basic hypothesis is that infants who are persistently task-directed, even if they are not highly skillful at tasks as infants (i.e., they do not have high developmental quotients), will catch up with and perhaps pass less task-directed peers who did have high infant test scores.

3. Research on mastery motivation is derived from an influential theory of motivation which has had little empirical verification (White, 1959, 1963). Although it is now well accepted that the infant is an active, striving being who attempts to master the environment and produce effects, there is still relatively little empirical knowledge about the motivation that guides the infant's early attempts to deal with the inanimate environment. Even less is known about individual differences in motivation and what factors enhance or inhibit the child's persistent attempts to solve problems and work at challenging tasks.

In his seminal paper, "Motivation Reconsidered: The Concept of Competence," White (1959) generated considerable discussion of issues concerning the origins of children's competence. He proposed that all children have relatively strong motivation to explore and play with toys and other objects. This "effectance motivation" helps the child learn about the world and acquire skills. Such motivation impels the child to become competent and is satisfied by a "feeling of efficacy." It is thought to be built into the child and not due to hunger, thirst, or previous reinforcement, as classical drive reduction and learning theories had held. Thus, White's theory postulated, through effectance motivation the very young child is curious, explores, and tries to master the environment even without approval from mother.

White's (1959, 1963) writings brought new insights to the topic of the motivation of childrens' behavior, but he did not attempt to operationalize the concept of competence motivation and he did not discuss individual differences in competence or effectance motivation. His aim apparently was to convince skeptics that such a motive existed.

Like White, Hunt's (1965) views about motivation were influenced by Piaget (1936/1952). Hunt suggested that there exists an intrinsic motivation to learn about objects and people through exploration. Purposive behavior begins when the neonate orients to and explores objects visually. As the infant's repertoire increases, new behavioral abilities are used to explore the environment. Thus, when the infant becomes capable of manipulative exploration, he or she employs this new capacity to learn about the environment, to have an impact on it, and to obtain feedback from it. According to Hunt, early in life the infant repeatedly explores objects until they become familiar. From such exploration, the child develops a generalized expectation that objects should be recognizable. Thereafter, novel stimuli elicit visual attention and exploration. Hunt argued that attention and exploration are most easily elicited by objects that have an optimal degree of discrepancy from familiar objects.

The importance of responsiveness to the infant's behavior has been emphasized by Watson (1972), who proposed that smiling and cooing result when an infant becomes aware of the contingency between its behavior and some change in the environment. This phenomenon is similar to White's discussion of a sense of efficacy and satisfaction derived from the mastery of a task. Goldberg (1977) pointed out the relevance of predictable contingent parental responses. Feelings of competence in the child can be decreased by an unresponsive and unpredictable parent.

The extensive research and writing of Harter (e.g., 1978, 1981a) on effectance and mastery motivation in both normal and retarded school-age children has had considerable effect on our research and especially on the maternal report questionnaire described later. Harter has been studying effectance motivation in school-age children for over a decade. Early in her research (Harter & Zigler, 1974), she found that mentally retarded (MR) children scored lower on effectance motivation tasks than nondelayed children even if the groups were matched for mental age. She also found that the MR children smiled less after they had solved a puzzle, perhaps indicating that they had less sense of efficacy and derived less pleasure from solving the problem (Harter, 1977).

More recently Harter (1981b) has developed a self-report instrument to assess intrinsic versus extrinsic motivation in the classroom in third-through ninth-grade children. She found a marked, and perhaps discouraging, decrease in intrinsic motivation in the older children. Ninth-graders were less likely than younger children to say they prefer a challenging task, like to figure things out themselves, or work at something just because it interests them. However, they were more confident about their own judgments and felt they knew when they were doing well or poorly without the teacher's telling them.

Scarr (1981b) has argued that a meaningful assessment of a child's competence must take into account motivational characteristics and adjustment as well as cognitive competence. Like several others (e.g., Yarrow & Pedersen, 1976; Ulvund, 1980), Scarr feels that in early childhood motivation and cognition, although not redundant, are interrelated sufficiently to form parts of a general competence factor. This issue will be discussed later. From the results of a recent longitudinal study, Scarr (1981a) demonstrated that IQ and motivational characteristics are mutually predictive from 2 to 4 years of age. The correlations between motivational and cognitive scores across these ages were just as high as the correlations between them at either age. Scarr states that this supports the importance of both motivational and cognitive characteristics in the child's development.

A major impetus for the research on mastery motivation described in this paper was a study by Yarrow, Rubenstein, and Pedersen (1975) of the environment and functioning of 70 predominantly lower income black families and their 5½-month-old infants. These investigators were impressed not only by the diversity of the infants' environments (for which the study is well known) but also by the motivational nature of many aspects of infant behavior during testing sessions. In a follow-up study of that sample, Yarrow, Klein, Lomonaco, and Morgan (1975) reported that these early-infancy cognitive-motivational scores, such as manipulation of novel objects, predicted Stanford-Binet IQ at 3½ years of age whereas the Bayley Mental Development Index did not.

These findings appeared to indicate the importance of understanding and measuring individual differences in infants' motivation. They also provided preliminary support for our hypothesis that measures of infants' motivation to master their environment will enhance the prediction of later competence.

Thus the authors and their colleagues began a series of studies of mastery motivation in infancy and early childhood, a series initiated by Leon Yarrow at the National Institute of Child Health and Human Development in the mid 1970s. In addition to continued research at NICHHD, programs of research on mastery motivation are now underway at the University of Colorado School of Medicine, Colorado State University, and the University of Pittsburgh. Table 1 summarizes 13 such studies by the authors and their colleagues. The table provides information about the objectives and methods of each study and where to find more information on the results.

The remainder of this chapter will selectively review the studies in Table 1, focusing on continuity and discontinuity in mastery motivation. The first section will describe the process of operationalizing the concept of mastery motivation, that is, the development and validation of the measures that have been used in most of the studies. The next section will discuss the current findings about continuity and discontinuity in mastery motivation. The final section will present a strategy to measure the mastery motivation of infants and toddlers relatively independent of their cognitive level. A key objective is to develop a standardized procedure and set of tasks to assess mastery motivation in children from approximately 12 to 36 months of age. It is hoped that the resulting procedure will be useful with a wide variety of children including those such as preterm, Down's syndrome, and physically handicapped children who are at risk for developmental problems.

MEASURING MASTERY MOTIVATION

Methodological Issues

The studies on which this review is based deal with the development of mastery motivation from infancy to the preschool period. Most of these studies have focused on methodology—on the development of methods for measuring mastery motivation at several ages and in several special populations. In developing these methods, a number of questions have been raised about basic measurement issues. How does one measure motivation in preverbal infants and very young children? Should one ask mothers, observe spontaneous play, or design structured test-like procedures? Can motivational constructs like "trying" to do something and "task-directedness" be scored reliably? Does there exist any long-term predictability? How does one assess the validity of a construct for which there are no established measures? Do such measures show short-term stability? This last point raises issues concerning continuity in behavior and the question of whether, in the course of early development, there are predictable transformations in infant behavior that index mastery motivation. Each of these questions will be considered in the following sections, and at least preliminary answers will be suggested.

Table 1. Studies of Mastery Motivation in Infancy and Early Childhood

Study	Objectives	Sample and methods	Citations
1.	Develop measures of mastery motivation for infants Study relationships of mastery motivation to: a. Cognitive development (Bayley MDI) b. Qualitative aspects of spontaneous play with toys c. Social and inanimate home environment at 6 months	44 normal, middle-class, 12- to 13-month infants 11 structured mastery tasks, mostly 2 minutes each. Three types: effect production, practicing emerging relational skills, and problem solving	Morgan, Harmon, Gaiter, Jennings, Gist, & Yarrow, 1977 Jennings, Harmon, Morgan, Gaiter, & Yarrow, 1979 Gaiter, Morgan, Jennings, Harmon, & Yarrow, 1982 Yarrow, Morgan, Jennings, Harmon, & Gaiter, 1982
2.	Study continuity in mastery motivation and cognitive (McCarthy) functioning from infancy to early childhood	35 normal 3½-year-old children who had participated in Study 1 4 mastery tasks of 2 types: Persistence at difficult problems, assessed by 2 problems (e.g., fitting wooden cutouts into a small box) Curiosity, assessed by letting the child play with a "curiosity box"	Jennings, Yarrow, & Martin, in press
3.	Develop measures of mastery motivation Study short-term stability over 2 weeks Study stability from 6 to 12 months Study relationship of mastery motivation to: a. Cognitive development (Bayley MDI) b. Parent's perception of temperament c. Parent–child interaction in home	75 normal, firstborn 6- and 12-month infants of middle-class background 12 structured mastery tasks, 3 minutes each (6 given at each of two sessions 2 weeks apart) 3 types of tasks: effect production, practicing emerging sensorimotor skills, and problem solving	Yarrow, McQuiston, MacTurk, McCarthy, Klein, & Vietze, 1983 McQuiston & Yarrow, 1982 McCarthy & McQuiston, 1983 Yarrow, Macturk, Vietze, Klein, & McQuiston, in press Vietze, Pasnak, Trembley, McCarthy, Klein, & Yarrow, 1981

4.	Study similarities and differences between infants with Down's syndrome and normal infants in mastery behavior Study similarities and differences in relationships between developmental status (Bayley) and mastery behavior	30 Down's syndrome 6-, 8-, and 12-month infants from middle-class backgrounds 12 structured mastery tasks as in Study 3	Vietze, McCarthy, McQuiston, Mac-Turk, & Yarrow, 1983 MacTurk & McQuiston, 1982 McCarthy & Vietze, 1982 MacTurk & Yarrow, 1983

5.	Develop measures of mastery motivation
	Study relationships of mastery motivation to:
	  a. Cognitive development (McCarthy)
	  b. Mastery behavior in developmental test situations
	  c. Behavior in free play
	  d. Mother–child interaction
	  e. Operant learning behavior
	Study relationships between mastery at 6 and 12 months and mastery at 2½ years
		53 normal 2½-year-old infants from Study 3
		6 structured mastery tasks, 5 to 10 minutes each
			Messer, Yarrow, & Vietze, 1982
			Messer, Ratchford, McCarthy, & Yarrow, 1983

6.	Study similarities and differences in mastery behavior between term and preterm infants
	Study the relationship between mastery behavior and:
	  a. Cognitive development (Bayley)
	  b. Affective development
	  c. Free play behavior
	  d. Attachment
		30 very low birth weight (less than 1500 grams) preterm infants at 12 months (corrected gestational age) and 30 full-term comparison infants
		8 structured mastery tasks, 3 minutes each
			Harmon & A. Culp, 1981
			Harmon, Glicken, & A. Culp, 1982
			Harmon & Glicken, 1981

(continued)

Table 1. (*Continued*)

Study	Objectives	Sample and methods	Citations
7.	Develop and refine measures of mastery motivation Develop maternal questionnaire (MOMM) to assess mastery motivation Study relationships of mastery motivation to: a. Task-directed free play b. Mothers' reports of mastery motivation (MOMM) c. Temperament (DOTS)	Approximately 25 normal 9-, 12-, and 24-month children 8 or 9 structured mastery tasks designed to assess persistence at challenging tasks MOMM questionnaire	David, Morgan, R. Culp, & Vance, 1982 Morgan, Busch-Rossnagel, R. Culp, Vance, & Fritz, 1982 Morgan & Jacobs, 1981a, 1981b Morgan, Harmon, Pipp, & Jennings, 1983
8.	Determine whether physically handicapped children differ from normal children in mastery behavior Study relationship to: a. Mother–child interaction b. Free play c. Intelligence (McCarthy) Study consistency in mastery behavior from age 3½ to age 4½	25 physically handicapped and 44 nonhandicapped 3½-year-old children, to be followed up at age 4½ Persistence at solving difficult problems (as in Study 2 above) Curiosity (as in Study 2) Preference for challenging tasks (e.g., asking which height block tower he or she would like to build) MOMM questionnaire	Jennings, Connors, Sankaranarayan, & Katz, 1982
9.	Examine effectiveness of "read your baby" intervention program on mastery motivation and cognitive functioning Compare mastery motivation of medium risk infants with non-risk infants Study relationships between perceptions of mastery motivation and mastery task scores	24 medium-risk 12-month-old (corrected for gestational age) infants (12 had intervention and 12 did not) and 12 non-risk 12-month-old infants 4 mastery tasks, 2 minutes each MOMM questionnaire	Harmon, Morgan, Jacobs, Glicken, A. Culp, Busch, & Butterfield, 1982 Harmon, Morgan, & Glicken, 1984 Butterfield & Miller, 1984

10.	Same as Study 9	Same sample as Study 9 tested at 24 months of age 6 Mastery Tasks (3 combinatorial and 3 means-end tasks), each designed to tap persistence at completing tasks.	Morgan, Harmon, Malpiede, A. Culp, & Renner, 1983
11.	To investigate whether persistence is a valid measure of mastery motivation or instead is an index of perseveration Compare 6- and 12-month-old infants on persistence and other behaviors vis-à-vis a contingent or noncontingent jack-in-the-box	MOMM questionnaire. 24 normal 6-month-old and 24 normal 12-month-old infants. Half at each age were randomly assigned to the infant contingent situation and half to the noncontingent situation. Each child was tested for 3 minutes during which he or she could press a cylinder which either did or did not activate a jack-in-the-box.	Caplovitz, Morgan, & Mardashti, 1982
12.	Examine the relative effectiveness of two methods of occupational therapy Compare mastery motivation of Down's syndrome and normally developing children of the same mental age	10 Down's syndrome 1½- to 5½-year-old children and 10 normal children matched for developmental age 4 mastery tasks: 2 used as pretest assessment and two used as a posttest	Flagle, 1982
13.	Check the construct validity of mastery task scores by relating them to mother and teacher ratings Test the feasibility of the procedure for selecting and presenting mastery tasks that is prescribed in this paper	Normal 1- to 3-year-old children from a university child care center A series of tasks selected so that those used for any given child are neither too hard nor too easy Teacher ratings of mastery motivation Mothers' ratings of mastery motivation (MOMM)	Morgan, Tolerton, Renner, & Harmon, 1983

Development of the Mastery Tasks

As we noted above, when the studies listed in Table 1 began, White's theory of effectance motivation had generated considerable theoretical discussion but little empirical research. Extensive discussion and observation of infants led us to concentrate on a test-like approach to the measurement of mastery motivation, as it came to be called. This decision was based on a preference for direct observation in contrast to maternal report and on the observation that young infants displayed much more task-directed behavior that we associated with mastery motivation during a structured test-like situation than during free play. In some studies, we have, however, also utilized free play and/or maternal reports to assess mastery motivation. A brief description of these types of measures is presented in the section on the validity of the mastery task measures.

In our research, mastery motivation has been operationally defined as the amount of task-directed behavior during the presentation of a set of toys that pose problems to be solved or completed. The procedure for the mastery motivation tasks has varied somewhat between the several studies in Table 1. In general, how to use the toy has been demonstrated; then it has been given to the child who has had the opportunity to play with it for a period of time (usually 3–5 minutes) with little encouragement from the experimenter or mother. The toys used for 6- and 12-month-old infants in Studies 1, 3, 4, 6, and 7 have posed tasks of three types. First, some toys gave the infant an opportunity to produce feedback or audiovisual effects from the toy by using a manipulandum such as a button, level, or dial. Second, other tasks required the infant to circumvent an obstacle such as a glass barrier or latch to obtain a goal object. Third, still other tasks offered the infant an opportunity to take apart or combine objects in an appropriate way such as putting pegs into holes or shapes into forms. These categories are, of course, somewhat arbitrary and overlapping; however, they have proven useful in part because they appear to form a hierarchy of difficulty and appropriateness for children in the second half of the first year. More discussion of this issue will be presented in the section on continuity. Some investigators (e.g., Messer, Yarrow, & Vietze, 1982) have argued that it is better to classify according to the types of behavior exhibited rather than by the characteristics of the tasks themselves. To some extent that is what has been done in Table 2.

As described below, the more recent studies and those with older children have used somewhat different procedures and categorization of tasks, but the approach to operationalizing the concept of mastery motivation has been very similar in all the studies in Table 1.

Table 2. A Hierarchary of Behaviors During the Mastery Motivation Tasks

Code	Behaviors
Other or non-task	Inattention to the test object, bids to mother or tester, and all other behaviors not described below
Visual attention	Looking at the object but not touching it
Exploration (only)[a]	Holding, touching, mouthing, examining, shaking, or manipulating the object
Task-directed behavior[b]	Behaviors which indicate an attempt to use the object in the appropriate way. These may or may not lead to success.
Success or solution[c]	Successful or appropriate use of the object or part of it. Solution of the problem.

[a]The exploration category has not included task-directed behavior.
[b]Some studies have made a distinction between task-related and goal-directed behavior. The former includes trying to solve the task, but in ways that will not lead to a solution (e.g., pushing on a barrier). Goal-directed behaviors are those which could actually result in a solution if done skillfully and/or persistently enough.
[c]In most studies, *persistence*, the key mastery motivation measure, has included the amount of time or number of intervals of both task-directed behavior and success or solution behavior.

Coding of the infant's behavior has also varied somewhat from study to study but has generally followed the scheme described in Table 2. More details on procedures and scoring are provided in Morgan, Harmon, Gaiter, Jennings, Gist, and Yarrow (1977) for Study 1, Vietze, Pasnak, Tremblay, McCarthy, Klein, and Yarrow (1981) for Studies 3 and 4, Harmon and Glicken (1981) for Study 6, and Morgan and Jacobs (1981b) for Study 7. The behaviors in Table 2 form a hierarchy with inattention, looking, or mouthing being lower than task-directed behaviors like trying to get a toy from behind a barrier or success behaviors like putting pieces in a puzzle. Both task-directed and success behaviors are indicators of mastery motivation.

Several measures have been derived to summarize the behaviors of infants during the mastery task session. These measures can be divided into three categories: (a) indices of mastery motivation, (b) causality pleasure, and (c) indices of competence.

The primary measure of mastery motivation has been called *persistence*. It is the amount of time or the number of intervals in which the infant exhibited task-directed or success behavior. That is, persistence is the amount of time the infant spends using or trying to use toys in an appropriate way. A second index of mastery motivation used in some studies has been *latency to task*, that is, the length of time it takes the infant to start engaging in task-directed behavior.

Causality pleasure has been coded in recent studies when positive affect is shown during or immediately after solving or completing some part of the task. We have assumed that this measure indicates that the child was experiencing pleasure in accomplishment (a sense of efficacy, White, 1963) and was aware of his or her own action (Kagan, 1981). Causality pleasure has not been recorded in all the studies because it is relatively infrequent, but it is considered by the authors to be an important measure conceptually related to mastery motivation.

Finally, several measures of *competence* have been derived. These are not considered to be indices of mastery motivation, but rather they indicate how well the child performed on the task. Later the reader will see how this score is important for selecting appropriately difficult tasks for a given child.

The above measures have varied slightly in some of the studies but are representative of what will be discussed below. It should be noted that the measure labeled *exploration* in Table 2 and later sections refers to exploration *only* (i.e., behavior that is not task-directed) and is considered to be a lower level or perhaps less mature measure of mastery motivation than persistence at the task.

Table 3 provides the authors' current conceptualization of the several types of mastery tasks that have been used in the studies on mastery motivation. This categorization is somewhat more complex than the one used by the NICHHD group. We feel that the tasks used with the 6- to 12-month-old infants in Studies 1, 3, 4, and 6 are mostly of the type which we labeled *practicing emerging skill* in Table 3. Such tasks expect the child to repeat or practice the solution once he or she uses the toy correctly or solves the problem. In general, these tasks involve mastering a limited range of maneuvers, such as reaching around a barrier or putting pegs in holes. This type of task appears especially appropriate for children around one year of age, in part because they may not yet be able to do tasks that involve a sequence or series of different maneuvers. As will be discussed later, it may be better to test children less than a year old using a different approach, perhaps one designed specifically to assess curiosity and exploration. That is why *exploration/curiosity* has been put first in Table 3; it is, of course, possible to measure curiosity at any age, and it may be indicative of one aspect of mastery motivation throughout. In fact, some of the tasks, especially those labeled *effect production*, might be better classified as curiosity.

In the authors' more recent work (i.e., Studies 7, 9, 10, and 13) with 1- to 3-year-olds, we have focused on measuring persistence at completing tasks that have several parts to put together or figure out so that the child is asked to do more than practice a skill. This type of task seems more appropriate for toddlers than simple practicing skills tasks. Of course, this distinction between persistence at practicing emerging skills and persistence at completing tasks is a relative rather than a clear-cut

Table 3. A Developmental Hierarchy of Mastery Motivation Tasks

1. *Exploration, or curiosity* (approximately 6 months and older). Because this kind of task should assess the child's interest in actively utilizing the object, it should not pose a major cognitive or fine motor challenge. The question is whether the child is motivated to explore all parts of the object and will maintain interest until he or she has done so. Measures of both the duration (persistence) and variety of behavior should be recorded. The object should be complex enough to promote sustained exploration. With appropriate objects these tasks could be used with children of almost any age.
2. *Persistence tasks*
   a. *Practicing an emerging skill* (approximately 9 months and older). This type of task can usually be completed (solved) rather quickly by the child but elicits, at least in children about 1 year of age, practicing or repeating of the maneuver that led to success. Three somewhat overlapping categories of persistence at emerging skills tasks have been used:
      i. *Effect production tasks*—e.g., simple cause-and-effect toys
      ii. *Combinatorial tasks* (called practicing emerging sensorimotor skills in studies 1–6)—e.g., putting pegs or balls in hole
      iii. *Means–end tasks* (called problem-solving tasks in studies 1–6)—e.g., getting a toy from behind a barrier
   b. *Completing a multipart task* (approximately 15 months and older). This type of task is sufficiently difficult or complex that it usually takes some time (i.e., at least 60 seconds) to complete. Two somewhat overlapping categories of persistence at completing a task item have been used:
      i. *Combinatorial tasks*—e.g., completing a shape sorter, peg board with different size holes, or a complex form board
      ii. *Means–end tasks*—e.g., appropriately using all parts of a surprise box, cash register, or lock board.
3. *Preference for challenging tasks* (approximately 3 years and older). In this type of task the child is presented with a choice of relatively hard and easy tasks and asked which one he or she wants to work at.
4. *Mastery for the sake of competence or self-initiated mastery* (4 years and older). In this type of task the extent of spontaneous grouping, order, or systematization is measured. This spontaneous behavior is compared to what the child *can* do when shown and/or given explicit instructions. The question is whether a child who can will in fact organize the materials, without specific instructions to do so.

one. For example, putting a series of pegs into holes involves repeating the same maneuver but also involves completing a task.

Jennings (Studies 2 and 8) has developed three types of tasks for 3½- and 4½-year-olds which she calls *curiosity, persistence,* and *preference for challenging tasks.* This scheme corresponds to that in Table 3. As mentioned above, curiosity/exploration and persistence tasks are also appropriate for infants and toddlers, but requiring a child to choose which of two or more tasks he or she would prefer to try does not appear to produce meaningful results in 2-year-olds (Morgan & Jacobs, 1981b) and may not be appropriate before 3 or even 4 years of age. Similarly, Harter and Zigler's (1974) *mastery for the sake of competence* tasks, listed as a fourth type in Table 3, do not appear to be appropriate for toddlers or perhaps even for preschoolers (Morgan & Jacobs, 1981b; Jennings, Yarrow, & Martin, in press).

Reliability of the Measures of Mastery Motivation

Although coding whether a young child's behavior is task-directed may seem very difficult, each of the studies in Table 1 has been able to define operationally task-directed behaviors so that excellent interobserver reliability has resulted. The authors' strong impression is that it is much easier to obtain reliability for children over a year than for younger infants, but available reports do not allow a formal test of such comparisons. Children under a year often display behaviors that are hard to interpret because they require considerable inference about whether the child is really trying to do the task. This observation is important for the discussion of discontinuity in mastery motivation in the latter part of the first year.

Validity of the Measures of Mastery Motivation

The validity of the mastery task measures has been confirmed in several ways. First, comparisons have been made between groups of children (such as those at risk or handicapped) who were predicted to score lower on mastery motivation measures than appropriate comparison groups. Second, we have examined the relationship between persistence at tasks and several other measures expected to reflect aspects of the concept of mastery motivation: (a) ratings of persistence during mental tests, (b) continuity of free play, and (c) mother and teacher ratings of components of mastery motivation. Finally, we have examined the validity of persistence as an index of mastery motivation by looking at differential relationships between responses of infants for whom persistence produced an effect (a jack-in-the-box) and those for whom the effect was not dependent on their behavior.

Differences between Groups. Harmon, Glicken, and Culp (1982) have compared the mastery motivation of very small (<1500 grams) preterm infants with that of a full-term sample when both groups were 12 months gestational age. The preterm infants showed less task-directed behavior (persistence) and causality pleasure than the full-term infants and more simple, non-task-directed manipulation or exploration. This lower amount of independent, task-directed behavior on the part of the preterm infants may be the result of the extensive involvement of their perhaps overly concerned mothers during the first year of life. The more intrusive behavior of the mothers of the preterm infants (demonstrated in part during a free-play session) may cause these infants to be lower on self-initiated behavior and less aware of their own impact on the environment. The higher scores of the full-term infants on causality pleasure may be a precursor of what Kagan (1981) has called self-awareness in somewhat older children. That is, the child may be aware that he or she was the cause of the effect or of the solution of the task.

Similar reasons may account for the lower mastery motivation scores found when groups of medium-risk, physically handicapped, and Down's syndrome children were compared with appropriate contrast groups. As in the study of very small preterm infants just described, Harmon, Morgan, and Glicken (1983) found lower persistence and causality pleasure in a group of infants who had been placed in a level II or medium-risk nursery after birth. Jennings, Connors, Sankaranarayan, and Katz (1982) found that physically handicapped preschool children scored lower on persistence tasks and on involvement in free play than nonhandicapped children; they did not, however, find differences in curiosity or preference for challenging tasks. The Down's syndrome infants studied by Vietze, McCarthy, McQuiston, MacTurk, and Yarrow (1983) had much longer latencies to task and thus less task-directed behavior, but they showed much the same pattern of behavior as the comparison sample after starting to use the toy.

Butterfield and Miller (1984) also found that an intervention designed to help the mothers interpret and adapt to their infants was effective. The nonintervention group scored lower on persistence and causality pleasure than the intervention group, which was no different from a low-risk, full-term comparison group.

Although we think that these are true differences in mastery motivation, it may be that they are due, at least in part, to differences in cognitive development, which have been found to be related to persistence in a number of studies (Harmon, Glicken, & Culp, 1982; Yarrow & Pedersen, 1976; Yarrow, Morgan, Jennings, Harmon, & Gaiter, 1982; Yarrow, McQuiston, MacTurk, McCarthy, Klein, & Vietze, 1983).

Relationship of Persistence at Tasks to Other Measures.   Yarrow *et al.* (1982, Study 1) found that individual differences in persistence on the mastery tasks were significantly correlated with an independent tester's ratings of persistence during a Bayley examination at another session. This finding and those that follow in this section provide evidence of the construct validity of the mastery task measures.

Although most investigators have focused on the cognitive or socioemotional aspects of free play, it can also be viewed as another reflection of the infant's early interest in mastering the environment. Jennings, Harmon, Morgan, Gaiter, and Yarrow (1979) found that although the quantitative aspects of play, such as total amount of exploration, showed no relationship to persistence, continuity of play, which is somewhat analogous to attention span or persistence, did show significant relationships to persistence at the mastery motivation tasks.

The authors have also examined the relationships of maternal ratings to persistence at the mastery tasks. We recently developed a questionnaire (Mother's Observation of Mastery Motivation or MOMM) to assess several aspects of a child's motivation in regard to playing with toys (Morgan, Harmon, Pipp, & Jennings, 1983). The questionnaire is, in part, a downward extension of Harter's (1981b) Intrinsic versus Extrinsic Orientation in the Classroom self-report scale for school-aged children. It is designed to obtain information on dimensions such as curiosity and autonomy as well as persistence and preference for challenging tasks, which correspond to those measured by the mastery tasks themselves. Data from about 200 mothers of normally developing, at-risk, and handicapped children from 6 months to 5 years have been collected. Some of the subjects have also been given the mastery tasks.

As predicted, we found a significant relationship between mothers' ratings of their children's persistence and the children's persistence score on a set of mastery tasks at 12 months of age (Morgan, Harmon, Pipp, & Jennings, 1983). Mother's rating of her infant's curiosity was related to the latency of the child to get involved in the task, and the rating of preference for challenging tasks was related to causality pleasure. Ratings of autonomy were not related to any of the mastery task measures, and how well the child actually did on the tasks (competence) was not related to mothers' ratings of any of the above dimensions of motivation. These correlations fit the predicted pattern and provide another source of validity.

Several studies have used the MOMM to compare maternal perceptions of the mastery motivation of children who are handicapped or at risk for developmental delays and comparison samples. Harmon, Morgan, Jacobs, Glicken, Culp, Busch, and Butterfield (1982) found that the mothers of the at-risk infants view them as significantly lower on persistence (e.g., gives up easily) and lower on preference for challenging tasks. Similarly, mother's of Down's syndrome children (Flagle, 1982) and physically handicapped children (Jennings et al., 1982) rate their infants as less persistent and less interested in challenge than mothers of normally developing children. These differences in maternal ratings, which are similar to those found for at-risk and handicapped children on the mastery tasks, provide support for the validity of the MOMM questionnaire as well as for the mastery tasks.

Jennings *et al.* (1982) looked at the relationships of persistence at the tasks to both preschool teacher and mother ratings. Time spent on a curiosity task was significantly positively related to teacher ratings of curiosity for both handicapped and nonhandicapped preschool children, but the so-called persistence task scores were related only to teacher ratings for the handicapped children. On the other hand, mothers' ratings were not significantly related to the children's task behavior or to the teacher ratings.

These later findings raise some questions about the validity of the measures but should be tempered by the fact that in the Jennings *et al.* study the mothers and teachers used different questionnaires and neither were directly matched to the types of tasks that these children were asked to do. Since both the MOMM questionnaire and mastery tasks differentiate risk and nonrisk groups and are meaningfully related to other variables, both may be valid indicators of mastery motivation but may be reflecting somewhat different aspects of the concept.

Another study provides evidence for the validity of persistence as an index of mastery motivation by looking, in part, at its relation to causality pleasure. Caplovitz, Morgan, and Mardashti (1982) found results consistent with the hypothesis that persistent cylinder-pressing by 6- and 12-month-old infants reflects mastery-oriented behavior rather than perseveration of response. Two groups of subjects participated in the study : (1) *infant contingent* subjects, whose cylinder pressing activated a jack-in-the-box, and (2) *infant noncontingent* subjects, who had a cylinder to press and who watched the jack-in-the-box emerge at the same rate as a yoked infant contingent subject, but whose behavior did not control the emergence of the jack-in-the-box. Infant contingent subjects (who did control the jack-in-the-box) smiled more if they pressed more, suggesting that they experienced pleasurable "feelings of efficacy." Infant noncontingent subjects also seemed to attempt to control the jack-in-the-box, in that they pressed the cylinder more frequently if the jack-in-the-box emerged more frequently. However, these infants, who lacked control over the jack-in-the-box, did not smile more upon pressing more; instead they frequently turned to view their caregivers with a serious look on their faces as if asking why they could not control the jack-in-the-box. These patterns of relationships between cylinder pressing and other behaviors differed, as predicted, for infant contingent and infant noncontingent subjects as a function of the success or failure of their attempts to control the jack-in-the-box.

These several methods of assessing the validity of the major indices of mastery motivation, *persistence at tasks* and *causality pleasure*, seem to provide such support. More work on naturalistic measures of mastery motivation, such as from home observations or day-care settings, is

desirable, but the preliminary results of relationships with teacher ratings reported above appear promising. Another type of validity, predictive, can be inferred from some of the findings in the next section concerning development and continuity in mastery motivation.

CONTINUITY AND CHANGE IN MASTERY MOTIVATION

A Transformation in the Second Half of the First Year of Life

David, Morgan, Culp, and Vance (1982) addressed two questions:
(1) do 9-month-old infants show as much task-related behavior as 12-
month-old infants on a given task that is difficult for both age groups,
and (2) are individual differences in persistence at a task stable over this
three-month period. At both 9 and 12 months of age, the infants were
very interested in the toy (a shape-sorter), attending to it more than 95%
of the allotted time. The 9-month-old infants, however, showed significantly more looking and simple exploration (e.g., shaking and banging
of the object) than the 12-month-olds. On the other hand, the 12-month-
olds showed significantly more task-related behavior than the 9-month-
olds, for example, taking a block out of a hole. There were very few
intervals at either age during which the child put a block in the correct
hole, and the difference between ages was not significant in this respect.
Since solution behaviors were excluded from the age comparisons of
task-related behavior, the differences that were significant appear to
indicate that 12-month-old infants were more *motivated* to try to solve the
task than 9-month-old infants. This conclusion can not be due merely to
increased maturation or skill because that was factored out, but rather it
must be due to the fact that the older infants focused their attention
more persistently on the task, with fewer digressions into simple exploration. However, there was little stability over the three months in individual differences in persistence. The results of this study are consistent
with the conclusion that there is a shift from more general exploration to
focused mastery attempts in the last quarter of the first year of life. This
transformation of mastery behavior would contribute to a lack of stability in individual differences during this period.

Yarrow *et al.* (1983) have examined change and stability between 6
and 12 months. Their cross-age correlations suggest that there are developmental transformations between 6 and 12 months in the child's mastery motivation. They feel that there is a developmental progression or
hierarchy in mastery behavior from producing effects to practicing
emerging combinatorial skills to problem solving (see Table 3). Furthermore, there is some evidence that mastery motivation is becoming less

global and more differentiated during this time. McQuiston and Yarrow (1982), in discussing the results of Study 3, note that from 6 to 12 months the organization of the infant's mastery behavior changes. Twelve-month-old infants spend significantly less time in effect production than they did at 6 months. Given the prominent role that secondary circular reactions play in the sensorimotor activity of the 6-month-old infant, it is not surprising that producing environmental effects is such an important part of the 6-month-old's mastery behavior. By 12 months, interest in producing effects is beginning to diminish and the infant is finding other ways to explore a task. Further analysis led to the conclusion that at 6 months a variety of exploratory behaviors, such as hitting, banging, shaking, mouthing, and focused exploration of objects, which were at first considered indices of a lower level of mastery behavior, were really central to the development of mastery motivation and were better predictors of both cognitive competence and later mastery than are the schemas originally identified as goal-directed. McQuiston and Yarrow also further examined the hierarchial model through longitudinal analysis of individual change and continuity. High cross-age stability of behavior was not predicted because they expected developmental transformations such that mastery on less mature components at 6 months would predict mastery of more mature components at 12 months. In fact, the strongest predictor of overall 12-month persistence was persistent effect production at 6 months. Persistent task-directed behavior at 12 months was also predicted by 6-month exploration on practicing skills tasks. These findings add further support to the view that exploration and the production of environmental effects are the hallmarks of the 6-month-old's motivation to understand the environment.

Messer et al. (1982) found only low positive correlations between the Bayley scores obtained during infancy and the McCarthy Scales of Children's Abilities at 30 months. Such results are in agreement with many other studies which have failed to find continuity in scores on developmental tests between infancy and even early childhood. Messer et al. did, however, find a number of significant relationships between mastery task scores at 6 month and 30-month McCarthy scores that were puzzling at first. Exploratory behavior such as mouthing, touching, hitting, and pushing the object (which they considered a lower-level, less directed form of mastery behavior) was significantly *positively* related to later McCarthy scores, especially for girls. On the other hand, what had been coded as persistent task-directed behavior at 6 months was significantly *negatively* related to childhood McCarthy scores. Thus, some of the apparently less task-directed forms of behavior (i.e., exploration) were positively correlated with later competence. This finding is consistent with the earlier discussion of a developmental transformation from exploration to more task-directed mastery motivation near the end of the first year. Exploratory behavior corresponds to the way in which Piaget (1952) describes how typical 6-month-old infants usually attempt to investigate an object. Thus the positive relationship between exploratory behavior at 6 months and later developmental competence becomes understandable when account is taken of the way in which 6-month-olds attempt to master objects. Messer *et al.* suspect that the behaviors scored at 6 months as task-directed behaviors could have included some gross actions which precede mastery attempts or could even have been perseverative.

Some data (Morgan, Harmon, Pipp, & Jennings, 1983) from age comparisons of mothers' ratings of their children's mastery motivation are relevant here. The mothers of children around 1 year of age differed from mothers of 2-, 3-, and 4-year-olds in rating their children as more interested in novel or unfamiliar objects. This may point out, in a somewhat different way, the salience of curiosity and exploration as key measures of the motivation of children under one year in contrast to the mastery motivation of older infants and young children, which appears to be more related to persisting at a task.

Finally, an early study by Yarrow, Klein, Lomonaco, and Morgan (1975) found that exploration of novel objects at 6 months of age was predictive of Stanford-Binet IQ at 3½ years. This finding is again consistent with the proposed transformation and in addition provides support for the importance of very early motivation for competence three years later.

McQuiston and Yarrow (1982) summarized not only their conclusion but ours in a useful way. They state that "at 6 months, the highly motivated child is one who is persistent in producing effects and whose practice of sensorimotor schemes involves the use of a variety of exploratory behaviors" (p. 9). They go on to point out that the highly motivated child becomes more goal- or task-directed by 12 months. They expect "that goal directedness in problem solving becomes a particularly relevant dimension of mastery motivation during the 12- to 18-month period" (p. 9). This statement, data presented in the validity section, and some of the data in the next section indicate that the transition in mastery motivation from exploration to persistence at trying to solve tasks may lap over into the second year of life or at least that some, perhaps especially high-risk, infants will not have completed the transition even by 12 months gestational age.

Continuity and Change during Early Childhood

Fewer studies have examined stability and change in the second and third year of life than in the first year, but there is some evidence of greater stability and only a few tentative hints about possible transformations in mastery motivation during the period from 1 to 4 years.

First, during this age span there is some evidence of short-term stability in mastery motivation scores. For example, Jennings *et al.* (1982) assessed the test–retest stability of their mastery task measures with 10 3½-year-old children over a two-month period. Persistence and curiosity were relatively stable ($r = .65$ and $r = .61$), but preference for challenging tasks was less stable ($r = .32$). Morgan, Harmon, Pipp, & Jennings (1983) have reported the stability of maternal ratings of the mastery motivation of 16 toddlers using the MOMM twice approximately four months apart. There was good stability for preference for challenging tasks ($r = .75$), but only moderate correlations ($r = .25–.73$) for other clusters.

Messer *et al.* (1982) found some significant relationships, especially for girls, between 12-month mastery motivation scores and 30-month McCarthy Scales of Children's Abilities. The direction of the correlations, in constrast to the findings for 6 to 30 months, was what one would expect if there were no major transformations during this period. Less task-directed behaviors such as looking and exploration were negatively related to 30-month competence while goal-directed behavior at 12 months was positively related to later competence. These findings indicate that some predictability exists between the one-year-old infant's mastery behavior and later cognitive functioning. They also suggest that there may not be a major transformation in mastery motivation between approximately 12 and 30 months. This latter conclusion gets additional support from the finding that most of the significant age differences on the Mother's Observation of Mastery Motivation (MOMM) questionnaire were between ages 2 and 3, not between 1 and 2 (Morgan, Harmon, Pipp, & Jennings, 1983).

In summary, it appears that by 12 months infants have a clearer recognition of the means to achieve an end than they do earlier. Consequently, peripheral and general exploration will truly reflect less directed forms of task involvement. Thus it is not surprising that exploration at 12 months was negatively related to McCarthy scores at 30 months and that goal-directed behavior was positively related to later competence. Because there was not a significant relationship between the 6-month or 12-month Bayley MDI and the McCarthy, these findings also suggest that an infant's attempts to master a problem may predict later cognitive development more effectively than his or her initial level of cognitive competence. The infants who attempted to master objects in a way appropriate for their level of cognitive functioning were the ones who later had higher scores on the McCarthy.

Jennings, Yarrow, and Martin (in press) followed up the infants from Study 1 at 3½ years of age. The results provide some evidence of continuity in mastery motivation for boys, even though the measures for the older children were quite different from those for the infants. Producing effects was related to all three motivation measures at 3½. It may be that the boys were somewhat less mature at one year and thus that producing effects, which one might describe as in between exploration and persistence at problem solving, was the most appropriate index of mastery motivation for them. Another finding from this study, which is consistent with those described earlier, is that visually directed reaching (an aspect of exploration) at 6 months was related to the McCarthy General Cognitive Index (GCI). For girls, who may have been a little more mature than boys but not all the way through the transition, persistence at the mastery tasks at one year was not predictive of motivation at 3½, but was predictive of the GCI of the McCarthy Scales. These results provide further support for the theory that competence develops out of early attempts first to explore and then to master the environment.

There is some evidence from the age comparisons of mothers' ratings of mastery motivation (MOMM) that certain aspects of the concept may change between 2 and 3 years of age (Morgan, Harmon, Pipp, & Jennings, 1983). Surprisingly, in this cross-sectional data mothers of normal 3- and 4-year-olds rated their children as significantly lower on persistence at tasks, preference for challenge, and solving problems for intrinsic reasons rather than to please adults than did the mothers of 1- and 2-year-olds. These findings are hard to interpret because it seems unlikely that the older children would actually score lower on persistence and preference for challenge. However, it may be that mothers are reporting the beginning of the shift from more intrinsic to more extrinsic motivation that Harter (1981b) has reported in school-age children.

Except for the data on short-term stability mentioned above, none of the observational mastery motivation studies in Table 1 were designed to test for longer-term continuity during the 18-month to 4-year period, that is, during the toddler stage and early childhood. The ongoing Study 8 by Jennings will provide mastery motivation data on 4½-year-olds who were also tested at 3½. As mentioned earlier, two other studies, not done within the mastery motivation framework, provide some findings relevant to this issue. Scarr (1981b) found moderate stability in motivational measures such as attention span between ages 2 and 4. These correlations were similar, but slightly lower, than those for cognitive variables (i.e., Stanford-Binet IQ). Bretherton and Bates (Chapter 10, this volume) have observed that cross-age stabilities are higher for language than for symbolic play.

These two studies and the data described previously appear to indicate only moderate stability for mastery motivation measures between 1 and 4 years. This may well be because transformations in appropriate indices of mastery motivation during this age period are only partially understood.

To summarize this entire section on continuity and change from early infancy to the preschool period, the findings suggest that mastery motivation in infants less than 1 year has a different meaning than after one year. General exploration and the production of effects appear to be the way in which the typical 6- to 12-month-old attempts to master his or her environment. Older infants and toddlers, on the other hand, are oriented toward trying to solve problems and complete tasks. In testing infants from 6 months to 3 years, it has become evident that after 12 months of age behavior becomes more focused and more easily classified as task-directed or not task-directed. There may also be a later transition at approximately 3 years of age, but this is less well documented and understood. These transitions, which may be as much related to the child's cognitive development as to motivation, no doubt contribute to the relative lack of stability found in some of the studies described above.

THE INTERRELATIONSHIP OF MASTERY MOTIVATION AND COGNITIVE DEVELOPMENT: PROBLEMS AND POSSIBLE SOLUTIONS

The studies described above have provided considerable evidence for the validity of the mastery motivation tasks and measures, in part through comparisons of full-term, normally developing children with those who are at risk or handicapped. These studies have sensitized us to problems caused by the interrelations of our measures of mastery motivation and measures of cognitive competence.

Several studies in Table 1 have found moderately high correlations between mastery motivation scores and cognitive or mental development. It is unclear to us to what extent these results reflect an inseparable intertwining of cognition and motivation in infancy, as several investigators have suggested (e.g., Yarrow & Pedersen, 1976; Ulvund, 1980; Yarrow & Messer, 1984). It is also possible that these correlations are at least partly an artifact of the way in which both mental development and mastery motivation have been measured. For example, a number of Bayley MDI items, such as how many cubes the child will put in a cup, have a clear motivational component. On the other hand, mastery motivation scores are bound to be affected, perhaps artifactually, by cognitive level. If a task is too difficult, the child may not understand how to attempt it. Even when the child understands what to do, if he discovers that he can not do it, prolonged attempts may be unreasonable. Similarly, it may not be appropriate to persist at doing a task that is so easy that it is well mastered.

The need to have a method to assess mastery motivation as independently as possible from cognitive level results, in part, from the authors' goal of developing a test of mastery motivation that could be used to compare children of different ages, much as is done for cognitive development with the Bayley Scales of Infant Development. We are also interested in being able to make reasonable comparisons of the motivation of children of the same chronological age but of different levels of ability.

Investigators have dealt in several ways with the problem of comparing samples that differ in cognitive abilities. For example, Harmon, Glicken, and Culp (1982) used analysis of covariance while Vietze et al. (1983) attempted to match samples on cognitive ability. We (Harmon, Malpiede, Culp, & Morgan, 1983; Morgan, Harmon, Malpiede, Culp, & Renner, 1983) have also carefully selected tasks which could be solved, at least in part, by children with a wide range of abilities. Our most recent strategy to untangle motivation and cognition (Morgan, Tolerton, Renner, & Harmon, 1983) deals well with the problem of controlling for cognitive differences between groups and also makes longitudinal analysis more meaningful. This strategy is to design sets of tasks that have several levels of difficulty. Each child is tested with the level that is appropriately difficult for him or her. Specifically, the level of the task is selected so that the child can and does successfully complete at least part of it, but does not finish all parts of the task too quickly (i.e., in less than 60 seconds). Thus, the level chosen for a given child is challenging but not so hard that partial completion is not possible.

A child's mastery motivation is assumed to be assessed most appropriately during tasks that are moderately difficult for him or her. It may also be necessary to examine persistence at very difficult and at easy tasks in order to get a complete picture. Perhaps the most highly motivated children are the ones who will try, appropriately, to do tasks that are very difficult for them.

On the other hand, persistence, if not appropriate to the task and/or if highly repetitive, may reflect perseveration rather than mastery motivation. This is especially a problem with easy tasks but is not necessarily restricted to them. Our research (e.g., Caplovitz et al., 1982) and Piaget's writing (e.g., 1936/1952) lead us to believe that repetition of a task in children one year old or younger is usually better characterized as mastery motivation than as perseveration. We are, however, more concerned about perseveration in older children and thus think that multipart tasks that require some time to complete are preferable to ones that are easily completed and elicit repetition.

SUMMARY

The motivation of infants and young children to master their environment has received much more theoretical than empirical attention. This chapter reviews the research that has led to operationalizing the concept of mastery motivation primarily in terms of persistence at age-appropriate tasks. Also summarized are studies that provide data about the reliability and validity of the measures of mastery motivation. Persistence at tasks has been found to be related to other measures that are assumed to reflect aspects of mastery motivation and to predict later competence. Groups of at-risk or handicapped children, predicted to score lower on mastery motivation, have in fact scored lower than appropriately matched comparison groups.

Studies described in this review also indicate that there is a developmental progression in the types of tasks appropriate for different age levels. Until at least 9 months of age, tasks that elicit curiosity and *exploration* appear to be most representative of the child's motivation to master the environment. From approximately 9 to 15 months, most tasks that have been used in these studies have in common the *practicing of emerging skills;* that is, mastery motivation is primarily the repetition of the same or highly similar task-directed behaviors. Within the category of practicing emerging skills, persistence at producing effects or feedback may be more important before one year of age, whereas attempting to master simple combinatorial and barrier problems may be more salient in the early part of the second year. For children approximately 15 months and older, tasks that are more complex and involve a sequence of behaviors to *complete* appear to be the most appropriate. For children over three years of age, two other types of problems, *preference for challenging tasks* and *self-initiated mastery* used by Harter with school-age children, may become appropriate, but little mastery motivation research has been done with preschool children.

We believe that the strategies for assessing mastery motivation described above have been fruitful and will continue to provide valuable data. The next steps are to develop a better understanding of mastery motivation during the toddler and preschool periods and to untangle motivation and cognition to the extent that that is possible.

Acknowledgments

This chapter is dedicated to Leon Yarrow, who was the inspiration for the series of studies on mastery motivation reviewed here. The authors would like to thank I. Bretherton, K. Caplovitz, R. Culp, K. Jennings, R. MacTurk, C. Maslin, D. Messer, and P. Vietze for their critical comments on an earlier version of the manuscript. The published and unpublished manuscripts of the colleagues named in Table 1 have been essential to this chapter.

REFERENCES

Butterfield, P. M., & Miller, L. Read your baby: A followup intervention program for parents with NICU infants. *Infant Mental Health*, 1984, in press.

Caplovitz, K., Morgan, G., & Mardashti, S. Mastery motivation in infancy: What does persistance index? *Program and Proceedings of the Developmental Psychobiology Research Group Second Biennial Retreat*, Estes Park, Colorado, 1982, 2, 10 (summary).

David, D. S., Morgan, G. A., Culp, R. E., & Vance, A. J. Changes in mastery motivation of 9- and 12-month-old infants. *Program and Proceedings of the Developmental Psychobiology Research Group Second Biennial Retreat*, Estes Park, Colorado, 1982, 2, 19–20 (summary).

Flagle, J. R. *The effects of occupational therapy intervention on task performance in Down's syndrome children.* Unpublished master's project, Colorado State University, 1982.

Gaiter, J. L., Morgan, G. A., Jennings, K. D., Harmon, R. J., & Yarrow, L. J. Variety of cognitively-oriented caregiver activities: Relationships to cognitive and motivational functioning at 1 and 3½ years of age. *Journal of Genetic Psychology*, 1982, 141, 49–56.

Goldberg, S. Social competence in infancy: A model of parent–infant interaction. *Merrill-Palmer Quarterly*, 1977, 23, 161–177.

Harmon, R. J., & Culp, A. M. The effects of premature birth on family functioning and infant development. In I. Berlin (Ed.), *Children and our future*. Albuquerque: University of New Mexico Press, 1981, 1–9.

Harmon, R. J., & Glicken, A. D. *Mastery motivation scoring manual for 12-month-old infants.* Unpublished manuscript, University of Colorado School of Medicine, 1981.

Harmon, R. J., Glicken, A. D., & Culp, A. M. *Assessment of mastery motivation in term and pre-term infants.* Presented at the International Conference on Infant Studies, Austin Texas, March, 1982.

Harmon, R. J., Morgan, G. A., Jacobs, S. E., Glicken, A. D., Culp, A. M., Busch, N. A., & Butterfield, P. M. Comparison of risk and low-risk infants' motivation on a maternal report questionnaire and mastery tasks. *Program and Proceedings of the Developmental Psychobiology Research Group Second Biennial Retreat*. 1982, 2, 25 (summary).

Harmon, R. J., Malpiede, D., Culp, A. M., & Morgan, G. A. *Manual for 12-month mastery motivation tasks.* Unpublished manuscript, University of Colorado School of Medicine, Denver, 1983.

Harmon, R. J., Morgan, G. A., & Glicken, A. D. Continuities and discontinuities in affective and cognitive-motivational development. *International Journal of Child Abuse and Neglect*, 1984, in press.

Harter, S. The effect of social reinforcement and task difficulty level on the pleasure derived by normal and retarded children from cognitive challenge and mastery. *Journal of Experimental Child Psychology*, 1977, *24*, 476–494.

Harter, S. Effectance motivation reconsidered: Toward a developmental model. *Human Development*, 1978, *21*, 34–64.

Harter, S. A model of intrinsic motivation in children: Individual differences and developmental change. In A. Collins (Ed.), *Minnesota Symposium on Child Psychology* (Vol. 14). Hillsdale, N.J.: Lawrence Erlbaum, 1981. (a)

Harter, S. A new self-report scale of intrinsic versus extrinsic orientation in the classroom: Motivational and informational components. *Developmental Psychology*, 1981, *17*, 330–312. (b)

Harter, S., & Zigler, E. The assessment of effectance motivation in normal and retarded children. *Developmental Psychology*, 1974, *10*, 169–180.

Hunt, J. Intrinsic motivation and its roles in psychological development. In D. Levine (Ed.), *Nebraska Symposium on Motivation* (Vol. 13). Lincoln: University of Nebraska Press, 1965.

Jennings, K., Harmon, R., Morgan, G., Gaiter, J., & Yarrow, L. Exploratory plays as an index of mastery motivation: Relationships to persistence, cognitive functioning and environmental measures. *Developmental Psychology*, 1979, *15*, 386–394.

Jennings, K. D., Connors, R. E., Sankaranarayan, P., & Katz, E. *Mastery motivation in physically handicapped and nonhandicapped preschool children*. Paper presented at the American Academy of Child Psychiatry Meeting, Washington, D.C., October, 1982.

Jennings, K., Yarrow, L., & Martin, P. Mastery motivation and cognitive development: A longitudinal study from infancy to three and one half years. *International Journal of Behavioral Development*, in press.

Kagan, J. *The second year: The emergence of self-awareness*. Cambridge, Mass.: Harvard University Press, 1981.

MacTurk, R. H., & McQuiston, S. *Social competence and mastery motivation in infants with Down syndrome*. Presented at the annual meeting of the American Psychological Association, Washington, D.C., August, 1982.

MacTurk, R. H., & Yarrow, L. J. *The structure of mastery motivation in Down and normal infants*. Presented at the biennial meeting of the Society for Research in Child Development, Detroit, April, 1983.

McCall, R. B. The development of intellectual functioning in infancy and the prediction of later IQ. In J. D. Osofsky (Ed.), *Handbook of infant development*. New York: Wiley, 1979.

McCarthy, M. E., & McQuiston, S. *Relationship of contingent parental behaviors to infant mastery motivation and competence*. Presented at the biennial meeting of the Society for Research in Child Development, Detroit, April, 1983.

McCarthy, M. E., & Vietze, P. M. *Development of mastery motivation across age in Down syndrome infants*. Presented at the annual meeting of the American Psychological Association, Washington, D.C., August, 1982.

McQuiston, S., & Yarrow, L. J. *Assessment of mastery motivation in the first year of life*. Presented at the annual meeting of the American Psychological Association, Washington, D.C., August, 1982.

Messer, D. J., Yarrow, L. J., & Vietze, P. M. *Mastery in infancy and competence in early childhood*. Presented at the annual meeting of the American Psychological Association, Washington, D.C., August, 1982.

Messer, D. J., Ratchford, D., McCarthy, M., & Yarrow, L. J. *An assessment of mastery behavior with 30-month children*. Presented at the biennial meeting of the Society for Research in Child Development, Detroit, April, 1983.

Morgan, G. A., & Jacobs, S. E. (Eds.) *Issues in measuring mastery/effectance motivation in infants and young children.* Fort Collins: Colorado State University, 1981. (ERIC Document Reproduction Service ED No. 281 359). (a)

Morgan, G. A., & Jacobs, S. E. *Manual for assessing mastery motivation in two-year-old children.* Unpublished manuscript, Colorado State University, 1981. (b)

Morgan, G. A., Harmon, R. J., Gaiter, J. L., Jennings, K. D., Gist, N. F., & Yarrow, L. J. A method for assessing mastery motivation in one-year-old infants. JSAS *Catalog of Selected Documents in Psychology,* 1977, 7, 68. (Ms. No. 1517, 41 pages.)

Morgan, G., Busch-Rossnagel, N., Culp, R., Vance, A., & Fritz, J. Infants' differential social responses to mother and experimenter: Relationships to maternal characteristics and quality of infant play. In R. N. Emde & R. J. Harmon (Eds.), *Attachment and affiliative systems: Neurobiological and psychobiological aspects.* New York: Plenum Press, 1982.

Morgan, G. A., Harmon, R. J., Malpiede, D. M., Culp, A. M., & Renner, S. *Manual for assessing mastery motivation in two-year-old children.* Unpublished manuscript, University of Colorado School of Medicine, Denver, 1983.

Morgan, G. A., Harmon, R. J., Pipp, S., & Jennings, K. D. *Assessing mothers' perceptions of mastery motivation: Utility of the MOMM questionnaire.* Unpublished manuscript, Colorado State University, Fort Collins, 1983.

Morgan, G. A., Tolerton, S., Renner, S. J., Harmon, R. J. *Mastery motivation tasks: Manual for one to three year-children.* Unpublished manuscript, Colorado State University, Fort Collins, 1983.

Piaget, J. *The origins of intelligence in children* (trans. Margaret Cook). New York: International Universities Press, 1952. (Originally published in 1936.)

Scarr, S. *On the development of competence and the indeterminate boundaries between cognition and motivation: A genotype–environment correlation theory.* Invited address to the Eastern Psychological Association, New York, May 1981. (a)

Scarr, S. Testing for children. *American Psychologist,* 1981, 36, 1159–1166. (b)

Ulvund, S. E. Cognition and motivation in early infancy: An interactionist approach. *Human Development,* 1980, 23, 17–32.

Vietze, P. M., Pasnak, C. F., Tremblay, A., McCarthy, M. E., Klein, R. P., & Yarrow, L. J. *A manual for assessing mastery motivation in 6- and 12-month-old infants.* Unpublished manual, NICHHD, Bethesda, Maryland, 1981.

Vietze, P., McCarthy, M., McQuiston, S., MacTurk, R., & Yarrow, L. Attention and exploratory behavior in infants with Down syndrome. In T. Field & A. Sostek (Eds.), *Infants born at risk: Perceptual and physiological processes.* New York: Grune & Stratton, 1983.

Watson, J. S. Smiling, cooing, and "The Game." *Merrill-Palmer Quarterly,* 1972, 18, 323–329.

White, R. W. Motivation reconsidered: The concept of competence. *Psychological Review,* 1959, 66, 297–333.

White, R. W. Ego and reality in psychoanalytic theory. *Psychological Issues,* 1963, 3 (11), 1–40.

Yarrow, L. J., & Messer, D. J. Motivation and cognition in infancy. In M. Lewis (Ed.), *Origins of intelligence* (2nd ed.). New York: Plenum Press, 1984, in press.

Yarrow, L. J., & Pedersen, F. The interplay between cognition and motivation in infancy. In M. Lewis (Ed.), *Origins of intelligence.* New York: Plenum Press, 1976.

Yarrow, L. J., Klein, R., Lomonaco, S., & Morgan, G. Cognitive and motivational development in early childhood. In B. Z. Friedlander, G. M. Sterritt, & G. E. Kirk (Eds.), *Exceptional infant 3: Assessment and intervention.* New York: Brunner/Mazel, 1975.

Yarrow, L. J., Rubenstein, J. L., & Pedersen, F. A. *Infant and environment: Early cognitive and motivational development.* Washington, D.C.: Hemisphere, Halsted, Wiley, 1975.

Yarrow, L. J., Morgan, G. A., Jennings, K. D., Harmon, R. J., & Gaiter, J. L. Infant's persistence at tasks: Relationships to cognitive functioning and early experience. *Infant Behavior and Development*, 1982, *5*, 131–142.

Yarrow, L. J., McQuiston, S., MacTurk, R. H., McCarthy, M. E., Klein, R. P., & Vietze, P. M. Assessment of mastery motivation during the first year of life. Contemporaneous and cross-age relationships, *Developmental Psychology*, 1983, *19*, 159–171.

Yarrow, L. J., MacTurk, R. H., Vietze, P. M., Klein, R. P., & McQuiston, S. The developmental course of parental stimulation and its relationship to mastery motivation during infancy. *Developmental Psychology*, 1984, in press.

CHAPTER 12

The Development of Cerebral Specialization in Infants

ELECTROPHYSIOLOGICAL AND BEHAVIORAL STUDIES

David W. Shucard, Janet L. Shucard, and David G. Thomas

In the various fields of investigation that come under the rubric of developmental psychobiology there are but a few basic theoretical issues that serve as the impetus for much of the research generated in these fields. The nature–nurture controversy, for example, has influenced virtually all of developmental psychobiology to the extent that we have now abandoned previous, simplistic, theoretical approaches in favor of more complex interactional conceptualizations. A second theoretical issue that occupies a similar position of significance is that of continuity versus discontinuity in development: Is development, or at least the developmental process one is studying, characterized by gradual change or is it marked by sudden reorganization? Another way to pose the question is whether development involves the same principles across the life span, or do different principles operate at different developmental levels (Lerner, 1976)?

Heinz Werner (1957) was perhaps the most influential theorist to elucidate the continuity/discontinuity issue. To Werner, discontinuity in

David W. Shucard, Janet L. Shucard, and David G. Thomas • Brain Sciences Laboratories, Department of Pediatrics, National Jewish Hospital and Research Center, and Department of Psychiatry, University of Colorado School of Medicine, Denver, Colorado 80206. The research reported here has been supported by National Institute of Child Health and Human Development grants HD 11747 and HD 15844, and by Social and Behavioral Sciences Research grant 12-83 from the March of Dimes Birth Defects Foundation.

development is characterized in part by abruptness or "gappiness," but more importantly discontinuity involves emergent properties that cannot be reduced to those that preceded them. An example might be seen in Piagetian theory (Piaget & Inhelder, 1969), wherein each stage emerges from the intercoordination of the schemes of the previous stage, but the schemes of the new stage are unique. They are not contained in the previous stage and cannot be predicted from it.

However, as Lerner (1976) pointed out, the categorizing of a developmental process as continuous or discontinuous depends on one's theoretical assumptions, especially the criteria one establishes as the definition of an emergent property. Moreover, we may be unable to deduce an advanced stage from an earlier one and thus assume discontinuity when actually our experimental methods are not sensitive enough to measure the variables needed for such a deduction.

When considering the ontogeny of cerebral specialization of function in the human brain in the context of continuity or discontinuity of development, one seems to be stalemated not so much by these shortcomings as by a lack of data with which to describe accurately the developmental processes involved. Thus, at this point, we do not have enough information to be able to answer the question of whether the development of cerebral specialization is a continuous or a discontinuous process within the traditional framework of this classic issue. Nevertheless, in this chapter we have formulated a definition of continuity–discontinuity which can be applied to the concept of cerebral specialization. First, we will review some of the background pertaining to the topic of cerebral specialization. This review will focus on the adult literature. Next we will discuss cerebral specialization within the context of development, including some of our own work with infants using electrophysiological methods. Throughout this chapter we will attempt to frame our discussion within the context of the continuity–discontinuity issue.

CEREBRAL SPECIALIZATION IN ADULTS

Clinical, Anatomical, and Behavioral Studies

It has become clear from both clinical and experimental evidence gathered over a century that certain homologous areas of the cerebral hemispheres[1] of man are not simply duplicates of each other (as is the

---

[1] In this chapter we will use the term *hemisphere* with the implication that is used here, that is, that certain *areas* of the cerebral hemispheres are relatively more or less involved in the performance of a given task. Further, we do not mean to imply that one cerebral hemisphere is exclusively involved in the performance of any task.

case with a number of other organs, such as the kidneys), but differ both structurally and functionally. Studies of behavioral deficits produced by unilateral cerebral lesions and cerebral disconnections, as well as anatomical, behavioral, and more recent electrophysiological studies, all indicate that homologous areas of the two cerebral hemispheres are specialized for different cognitive functions.

Reports by Broca (1865) and Wernicke (1874) based on different types of aphasic patients indicated that the posterior section of the inferior frontal gyrus (Broca's area) and the posterior section of the superior temporal gyrus (Wernicke's area) of the left cerebral hemisphere play a major role in language production and comprehension respectively. Anatomical evidence presented by Geschwind and Levitsky (1968), Teszner, Tzavaras, Gruner, and Hécaen (1972), Wada, Clark, and Hamm (1975), and Witelson and Pallie (1973) has shown that indeed Wernicke's area differs anatomically from the homologous area of the right hemisphere. Wada and Rasmussen (1960) and Gordon and Bogen (1974), using intracarotid injections of sodium amytal to inhibit speech, have further substantiated the importance of the left hemisphere for language in most individuals.

Studies of commissurotomy ("split brain") patients by R. W. Sperry and his colleagues (Sperry, 1974, 1982) done in the last two decades have produced evidence that when the two hemispheres are surgically disconnected they each appear conscious, that is, "two separate conscious minds in one head" (Galin, 1974, p. 572). Sperry and his collaborators showed that in split-brain patients the left hemisphere is capable of speech, writing, and calculation but is limited in its ability to solve problems involving spatial relationships and novel figures. The right hemisphere, on the other hand, appears better able to perform tasks involving complex spatial and musical patterns, although it can perform simple addition and has the use of a few words (see Galin, 1974, for a review).

Many studies of cerebral function in noncommissurotomy subjects using various methods such as dichotic listening (Kimura, 1973), tachistoscopic techniques (Kimura & Durnford, 1974), and recordings of the direction of eye movements (Galin & Ornstein, 1974; Kinsbourne, 1972), as well as the previously noted intracarotid amytal injection studies, have lent further support to the findings of Sperry and his colleagues.

Electrophysiological Studies

Recent investigations have attempted to relate cerebral specialization of function to changes in electrophysiological measures (the electroencephalogram and evoked potential) recorded from the surface of the scalp in normal individuals. These measures are appealing because they are noninvasive, do not have to be justified in terms of an individual's medical needs, are readily quantified, but yet with appropriate controls tap cortical functioning more directly than many of the other methods previously used. Consequently, normal, intact subjects can be used to study cerebral specialization with this methodological approach.

The findings of investigations using electrophysiological methods tend to support those using other techniques showing differential functioning between the two hemispheres. For example, Morrell and Salamy (1971), in studying adults, found that auditory evoked potentials (AEPs) produced by speech sounds (e.g., "pa," "pi") were significantly higher in amplitude in the left temporal area than the right. Similar results were reported by Molfese, Freeman, and Palermo (1975), who also showed that greater right-hemisphere evoked potential amplitudes occurred in response to musical chords. Galin and his associates, using the electroencephalogram (EEG) (Doyle, Ornstein, & Galin, 1974; Galin & Ornstein, 1972) and visual evoked potential (VEP) measures (Galin & Ellis, 1975), also showed differences between left and right hemisphere responses that were dependent on the tasks the adult subjects were performing. For example, when subjects were asked to write from memory (presumably more of a left hemisphere task), the EEG and VEP measures indicated that greater activation was occurring in the left hemisphere. When these same subjects were asked to manipulate blocks (presumably more of a right hemisphere task), the EEG and VEP measures, as predicted, were indicative of greater right hemisphere activation (Galin & Ellis, 1975). McKee, Humphrey, and McAdam (1973), using EEG alpha activity, reported similar results for linguistic and musical tasks.

Not all previous results from electrophysiological studies of hemispheric specialization have been positive. Researchers at UCLA (Taub, Tanguay, Doubleday, & Clarkson, 1976) presented musical chords and consonant–vowel sounds to adult subjects and measured differences in evoked potentials over the right and left hemisphere. Although evidence of differentiation of function between the right and left hemisphere was obtained for musical stimuli, it was not obtained for speech sounds. Further, Tanguay, Taub, Doubleday, and Clarkson (1977) failed to produce "hemispheric effects" in evoked potential latency or amplitude to consonant–vowel stimuli presented to adults. Friedman, Simson, Ritter, and Rapin (1975), focussing on the P300 wave, found inconsistent evidence for AEPs to be different over right and left hemispheres to words and human sounds. These investigators concluded from a review of the literature that flaws in design and in statistical techniques, together with "inconsistencies in reported findings" indicate that the relationship between the evoked potential and hemispheric functioning "is marginal at best" (1975, p. 18). Although Galambos, Benson, Smith, Schulman-Galambos, and Osier (1975) showed evidence of cerebral specialization with AEP measures to speech and tones, they stated that "either the evoked response method is virtually blind to the crucial events we believe must be there, or the hemispheric differences are barely present in the conditions under which the measurements are currently being made" (1975, p. 282). Shipley (1977), in discussing the inconsistencies of previous results, suggested that the reason for these inconsistencies is that "the meaning of asymmetries in the evoked cortical potential is not understood, even though the search for correlates between them and stimulus or organismic variables is presently quite active" (1977, p. 138).

Tanguay et al. (1977) have also discussed the inconsistency of previous electrophysiological research and have persuasively argued that the reasons for inconsistency and/or weaknesses of results in other investigations are that the use of "simple AER [auditory evoked response] methodology" may be incapable of reflecting complex neurophysiological events which subserve language; in addition, factors such as set, attention, and boredom may be crucial in accounting for these discrepancies in findings. We agreed with Tanguay et al. (1977) and also believed that more robust eliciting conditions had to be developed in order to amplify right-left hemispheric differences, and to this end we have explored a novel technique which addresses itself to many of these problems (Shucard, Shucard, & Thomas, 1977). This technique differed fundamentally from previous evoked potential studies. In those previous studies, evoked potentials were elicited by discrete, transient stimuli such as a single word or musical chord repeated several times. We reasoned that such procedures may not produce optimal differential hemispheric processing because the subject was not involved in active, natural verbal or musical information processing. Rather than becoming involved in the phonemic, syntactic, and semantic processing required to comprehend a narrative, subjects in previous studies typically heard single speech sounds removed from such a rich linguistic context. Similarly, we felt that the presentations of a single musical chord did not require on the part of the subject the holistic processing of the interrelationships among sequential notes and chords which is thought to involve more right cerebral hemisphere functions than left. Our solution was to present to subjects tasks in which they had to listen actively to ongoing narrative speech and classical music while evoked potentials were recorded to *task-irrelevant tones* superimposed on each type of ongoing stimulus.

Briefly, in this 1977 study we found that when subjects were processing language information (i.e., listening for content as well as attempting to detect specific words in verbal passages), they produced higher-amplitude bipolar AEPs from the left hemisphere to pairs of auditory tone pips that were irrelevant to the task and superimposed on the language stimuli being presented. Conversely, when these same subjects were processing musical information (i.e., attempting to recognize specified melodies in musical selections), they produced higher-amplitude bipolar AEPs from the right hemisphere to the tone pips (Shucard *et al.*, 1977). The effects were most salient for the second of the paired tones. The results of this study and a replication of it (Shucard, Cummins, Thomas, & Shucard, 1981) showed that electrophysiological measures are indeed sensitive to differential hemispheric functioning in the normal, intact human.

CEREBRAL SPECIALIZATION AND DEVELOPMENT

There is increasing interest in questions pertaining to the *development* of cerebral specialization of brain functioning in humans. The question being frequently asked (e.g., Kinsbourne, 1975; Molfese *et al.*, 1975) is whether cerebral specialization for language and other functions is present at birth, or whether areas of the brain come to be specialized with development. The way that this question is asked invariably pits Lenneberg (1966), who believed that cerebral specialization is not complete until late in childhood, against others (e.g., Entus, 1977; Kinsbourne, 1975; Molfese *et al.*, 1975), who believe that cerebral specialization is present very early in life, possibly even at birth. The strongest proponent of this latter view is probably Kinsbourne (1975; Kinsbourne & Hiscock, 1983). Kinsbourne maintains that the brain is organized in a lateralized fashion from birth or before and that fundamental aspects of brain organization do not develop postnatally. Kinsbourne concedes, however, that functions which rely on this lateralized organization may develop, such as face recognition for which asymmetrical responding is not found in young children but is in older children (Kinsbourne & Hiscock, 1983). To Kinsbourne, the behavioral organization of the skill changes with age, not the neural substrates.

We feel that the questions being asked about the ontogeny of cerebral specialization in the infant are important, but we particularly regret the tendency of many investigators to treat cerebral specialization as a unitary construct. To be specific, the development of cerebral specialization can be divided into numerous subsidiary categories. We would first propose differentiation of the concept of cerebral specialization into *ana-*

tomical versus *functional* categories. Regarding possible innate hemispheric differences in anatomical structure, we point to the studies by Geschwind and Levitsky (1968), Teszner *et al.* (1972), Wada *et al.* (1975), and Witelson and Pallie (1973), who reported finding anatomical differences between homologous areas of the two hemispheres of infants. In particular, it has been found that the planum temporale (part of the classical area of Wernicke) was significantly larger than the homologous area on the right in both adults and newborns. However, the presence of anatomical signs of lateralization at birth may not necessarily indicate that *functional* specialization is simultaneously present. A distinction between anatomical lateralization and functional lateralization is particularly relevant when one realizes that considerable microanatomical and biochemical development takes place after birth. For example, dendritic arborization in the cortex does not reach an advanced stage of density until well after infancy and does not approach completion until the fourth year of life (Lenneberg, 1974).

Lenneberg (1966) also alluded to the importance of the chemical composition of the brain as providing a physiological basis for the development of language. Furthermore, a number of experiments reviewed by Szentagothai (1974) indicated the importance of the interaction between environmental inputs and central nervous system development. Thus, if an inborn anatomical structure is not fully functional at birth due to lack of the necessary biochemical and microanatomical development, as well as environmental stimulation, then there may be limited functional specialization at that particular developmental period.

Even within the functional domain, cerebral specialization as a unitary concept may be inadequate. Cerebral specialization as a unitary concept is the basic assumption underlying theories that, for example, characterize left hemisphere functioning as analytic and linear and right as holistic and gestalt (Galin, 1974). Such characterizations depict each hemisphere as having a single, intrinsic *modus operandi* presumably inherent in its neural organization. For example, a particular functional process such as face recognition is considered to be within the domain of the right hemisphere because that skill requires holistic, gestalt processing. Thus, the hemispheric specialization of any given task can be derived and predicted on the basis of which *modus operandi* is most appropriate for that task.

The antithesis of this unitary concept is one which recognizes lateralized functions but which does not recognize a *single* organizational principle for the functioning of each hemisphere. Here an ability is not necessarily related to other abilities localized in the same hemisphere. The localization of a skill is not predictable from a single organizational principle, such as analytic or holistic processing. This issue of whether cerebral specialization is a unitary phenomenon or in actuality involves several independent processes leads us to our own definition of the continuity–discontinuity question, which is somewhat different from that of Werner (1957) and may be more quantifiable. If cerebral specialization is a unitary phenomenon, then functions lateralized within a cerebral hemisphere are not independent and should then not develop independently of each other. Such interdependence of asymmetric functions can be thought of in terms of the development of cerebral specialization being a *continuous* process that is, emanating from a common neural organization. However, if these functions all arise independently, there is no continuity, no common neural organization, and the developmental process of cerebral specialization would then appear to be *discontinuous*. Of course, there is a less extreme position than these two: there may be clusters of functions which develop in concordance but are independent from other clusters. Thus, we can use measures of relationship such as a correlation coefficient to represent our definition: Two lateralized functions that correlate highly within or across time show continuity (interdependence). For example, the degree of asymmetry in musical chord recognition in a dichotic listening paradigm at a young age may correlate with spatial abilities later in development. Those lateralized functions with correlations approaching zero show discontinuity (independence).

Next we will review the literature pertaining to cerebral specialization of function in infancy. In particular, we will focus on developmental studies of dichotic ear asymmetries, hand preference, and electrophysiological indices of cerebral specialization in infancy.

Dichotic Listening Studies

It is well established that when adults are simultaneously presented with verbal stimuli to the two ears, there is a right ear advantage (REA) in recalling those stimuli (Shankweiler & Studdert-Kennedy, 1975). The REA is attributed to the more direct access of the right ear to the left cerebral hemisphere, which is typically the more specialized for linguistic processing. Similarly, a left ear advantage (LEA) has been described for musical patterns (Kimura, 1964). Entus (1977) has ingeniously adapted this dichotic listening paradigm for use with young infants by measuring the habituation and dishabituation of high-amplitude, nonnutritive sucking to presentations of verbal (consonant–vowel syllables) or musical stimuli (single notes played by one instrument). Entus found a REA for the verbal stimuli and a LEA for the musical stimuli in infants approximately 50 and 100 days old; age did not appear to influence the results significantly.

The findings of Entus (1977) suggest that infants as early as 2 months of age can distinguish certain phonetic changes better with the right ear (presumably the left hemisphere) and certain timbre changes better with the left ear (presumably the right hemisphere). These abilities, moreover, do not appear to show substantial developmental differences between 2 and 5 months of postnatal age. Although a replication of Entus's findings by Vargha-Khadem and Corballis (1979) failed, an attempt by Glanville, Best, and Levinson (1977) using heart rate deceleration rather than sucking substantiated Entus's (1977) findings with a 3 month-old sample. Best, Hoffman, and Glanville (1982) again replicated both the REA for verbal stimuli and the LEA for musical stimuli using heart rate deceleration. However, Best *et al.* discovered that, whereas the LEA for musical stimuli was present at 2 months, the REA for verbal stimuli was not seen until 3 months of age. Consequently it appears that certain lateralized functions are indeed present very early in the human, but there does seem to be a period of development at least for the function or functions tapped in the dichotic listening paradigm. It is unclear from these reports how significant the role of gender was in the findings.

Hand Preference Studies

Unimanual and Bimanual Handedness.   Hand preference is perhaps the most obvious of lateralized functions. Consequently the scientific study of the development of hand preference in infancy has a history almost 50 years old (Young, 1977). Young reviewed much of the research in this area and found that hand preference is inconsistent in the infancy period: Either no lateral preference is present or vacillating preference occurs, the latter often being characterized by an initial left hand preference that later gives way to the right hand in most cases. In that review chapter, Young (1977) called for more carefully controlled, longitudinally based research on the topic of hand preference and its development in infancy. In a series of well-integrated studies, Ramsay (1979, 1980a, 1980b, 1984; Ramsay, Campos, & Fenson, 1979) has investigated the development of unimanual handedness, bimanual handedness, the relationship of the two to each other, and, as we shall later discuss, the relationship of each to language development.

Unimanual handedness was measured by Ramsay (1980b) as the hand preference shown by an infant in an attempt to manipulate the movable portion of a number of toys. In a cross-sectional sample of 5-, 7-, and 9-month-olds, the 7- and 9-month-olds showed a distinct hand preference (usually the right), whereas the 5-month-olds as a group demonstrated no preference. However, when this youngest group of infants was tested again at 9 months, clear right hand preference was observed. These results were partially replicated in a later longitudinal study (Ramsay, in press) in that unimanual handedness was established by about 7 months of age. However, two weeks following the onset of hand preference it began to disappear, only to reemerge five weeks later.

Thus, Ramsay's studies of unimanual handedness tend to concur with those studies reviewed by Young (1977) in that handedness during the first year of life is unstable. Bimanual handedness—measured as the preferred hand for the manipulation of objects that require bimanual coordination, one hand to hold the object and the other to do the manipulating—appears later in development and does show greater stability (Ramsay, 1980a; Ramsay *et al.*, 1979). In both cross-sectional and longitudinal samples, bimanual hand preference appeared at about 12 months of age. Ramsay *et al.* (1979) studied the stability of bimanual hand preference for five months following its onset in a group of 23 infants. It was found that the established bimanual preference, be it for the left or right hand, wavered little over that five-month period.

These data from Ramsay's studies suggest that the unimanual handedness seen in the first year of life is not stable but that bimanual handedness, once it develops, tends to show little variability. However, even though unimanual preference vacillates, it appears to be predictive of later bimanual preference. Ramsay (1980b) assessed unimanual handedness at 7 or 9 months and then at 13 months measured bimanual preference in 28 infants; of 22 infants found to show a bimanual right hand preference at 13 months, 18 had shown a unimanual right hand preference when they had been tested at 7 or 9 months. This same correspondence held true for 5 of 6 left-handers. In these studies, no significant gender effects were related to unimanual handedness.

Handedness and Language Development. There is a fairly large literature that addresses the relationship between handedness and hemispheric dominance for language in adults. In fact, Broca, as early as 1865 (see Hécaen & Sauguet, 1971) postulated different patterns of hemispheric organization for sinistrals versus dextrals. More recently, there have been speculations that the development of language is related to the development of handedness. Steffen (1974), for example, described the developmental courses of handedness and language over the first few years of life. Noteworthy is Steffen's description of how, in general, children show coincidental developments in the two domains. However, such coincidental emergence is only suggestive of continuity. To lend further support for continuity between handedness and language, both domains should be measured in the same subjects.

Ramsay (1980a, 1984) has investigated the relationship of certain language-related developments in studies of the ontogenesis of handedness. In his study of unimanual handedness, Ramsay (1984) followed infants over several weeks and found that individual infants began to show hand preference during the same week that they began to babble with duplicated syllables such as "mama." Ramsay (1980a) has also found that bimanual handedness, which emerges at about 12 months, develops simultaneously with the appearance of multisyllabic expressions (such as "mommy" rather than "mama"). In a longitudinal sample of 12 infants he found that 11 used their first multisyllabic expression in the same month as or following the onset of bimanual handedness.

Infant Electrophysiological Studies: EEG and AEP Discrete Stimulus Paradigms

To date, there have been few electrophysiological studies of cerebral specialization in infancy despite the likelihood that the development of specialized functions is occurring during this time period. The sparsity of these investigations is perhaps due to the difficulties in measurement that arise with applying electrophysiological techniques to waking infants. Nevertheless, several studies have attempted to investigate the development of the brain in normal infants using electrophysiological methods. For instance, Crowell, Jones, Kapuniai, and Nakagawa (1973) in studying cortical driving to light stimulation in newborns found that there were differences in maturation between the right and left hemispheres, which they felt contradicted studies on early lesions. These investigators reported that only a subset of neonates showed photic driving, that is, the EEG spectral frequency matched that of a 3 Hz light flash. However, of those infants showing only unilateral driving, 16 of 18 showed greater driving in the right hemisphere. Since photic driving is a property of the mature brain, Crowell et al. proposed that, at least for the visual cortex, the left hemisphere is slower to mature and respond than the right. Other workers have suggested that the left hemisphere matures first (Woods & Teuber, 1973). Crowell, Kapuniai, and Garbanati (1977) replicated the findings of Crowell et al. (1973) pertaining to unilateral photic driving and also investigated the development of bilateral driving. In their sample of 217 neonates, only 12% showed bilateral photic driving; at 30 days of age, this proportion had increased to 48%.

The development of linguistic functions and their asymmetrical organization in infancy has also been studied using AEP techniques. Molfese and his associates (Molfese, 1977; Molfese & Molfese, 1979; Molfese et al., 1975; Molfese, Nunez, Seibert, & Ramanaiah, 1976), in a series of studies, investigated the effects of various linguistic and nonlinguistic stimuli on evoked potential amplitude asymmetry. Independent variables such as formant structure, bandwidth, and voice onset time were studied, as well as musical chords, in neonates, infants, children, and adults. The findings from these studies in which AEPs were recorded to discrete stimuli tend to support the viewpoint that functional specialization is present at birth for grosser levels of speech processing, such as the discrimination of acoustic bandwidth. More refined lateralized processing, such as the discrimination of phonetic cues like formant structure, appears to develop later.

Infant Electrophysiological Studies: AEP Probe Paradigm

In the research discussed below we will focus on the results obtained in our laboratory with infants whose auditory evoked potentials were recorded while they were awake and receiving continuous auditory stimuli (e.g., speech, music). In these studies we attempted to investigate cerebral specialization in infants through the use of pairs of tones which acted as probes and which were superimposed on the continuous auditory stimuli. The AEPs were recorded to the probe tones, not to discrete linguistic or nonlinguistic stimuli as has been typical of research using evoked potential methodology. This procedure, as discussed above, appeared to us to be sensitive to differential hemispheric involvement in information processing in adults (Shucard *et al.*, 1977; Shucard *et al.*, 1981). Further, in these studies particular attention was given to methodological problems that have plagued earlier electrophysiological work with infants by carefully controlling for level of arousal, equating stimulus intensity to both ears, and studying infants within narrowly delineated age ranges (see Shucard, Shucard, Cummins, & Campos, 1981; Shucard, Shucard, & Thomas, 1982).

Methods. In order to examine the development of cerebral specialization of function in infants as well as possible sex-related differences in cerebral specialization, both 3-month-old and 6-month-old male and female infants were studied. The 3-month-old group consisted of 8 male and 8 female infants ranging in age from 10 weeks, 6 days, to 13 weeks, 2 days; the 6-month-old group consisted of 10 male and 10 female infants ranging in age from 24 weeks, 0 days, to 26 weeks, 5 days. Each infant had a normal gestation time, normal birth weight, and Apgar scores of 8 or above. All of the parents were right-handed.

The AEPs to pairs of tone pips were recorded between T_4–C_z (right hemisphere) and T_3–C_z (left hemisphere) according to the International 10–20 System, with a ground electrode located on the forehead. Two polygraph amplifier channels were used to record the EEG. The AEPs for Tone 1 and Tone 2 were averaged and printed out separately. The tone pips (600 Hz, 100 msec) were superimposed in pairs on white noise, verbal passages, and musical selections. The interval between the tones of each pair was 2.0 seconds, and a minimum of 4.0 seconds occurred between the presentation of tone pairs. All auditory stimuli were presented over lightweight headphones placed on the infants ears. The white noise, verbal passages, and musical selections were pre-recorded on audio tape. The verbal and musical selections presented to the 3-month-old group of infants were the same as those used in a previous study with adults (Shucard *et al.*, 1977). The 6-month-old group was presented with verbal passages taken from a standardized second-grade reader and recorded on tape by a female voice. The musi-cal selections were from "Adventures in Music," a collection of classical music for children. Tone pip intensity was approximately 70 db sound pressure level. The average intensity of the verbal passages was 68 db and the average intensity of the musical selections was 70 db.

Three-month-old Group.   Each infant participated in 2 recording ses-sions separated by a period of 2 to 7 days. One experimental condition (verbal passages or musical selections) was presented during each ses-sion and the order of conditions was counterbalanced across infants.

The infant came to the laboratory at a regular feeding time and at a time when the mother felt the infant was most likely to remain alert. During each recording session the infant sat on his or her mother's lap in a sound-attenuated, electrically shielded room and was bottle- or breast-fed while the stimuli were presented. Occasionally, a pacifier was used if the infant finished feeding before the completion of the recording session.

A closed circuit video system, continuous EEG recording, and the mother's observations were used to monitor the infant's level of arousal throughout each session. Only data from infants who were alert with their eyes open for approximately 95% of each recording session were used in the analyses.

During the verbal condition, infants were presented with verbal passages. Tone pairs were superimposed on the passages. The mean length of time that the passages were presented was 11.4 minutes ($SD =$ 1.2). The total number of tone pairs presented during the verbal condi-tion ranged from 30 to 70 across all infants studied (mean = 58, $SD =$ 12). During the music condition, the paired tone pips were superim-posed on musical selections. The tones were presented in the same manner as during the verbal condition. The musical selections were presented for a mean of 11.3 minutes ($SD = 2.1$) and the number of tone pairs ranged from 40 to 80 across all infants studied (mean = 59, $SD =$ 12). Verbal and musical stimuli were presented until enough tone pairs were delivered to product reliable AEPs. None of the infants had a difference greater than 10 between the number of tone pairs presented during the verbal and music conditions. For both conditions, tone pairs were presented only when the infant was alert and quiet. Breaks were taken when needed (e.g., if the baby became fussy or had to be burped).

A separate group of 3-month-old infants, with similar age, sex, parental handedness, gestation time, normalcy of birth weight, and Apgar scores as the above group was studied in an identical manner, except that the tone pips were presented against a background of a hissing sound of 55–60 db, approximating the frequency spectrum of white noise (baseline condition). This baseline group was studied in order to examine the responses to the tone probes when they were not accompanied by complex auditory stimuli (verbal or musical). The infants in this group received between 30 and 70 tone pairs (mean = 56, SD = 12), approximating the number used for the experimental group.

Six-month-old Group. The infants in this group were tested in a similar manner as the 3-month-olds. Each 6-month-old infant, however, participated in all three conditions (baseline, verbal, and music) on separate days. For the baseline condition between 40 and 80 tone pairs were presented (mean = 57, SD = 10), for the verbal condition between 40 and 70 pairs were presented (mean = 55, SD = 7), and for the music condition between 50 and 70 pairs were presented (mean = 58, SD = 8). The baseline condition was always presented on the first day of testing and the verbal and music conditions were counterbalanced for order across the other two testing days. All three testing sessions occurred within a 12-day period.

Results. Four AEPs were obtained for each infant during each condition (right and left hemisphere for both Tone 1 and Tone 2) and for each AEP obtained, four reliable peak-to-trough and trough-to-peak components were identified and designated as follows: Peak 1 (P_1–N_1), Peak 2 (N_1–P_2), Peak 3 (P_2–N_2), and Peak 4 (N_2–P_3). The AEPs recorded for one 3-month-old subject during the baseline condition and another 3-month-old subject during both the verbal and music conditions are illustrated in Figure 1 and these components are identified. The mean latencies in milliseconds of these peaks (maximum positive voltages) and troughs (maximum negative voltages) for the 3-month-olds collapsed across condition, tone, and hemisphere were as follows: P_1 = 77.3 msec, N_1 = 125.3 msec. P_2 = 205.5 msec, N_2 = 309.7 msec, P_3 = 416.1 msec. For the 6-month-olds these mean latencies were: P_1 = 74.3 msec, N_1 = 121.5 msec, P_2 = 198.5 msec, N_2 = 285.5 msec, P_3 = 398.0 msec.

Three-month-old Group. Analyses of the data indicated that females had higher-amplitude left-hemisphere AEPs as compared to the right, whereas males showed higher-amplitude right-hemisphere responses

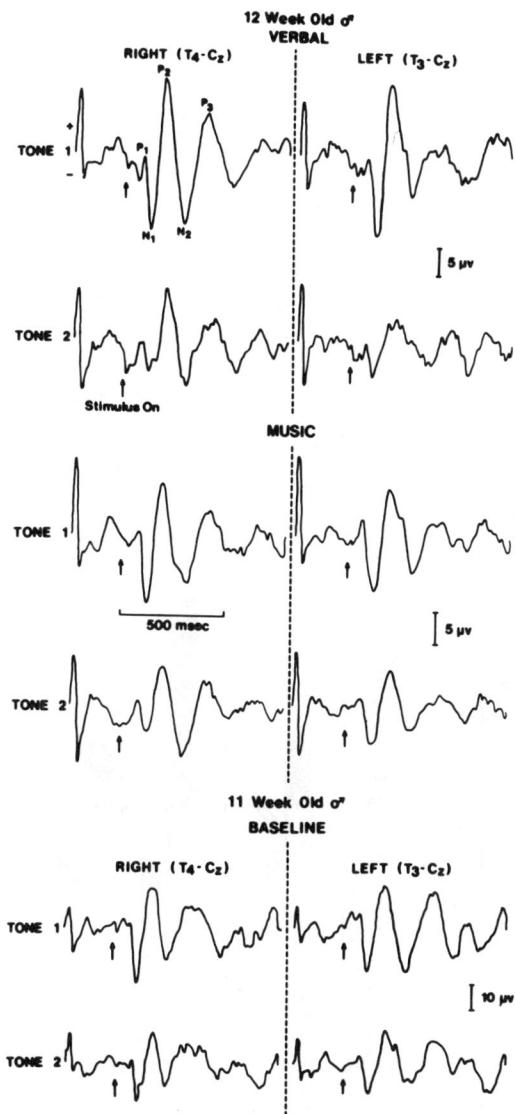

Figure 1. AEPs to pairs of tone pips during the verbal, music, and baseline conditions. The vertical arrows indicate stimulus onset. The positive deflection at the beginning of the tracing (before the stimulus onset) is the calibration signal. Peaks 1, 2, 3, and 4 as defined in the text refer respectively to the negative-going peak from P_1 to N_1, the positive-going peak from N_1 to P_2, the negative-going peak from P_2 to N_2, and the positive-going peak from N_2 to P_3. Positivity at C_z with respect to T_4 or T_3 is up. (From "Auditory Evoked Potentials and Sex Related Differences in Brain Development" by J. L. Shucard, D. W. Shucard, K. R. Cummins, and J. J. Campos, *Brain and Language*, 1981, *13*, 91–102. Copyright 1981 by Academic Press, Inc. Reprinted by permission.)

compared to the left, irrespective of the experimental condition. This effect was seen for AEP components occurring later than 280 msec (Peaks 3 and 4). Thus, the pattern of AEP amplitude asymmetry for males differed significantly from that for females irrespective of the condition (verbal and music) or the tone (Tone 1 or Tone 2). This relationship between males and females across conditions and tones is illustrated in Figure 2. The robustness of this effect is further demonstrated by the number of individual infants who showed this relationship. The mean left-hemisphere AEP amplitude collapsed across tone and condition was higher than the right for 7 out of 8 females for Peak 3 and 8 out of 8 females for Peak 4, whereas the right-hemisphere response was

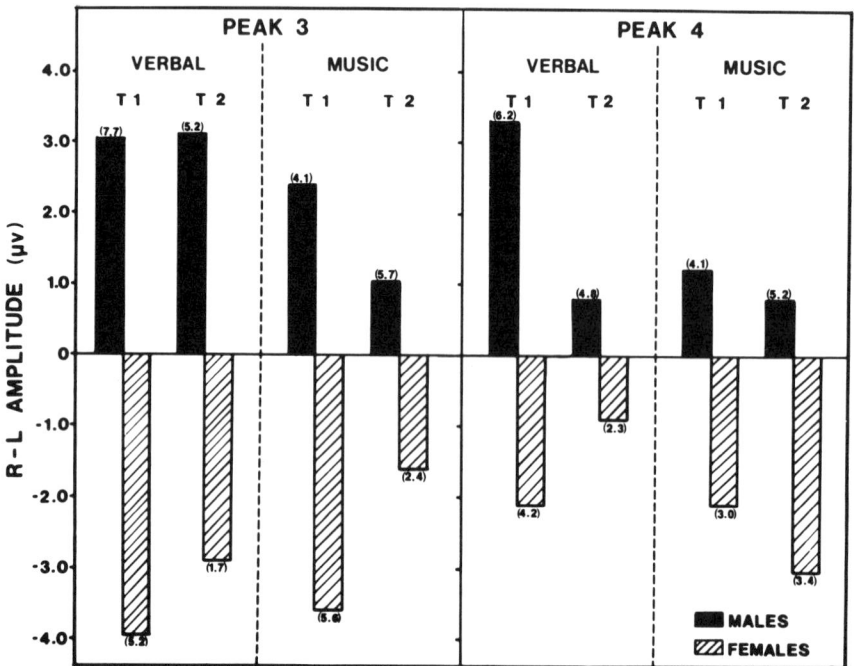

Figure 2. Mean amplitude right minus left (R − L) difference scores for the 3-month-old infants for AEP Peaks 3 and 4 obtained during the verbal and music conditions. T1 refers to Tone 1 and T2 to Tone 2. A positive value indicates that the mean right-hemisphere AEP amplitude was greater than the left, whereas a negative value indicates that the mean left-hemisphere AEP amplitude was greater than the right. Numbers on the bars in parentheses indicate standard deviations. (From "Auditory Evoked Potentials and Sex Related Differences in Brain Development" by J. L. Shucard, D. W. Shucard, K. R. Cummins, and J. J. Campos, *Brain and Language*, 1981, *13*, 91–102. Copyright 1981 by Academic Press, Inc. Reprinted by permission.)

higher than the left in 7 out of 8 males for Peak 3 and 5 out of 8 males for Peak 4.

The results for the baseline group indicated that there were no significant hemispheric AEP amplitude differences in the pattern of responding between males and females in this condition for any of the AEP components. Thus, AEP amplitude responses to tone pairs not accompanied by auditory stimuli containing complex information (verbal and musical) did not differ significantly between males and females.

Six-month-old Group. Analyses of the data from 6-month-olds indicated that the males had higher-amplitude right-hemisphere AEPs as compared to the left, regardless of the condition (verbal or music). This effect was most prevalent for Peak 2 and was comparable to the findings for 3-month-old males. The 6-month-old females, on the other hand, generally showed higher-amplitude left-hemisphere AEPs for the verbal condition and higher right-hemisphere AEPs for the music condition for both Peak 2 and Peak 3. The effect is in contrast to the 3-month-old females, who had greater left-hemisphere AEP amplitudes regardless of condition. Figure 3 illustrates the results for the 6-month-old group.

Using this probe technique, we have uncovered a pattern of hemispheric development that differs between male and female infants. For 3-month-old infants, speech and musical stimuli produced a pattern of asymmetry that was more dependent on the sex of the infant than on the stimuli themselves. For females, larger left-hemisphere AEPs suggest that it is this area of the brain that was more involved in the processing of both types of stimuli. For males, it appears that the right hemisphere was more involved in processing both types of stimuli.

Six-month-old males showed essentially the same pattern of asymmetry as 3-month-old males. However, 6-month-old females tended to show the pattern found in adults: greater left-hemisphere AEP amplitude to the probe tones during the verbal condition and greater right-hemisphere amplitude during the music condition (Shucard et al., 1977). The findings indicate that 6 months of age may be a transitional period in the development of the cerebral cortex in female infants.

Handedness and Auditory Evoked Potential Asymmetries. On the basis of the potential relationship between handedness and language development described above, we attempted to investigate the relationship between handedness and electrophysiological patterns of asymmetry in our 6-month-old infants. At the end of both the second and third recording sessions (verbal and music), after the electrodes were removed, infants were given a behavioral rattle-reaching task. Each infant sat upright in his or her mother's lap facing away from her; the mother held the infant's sides. The experimenter sat directly facing the mother and infant. The experimenter held the rattle directly in front

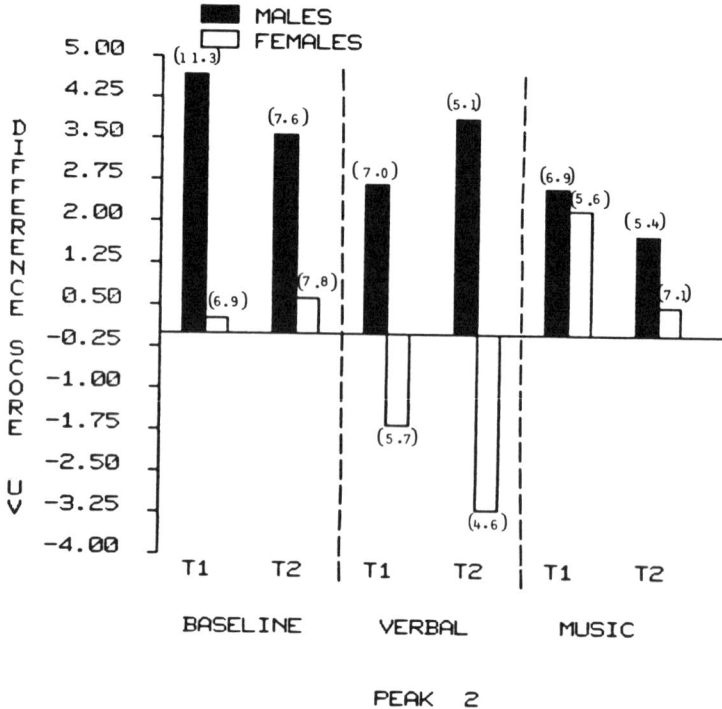

Figure 3. Mean amplitude right minus left (R − L) difference scores for the 6-month-old infants for AEP Peak 2 obtained during the baseline, verbal, and music conditions. T1 refers to Tone 1 and T2 to Tone 2. A positive value indicates that the mean right-hemisphere AEP amplitude was greater than the left, whereas a negative value indicates that the mean left-hemisphere AEP amplitude was greater than the right. Numbers on the bars in parentheses indicate standard deviations.

of the infant within reaching range and with the top of the rattle at the infant's shoulder level. The rattle was held as close to the infant's midline as possible to prevent biasing the left or right side.

The hand with which the infant reached and grabbed the rattle was recorded. This procedure was repeated for a total of six trials for each of the two sessions. Each trial was scored as to how the infant reached. If the infant reached for and grabbed the rattle with his or her right hand, the score was +1.0. If the infant reached first with the right hand and then grabbed the rattle with both hands, the score was +0.5. If the infant reached for and grabbed the rattle with both hands simultaneously, the score was 0. Conversely, if the infant reached with his or her left hand, the score was −1.0; for the left then both, the score was −0.5. By summing the scores for each trial within each session, an infant who always reached and grabbed the rattle with the right hand would receive a score of +6.0, whereas an infant who always reached and grabbed with the left hand would receive a score of −6.0. This hand preference measure appeared to be quite reliable over the two sessions. A Pearson product–moment correlation calculated between the two testing sessions was highly significant ($r = .65, p < .001$).

We were particularly interested in how this measure of handedness was related to our electrophysiological measures of cerebral asymmetry. We assumed that greater AEP amplitude differences between the two cerebral hemispheres in the verbal or music condition signified a greater degree of functional asymmetry. In order to assess the relationship of interest, we calculated right minus left (R − L) difference scores for AEP amplitudes and correlated these R − L scores with measures of hand preference. The findings were quite interesting. For the females, higher right-hand preference scores (Session 3) were correlated with AEP indices of greater right-hemisphere activation (positive R − L score) for Peak 2 during the *music condition* ($r = .64, p < .02$). For the males, higher right-hand preference scores (Session 3) were correlated with AEP indices of greater left-hemisphere activation (negative R − L score) for Peak 2 during the *verbal condition* ($r = −.77, p < .004$).

At this point we can only speculate as to why the two genders showed significant correlations between hand preference and electrophysiological measures for the two conditions. We hypothesize that males have a more slowly developing left hemisphere than right and that females have the reverse pattern of development, that is, a more rapidly developing left hemisphere. This hypothesis was derived from the patterns of asymmetry we obtained with the 3-month-olds. At 6 months of age some males may be beginning to show increased functioning of the left hemisphere. Since language processing is believed to be more the domain of the left hemisphere, the significant negative correlation between right-hand preference and R − L AEP amplitude during the verbal condition may reflect increasing functional organization of the left hemisphere in 6-month-old males. The females, on the other hand, showed a significant positive correlation between right-hand preference and R − L AEP amplitude during the music condition. This relationship may reflect increasing functional organization of the right hemisphere and increasing asymmetry of function in females, which allows hand preference to become established.

CONCLUSION

Thus, our data, as well as those of others, indicate that contrary to theories of cerebral specialization that posit fixed lateralization at birth, there are developmental patterns for at least some lateralized functions. Further, we conclude that the development of cerebral specialization has continuous aspects to it in that different lateralized functions appear to develop interdependently. Although we cannot conclude that cerebral specialization is a unitary concept that shows total continuity in neural organization, we can conclude that the existence of a totally discontinuous development of the many lateralized functions in humans appears to be unlikely.

REFERENCES

Best, C. T., Hoffman, H., & Glanville, B. B. Development of infant ear asymmetries for speech and music. *Perception and Psychophysics*, 1982, *31*, 75–85.

Broca, P. Sur la faculté du langage articulé. *Bulletin de la Société d'Anthropologie de Paris*, 1865, *6*, 377–393.

Crowell, D. H., Jones, R. H., Kapuniai, L. E., & Nakagawa, J. K. Unilateral cortical activity in newborn humans. An early index of cerebral dominance? *Science*, 1973, *180*, 205–208.

Crowell, D. H., Kapuniai, L. E., & Garbanati, J. A. Hemispheric differences in the rhythmic responses to photic stimulation in human infants. In J. E. Desmedt (Ed.), *Visual evoked potentials in man: New developments*. Oxford: Clarendon Press, 1977.

Doyle, J. C., Ornstein, R., & Galin, D. Lateral specialization of cognitive mode: II. EEG frequency analysis. *Psychophysiology*, 1974, *11*, 567–578.

Entus, A. K. Hemispheric asymmetry in processing of dichotically presented speech and nonspeech stimuli by infants. In S. J. Segalowitz & F. A. Gruber (Eds.), *Language development and neurological theory*. New York: Academic Press, 1977.

Freidman, D., Simson, R., Ritter, W., & Rapin, I. Cortical evoked potentials elicited by real speech words and human sounds. *Electroencephalography and Clinical Neurophysiology*, 1975, *38*, 13–19.

Galambos, R., Benson, P., Smith, T. S., Schulman-Galambos, C., & Osier, H. On hemispheric differences in evoked potentials to speech stimuli. *Electroencephalography and Clinical Neurophysiology*, 1975, *39*, 279–283.

Galin, D. Implications for psychiatry of left and right cerebral specialization. *Archives of General Psychiatry*, 1974, *31*, 572–583.

Galin, D., & Ellis, R. R. Asymmetry in evoked potentials as an index of lateralized cognitive processes: Relation to EEG alpha asymmetry. *Neuropsychologia*, 1975, *13*, 45–50.

Galin, D., & Ornstein, R. Lateral specialization of cognitive mode: An EEG study. *Psychophysiology*, 1972, *9*, 412–418.

Geschwind, N., & Levitsky, W. Human brain: Left-right asymmetries in temporal speech region. *Science*, 1968, *161*, 186–187.

Glanville, B. B., Best, C. T., & Levenson, R. A cardiac measure of cerebral asymmetries in infant auditory perception. *Developmental Psychology*, 1977, *13*, 54–59.

Gordon, H. W., & Bogen, J. E. Hemispheric lateralization of singing after intracarotid sodium amylobarbitone. *Journal of Neurology, Neurosurgery, and Psychiatry*, 1974, *37*, 727–738.

Hécaen, H., & Sauguet, J. Cerebral dominance in left-hand subjects. *Cortex*, 1971, *7*, 19–48.

Kimura, D. Left-right differences in the perception of melodies. *Quarterly Journal of Experimental Psychology*, 1964, *14*, 355–358.

Kimura, D. The asymmetry of the human brain. *Scientific American*, March 1973, 70–78.

Kimura, D., & Durnford, M. Normal studies on the function of the right hemisphere in vision. In S. J. Dimond & J. G. Beaumont (Eds.), *Hemisphere function in the human brain.* New York: Halstead Press, 1974.

Kinsbourne, M. Eye and head turning indicates cerebral lateralization. *Science,* 1972, *176,* 539–541.

Kinsbourne, M. The ontogeny of cerebral dominance. *Annals of the New York Academy of Sciences,* 1975, *263,* 244–250.

Kinsbourne, M., & Hiscock, M. Functional lateralization of the brain: Implications for normal and deviant development. In M. M. Haith & J. J. Campos (Eds.), *Handbook of child psychology (vol. 2): Infancy and developmental psychobiology.* New York: Wiley, 1983.

Lenneberg, E. H. Speech development: Its anatomical and physiological concomitants. In E. C. Carterette (Ed.), *Brain function III: Speech, language and communication.* Berkeley: University of California Press, 1966.

Lenneberg, E. H. Maturation of the CNS and language. *Neurosciences Research Program Bulletin,* 1974, *12,* 619–636.

Lerner, R. M. *Concepts and theories of human development.* Menlo Park, Calif.: Addison-Wesley, 1976.

McKee, G., Humphrey, B., & McAdam, D. W. Scaled lateralization of alpha activity during linguistic and musical tasks. *Psychophysiology,* 1973, *10,* 441–443.

Molfese, D. L. Infant cerebral asymmetry. In S. J. Segalowitz & F. A. Gruber (Eds.), *Language development and neurological theory.* New York: Academic Press, 1977.

Molfese, D. L., & Molfese, V. J. Hemisphere and stimulus differences as reflected in the cortical responses of newborn infants to speech stimuli. *Developmental Psychology,* 1979, *15,* 505–511.

Molfese, D., Freeman, R., & Palermo, D. The ontogeny of brain lateralization for speech and nonspeech stimuli. *Brain and Language,* 1975, *2,* 356–368.

Molfese, D. L., Nunez, V., Seibert, S. M., & Ramanaiah, N. V. Cerebral asymmetry: Changes in factors affecting its development. *Annals of the New York Academy of Sciences,* 1976, *280,* 821–833.

Morrell, L., & Salamy, J. Hemispheric asymmetry of electrocortical responses to speech stimuli. *Science,* 1971, *174,* 164–166.

Piaget, J., & Inhelder, B. *The psychology of the child.* New York: Basic Books, 1969.

Ramsay, D. S. Manual preference for tapping in infants. *Developmental Psychology,* 1979, *15,* 437–442.

Ramsay, D. S. Beginnings of bimanual handedness and speech in infants. *Infant Behavior and Development,* 1980, *3,* 67–77. (a)

Ramsay, D. S. Onset of unimanual handedness in infants. *Infant Behavior and Development,* 1980, *3,* 377–385. (b)

Ramsay, D. S. Onset of duplicated syllable babbling and unimanual handedness in infancy: Evidence for developmental change in hemispheric specialization? *Developmental Psychology,* 1984, *20,* 64–71.

Ramsay, D. S., Campos, J. J., & Fenson, L. Onset of bimanual handedness in infants. *Infant Behavior and Development,* 1979, *2,* 69–76.

Shankweiler, D., & Studdert-Kennedy, M. A continuum of lateralization for speech perception. *Brain and Language.* 1975, *2,* 212–225.

Shipley, T. Interhemispherical differences in the evoked cortical potential to intersensory and repetitive stimulation: Hypotheses, methods and appraisals. *Neuropsychologia,* 1977, *15,* 133–141.

Shucard, D. W., Shucard, J. L., & Thomas, D. G. Auditory evoked potentials as probes of hemispheric differences in cognitive processing. *Science,* 1977, *197,* 1295–1298.

Shucard, D. W., Cummins, K. R., Thomas, D. G., & Shucard, J. L. Evoked potentials to auditory probes as indices of cerebral specialization of function—Replication and extension. *Electroencephalography and Clinical Neurophysiology,* 1981, *52,* 389–393.

Shucard, J. L., Shucard, D. W., Cummins, K. R., & Campos, J. J. Auditory evoked potentials and sex related differences in brain development. *Brain and Language,* 1981, *13,* 91–102.

Shucard, J. L., Shucard, D. W., & Thomas, D. G. *Relationships between hand preference and evoked potential asymmetries in 6-month-old infants.* Paper presented to the Second Biannual Retreat of the Developmental Psychobiology Research Group, Estes Park, Colorado, June, 1982.

Sperry, R. W. Lateral specialization in the surgically separated hemispheres. In F. O. Schmitt & F. G. Worden (Eds.), *The neurosciences: Third study program.* Cambridge: MIT Press, 1974.

Sperry, R. W. Some effects of disconnecting the cerebral hemispheres. *Science,* 1982, *217,* 1223–1226.

Steffen, H. Cerebral dominance: The development of handedness and speech. *Acta Paedopsychiatra,* 1974, *41,* 223–235.

Szentagothai, J. Plasticity in the central nervous system: Its possible significance in language mechanisms. *Neurosciences Research Program Bulletin,* 1974, *12,* 534–536.

Tanguay, P., Taub, J., Doubleday, C., & Clarkson, D. An interhemispheric comparison of auditory evoked responses to consonant–vowel stimuli. *Neuropsychologia,* 1977, *15,* 123–132.

Taub, J., Tanguay, P., Doubleday, C., & Clarkson, D. Hemisphere- and ear-asymmetry in the auditory evoked response to musical chord stimuli. *Physiological Psychology,* 1976, *4,* 11–17.

Teszner, D., Tzavaras, A., Gruner, J., & Hécaen, H. L'asymétrie droit-gauche du planum temporale. A propos de l'étude anatomique de 100 cerveaux. *Revue Neurologique,* 1972, *126,* 444–449.

Vargha-Khadem, F., & Corballis, M. Cerebral asymmetry in infants. *Brain and Language,* 1979, *8,* 1–9.

Wada, J., & Rasmussen, T. Intracarotid injection of sodium amytal for the lateralization of cerebral speech dominance. *Journal of Neurosurgery,* 1960, *17,* 266–282.

Wada, J., Clark, R., & Hamm, A. Cerebral hemispheric asymmetry in humans. *Archives of Neurology,* 1975, *32,* 239–246.

Werner, H. The concept of development from a comparative and organismic point of view. In D. B. Harris (Ed.), *The concept of development.* Minneapolis: University of Minnesota Press, 1957.

Wernicke, C. *Der Aphasische Symptomencomplex—Eine Psychologische und Anatomische Basis.* Breslau: Max Cohn and Weigert, 1874.

Witelson, S. F., & Pallie, W. Left hemisphere specialization for language in the newborn. *Brain,* 1973, *96,* 641–646.

Woods, B. T., & Teuber, H. Early onset of complimentary specialization of cerebral hemispheres in man. *Transactions of the American Neurological Association,* 1973, *98,* 113–116.

Young, G. Manual specialization in infancy: Implications for lateralization of brain function. In S. J. Segalowitz & F. A. Gruber (Eds.), *Language development and neurological theory.* New York: Academic Press, 1977.

CHAPTER 13

Continuity and Change in Socioemotional Development during the Second Year

Ross A. Thompson and Michael E. Lamb

Are there meaningful individual differences in behavior which are stable over time? Which conditions foster continuity or change in these differences? How well can we predict a child's later development on the basis of early assessments?

Such questions are at the heart of the developmental sciences and have thus generated considerable interest among students of child development, particularly those concerned with infancy and the importance of early experiences. Much of the current interest in this area derives from an apparent dissonance between theoretical propositions and research findings concerning the continuity of individual differences. Whereas much of the empirical literature demonstrates the flexibility and plasticity of early development, most developmental theories (particularly those emphasizing individual differences) assume that there is an underlying continuity to behavioral development. The belief that early experiences, influences, or traits are likely to have consistent,

Ross A. Thompson • Department of Psychology, University of Nebraska, Lincoln, Nebraska 68588.   Michael E. Lamb • Department of Psychology, University of Utah, Salt Lake City, Utah 84112. The research described in this chapter was based on a dissertation conducted by Ross A. Thompson in partial fulfillment of the requirements for the Ph.D. degree at the University of Michigan. It was supported by a grant from the Riksbankens Jubileumsfond of Sweden to Michael E. Lamb, and by a dissertation grant from the Rackham Graduate School to Thompson. While the research was conducted, Thompson was a predoctoral fellow in the Bush Program in Child Development and Social Policy at the University of Michigan and a National Science Foundation predoctoral fellow.

long-term implications has guided not only theoretical inquiry but also the design and implementation of a variety of early intervention programs.

The empirical evidence regarding behavioral continuity is unimpressive, however, especially when one attempts to make predictions from infancy. Whether one considers the consistency of individual differences on measures of intelligence (Bayley, 1970), temperament (Thomas & Chess, 1977), or levels of attentiveness (Kagan, 1971), the behavior of infants appears to tell us little about their likely behavior in childhood. Long-term predictions from infancy have been especially poor in some of the more ambitious longitudinal studies involving several decades of data-gathering (e.g., the Berkeley Growth Study: Bayley, 1964; Schaefer & Bayley, 1963; the Menninger studies: Escalona, 1968; Escalona & Heider, 1959; Murphy & Moriarty, 1976; the Fels project: Kagan & Moss, 1962; Moss & Kagan, 1964). Furthermore, infants and young children who have been subjected to early deprivation or trauma frequently do not show the long-term effects predicted by theorists like Bowlby (1951) (see Kagan, Chapter 2, this volume; Rutter, 1979). Such findings lead many researchers to ask: Why is there not greater evidence for developmental continuity?

Responses to this question have varied. Some have suggested that continuity exists, but that infancy is a period of such rapid change that it becomes difficult to detect (Lewis & Starr, 1979; Rutter, 1970). Qualitative changes in the nature of intellectual functioning, cognition, and other processes with development may also make the detection of behavioral continuities elusive (Beckwith, 1979; Sameroff, 1975). And others have argued that the stability of individual differences cannot be argued on empirical bases but must remain a theoretical assumption (Lewis & Starr, 1979).

On the other hand, some developmentalists have suggested that it is naive for theorists to expect substantial continuity in behavior (Clarke & Clarke, 1977; Kagan, Kearsley, & Zelazo, 1978). The leading proponent of this perspective, Jerome Kagan, has criticized "tape recorder theories" of child development for portraying the effects of early experiences too simplistically (Kagan et al., 1978). Kagan argues that early influences may later be difficult to detect because of the effects of intervening experiences, the complexity of maturational processes, or discontinuities in central nervous system maturation (Kagan, Chapter 2, this volume). From these perspectives, therefore, lack of evidence for developmental continuity is readily understandable.

These arguments assume, of course, that the empirical evidence fairly demonstrates the absence of developmental stability or continuity. It could be argued, however, that the case remains unproven because researchers have tended to explore simple, straightforward relationships between early and later assessments of behavior, without seriously considering a number of potential mediating influences. When one considers these, substantial within-individual changes over time, which initially appeared neither meaningful nor interpretable, become more understandable, and the conditions which influence developmental continuity can be better appreciated. From this perspective, understanding the origins of behavioral instability is as informative as detecting continuities, because it enables us to predict the conditions in which continuity should reasonably be expected, and when it should not. Two kinds of mediating influences may be especially important: (a) changes in family circumstances, caregiving arrangements, or other aspects of the child's interpersonal environment and (b) patterns of stability or change in variables which are strongly related to the behaviors under examination.

Concerning the first, researchers have typically failed to view the child in an appropriate social context. Quite often this has meant studying developmental processes without regard to caregiving arrangements, and usually it has meant ignoring the influences of other family members, peers, or broader socioeconomic factors affecting the family as a whole. Moreover, few studies of developmental continuity have taken into account concurrent stability or change in the child's social circumstances. The social context, however, constitutes an important aspect of the child's experiences. Consistency over time in some aspects of behavioral development may depend on stability in the baby's social relationships, especially when the behaviors of interest are in the socioemotional realm. Thus, for example, Rutter (1979) noted that changing family circumstances influence whether trauma like parent–child separation have long-term effects. In general, developmental continuity is less likely to occur when social circumstances change markedly.

Concerning the second influence, researchers have also tended to examine individual variables in isolation, instead of considering meaningful interrelationships among variables which may influence continuity. When two (or more) variables are strongly related in contemporaneous assessments, change in one may be highly related to changes in the others. Researchers will find it easier to interpret evidence of continuity or discontinuity if they take into account patterns of change and stability on related dimensions. As a result, both stability and change become more meaningful in light of the organization of behavior.

In sum, it seems premature to conclude that continuity does not occur on the basis of existing research evidence. Clearly, more sophisticated approaches are necessary which take into account mediating influences in future studies. If behavioral continuities do exist, they are unlikely to occur in all developmental domains, independent of intervening circumstances. Consequently, the key task for developmentalists interested in this issue is to specify those conditions in which continuity may reasonably be expected—and when it may not—in light of these intervening circumstances and behavioral interrelationships. In other words, the question is not *whether* continuity exists, but *under what conditions* it occurs. Furthermore, we suggest that when researchers study continuity and change by means of short-term longitudinal studies, these and other kinds of mediating influences can be more reliably detected and applied to longer-term developmental studies.

THE ORGANIZATION OF THIS RESEARCH

In this chapter, we describe a research study which was designed with these considerations in mind. Our objective was to examine the relationships among a set of socioemotional variables assessed on two occasions during the second year (at 12½ and 19½ months of age), with special focus on their stability over time in light of their interrelationships at each age. The variables we studied were (a) the security of infant–mother attachment, (b) social responsiveness toward an adult stranger, (c) separation distress and recovery, (d) maternal ratings of infant temperament, and (e) an inventory of family circumstances and caregiving arrangements.

Why were these variables selected for analysis? Variables a through d were chosen because they reflect significant dimensions of socioemotional development during the second year, including measures of social responsiveness, emotionality, and the child's behavioral style. Because of this, we expected these variables to yield informative interrelationships at each age, and we were interested in how these associations would be related to patterns of stability and change in each variable over time. Family circumstances and changes therein were also studied in order to explore their association with consistency in the other variables.

Two broad questions guided our inquiry. First, *how consistent is the security of attachment in the second year, and what kinds of influences are associated with changes in the baby's attachment status in a middle-class sample?* In recent years, researchers have focused extensively on individual differences in the quality or security of mother–infant attachment (Ainsworth, 1973; Ainsworth, Blehar, Waters, & Wall, 1978; Bowlby, 1969). According to attachment theorists, differences in attachment security develop from the history of interactions shared by the infant and adult. Infants who are securely attached are thought to have developed out of these interactions a confident expectation that the adult will be available when needed. Thus they learn that they can depend on the adult to be responsive in future instances of distress or alarm. By contrast, infants who are insecurely attached are thought to have less confidence in the adult's availability and helpfulness and thus respond with more negative emotions than do securely attached infants. Differences in the security of attachment are thus thought to reflect differences in an infant's expectations of the adult's behavior which are most clearly revealed in the Strange Situation procedure (Ainsworth *et al.*, 1978) described below. Research on the security of attachment using the Strange Situation has provided limited empirical support for these formulations, and more work remains to be done (see Lamb, Thompson, Gardner, Charnov, & Estes, 1984).

The stability of individual differences in the security of attachment in the second year has also been studied by several investigators. On one hand, Waters (1978) observed 50 middle-class infants and mothers in the Strange Situation at 12 and 18 months and reported that 96% of these infants received the same classification each time. On the other hand, Vaughn, Egeland, Sroufe, and Waters (1979) performed an identical test–retest study with 100 socioeconomically disadvantaged families and found that only 62% of the sample obtained the same classification at each age. Furthermore, Vaughn and his associates indicated that Waters's families were highly stable, with consistent paternal employment and almost no residential changes. In view of the differences between Waters's and Vaughn's findings, we were interested in comparing their estimates of stability with one obtained in a different middle-class sample in which family circumstances might change more frequently than in Waters' sample.[1]

Vaughn and his colleagues also found a relationship between changes in the baby's attachment classification between 12 and 18 months and concurrent changes in the family's circumstances. Specifically, the mothers of infants who changed from securely to insecurely

---

[1] In this review of stability studies, we have not included two studies which are sometimes cited elsewhere. Connell's (1976) dissertation, in which he found that 81% of his middle-class sample obtained the same attachment classification at 12 and 18 months, did not use Ainsworth's classification system. Rather, Connell employed a set of weighted equations derived from Ainsworth's data, which he used to classify infants in his own sample. In calculating these equations, Connell eliminated from Ainsworth's sample those infants in two of the four secure-classification subgroups (i.e., subgroups B_1 and B_4) because he found them difficult to classify accurately. Main and Weston (1981) found 73% stability from 12 to 18 months in a small ($N = 15$) sample. They used an additional classification category ("unclassifiable") and stringent sample selection criteria which probably resulted in an unusual group of families. For these reasons, both stability estimates are difficult to compare validly with the estimates of Waters, Vaughn and colleagues, and the results of this study.

attached reported a greater number of stressful events between the two assessments than did the mothers of infants who were securely attached on both occasions. In our study, we were also interested in the influence of family events on attachment stability, although we were more interested in determining what *kinds* of family events were associated with changes in attachment status between 12½ and 19½ months. Would single, critical events, such as a major separation between mother and baby, be more influential than experiences with regular, recurrent effects (such as maternal employment) or other kinds of family changes? Answering such a question would help us determine the kinds of events that may affect the infant expectations possibly underlying differences in the security of attachment.

Our second question was: *What is the relationship between stability and change in attachment status and the temporal consistency of other socioemotional variables?* Because of the importance of the mother–infant relationship in early development, we expected that the temporal consistency of behaviors which were strongly related to the security of attachment would be greater when the baby's attachment status was stable over the same period than when it was not. Other variables which were weakly associated with the security of attachment should show patterns of stability or change which are independent of changes in the baby's attachment status. Thus we were interested in how attachment stability was related to the temporal consistency of the other variables in light of their contemporaneous associations. This would enable us to interpret changes as well as stabilities meaningfully in light of the interrelationships among socioemotional behaviors.

In sum, our goal was to identify the conditions in which continuity in socioemotional development could be observed during the second year. To this end, we sought (a) to examine stability and change in the security of mother–infant attachment in light of changing family conditions and (b) to understand how changes in attachment status might be associated with stability in other aspects of socioemotional development. Our purpose was not to demonstrate whether continuity in development does or does not exist over this period, but to explore the conditions which influence continuity in behavior over a relatively short span of time.

METHOD AND MEASURES

The sample consisted of 43 infants (21 boys, 22 girls) and their mothers. Fifteen children were firstborns. These infants were tested when they were 12½ months (±2 weeks) and 19½ months (±2 weeks) of age.[2]

Our subject recruitment procedures yielded a relatively heterogeneous middle-class sample. Occupations of the major breadwinners ranged from college professors and engineers to assembly workers at local automotive plants. Educational level ranged from graduate degrees to partial high-school experience. On Hollingshead's (1975) Four Factor Index of Social Status, 14 families were in Hollingshead's Class I (major professional and business), 12 were in Class II (minor professional and technical), 13 were in Class III (skilled craftsmen, clerical, sales workers), and 4 were in Class IV (machine operators and semiskilled workers).

Identical laboratory assessments were conducted at each age in a large carpeted playroom (5.75 m × 4.5 m) containing two chairs (one for mother and one for stranger), a table, decorative wall hangings, and an assortment of age-appropriate toys for the baby in the middle of the room. Each assessment began with the stranger sociability procedure. Following a short interruption, the Strange Situation procedure was then conducted, with a different research assistant serving as stranger. The assessment of stranger sociability always preceded the Strange Situation because the latter usually involves marked distress for the baby, whereas the sociability procedure is brief and pleasant. As a result, no infant began either assessment in a distressed state. All sessions were videotaped. Following the Strange Situation, mothers were given a questionnaire to complete at home and return by mail. The questionnaire included inquiries concerning demographic characteristics of the family, the nature of the family and caregiving circumstances, and the infant's perceived temperament.

Stranger sociability was assessed using a modified version of a procedure originally developed by Stevenson and Lamb (1979). The procedure involved social overtures of gradually increasing intrusiveness conducted by a female stranger in the mother's presence. The initial social bids occurred while the infant was on mother's lap, with the stranger offering the baby a toy and then initiating a give-and-take exchange. Later the baby was placed on the floor, and a similar series of social initiatives was conducted by the stranger. Finally, the stranger attempted to pick up the baby before leaving the room. The entire session typically lasted about five or six minutes. Further procedural details are provided by Lamb (1982) and Thompson and Lamb (1982a).

[2] There were two exceptions: One infant was initially tested three weeks before his first birthday; another was seen for the second time one week following her 20-month birthday. Because inspection of the data revealed that these infants did not differ from others in relevant behaviors, their data were included.

From videotaped records, the infant's initial response to each of the eight steps of the procedure was scored on a 1 (withdrawal, distress) to 5 (outgoing, friendly) point scale. In addition, a summary rating of the baby's sociability was scored on a 9-point scale. The sum of these ratings constituted the infant's score for stranger sociability.

The *Strange Situation* is a 21-minute procedure designed to appraise the security of the infant–mother attachment relationship. It involves seven 3-minute episodes, with each episode entailing changes in the social setting of the playroom (see Ainsworth *et al.*, 1978, for further details). By design, the Strange Situation creates a condition of gradually escalating stress for the baby (i.e., interaction with an unfamiliar adult, then separation from the mother in the company of the stranger, then reunion with the mother, and finally being left alone) so that researchers can examine changes in the organization of the infant's behavior toward the parent as the attachment behavioral system is likely to become activated (Ainsworth *et al.*, 1978; Bowlby, 1969).

In assessing the security of attachment, particular attention is devoted to the baby's behavior during the two reunion episodes. Most common are *securely attached* (Group B) infants, who characteristically greet their mothers positively either through distal or proximal interaction (subgroups B_1 and B_2) or through seeking and maintaining contact (subgroups B_3 and B_4). By contrast, insecurely attached infants show more negative responses upon reunion. *Avoidant* (Group A) infants avoid or ignore the mother during reunions, typically by means of gaze-avoidance, moving away from her, or ignoring her social initiatives. *Resistant* (Group C) infants combine contact-seeking with conspicuously angry or resistant behaviors such as pushing away, slapping their mothers, or persistently rejecting the toys offered during reunion. Two subgroups within each of these A and C groups reflect variations in these reunion responses.

The reason for Ainsworth's emphasis on reunion behaviors in assessing the security of attachment is that they are likely to reveal the baby's expectations for the mother under stress (Ainsworth *et al.*, 1978). In contrast, variations in separation distress are thought to be less informative concerning infant expectations because of the absence of the attachment figure. Thus they figure minimally in assessing the security of attachment. For this reason, although our measures of *separation distress and recovery* were derived from infants' Strange Situation behaviors, they were not highly confounded with the criteria used in determining the security of attachment. From videotaped records, these assessments were conducted through a microanalysis of behavior in the two separation episodes (Episodes 3 and 5) and the three recovery episodes (Episodes 4, 6, and 7) of the Strange Situation.[3] Each episode was divided into consecutive 15-second scoring intervals, in each of which summary ratings were made of the baby's (a) facial expression and (b) vocalizations.

The facial expression measure was a 6-point measure drawing primarily upon information in the mouth region and in the eyes–forehead region of the face. Scoring criteria were broadly defined because infants were freely mobile throughout the procedure, making a more discrete analysis of facial components (e.g., Hiatt, Campos, & Emde, 1979) impossible. The points on this measure were: unrateable, smiling, pleasant expression, neutral, sober, and cryface. Since infants were sometimes turned away from the camera, raters were instructed to make an inference about the baby's facial expression when the face could not be observed directly.

The vocalization measure was a 13-point measure for assessing variations in infant vocal activity, with an emphasis upon distress vocalizations. Variations in vocalizations were delineated in terms of sensory dimensions such as pitch, intensity, and rhythmicity. The points on this measure were: no vocalization, pleasure vocalization, neutral vocalization, mild distress, calling, distress gasps, fussing or whimpering, whining, protest, sobbing, screaming, panic cry, and hyperventilated cry. This measure was later collapsed into six levels denoting clear distinctions in distress intensity. (Copies of these measures are available from the first author on request.)

From the raw data two kinds of summary measures were calculated. The peak *distress intensity* during each separation episode was calculated as the highest rating assigned in any scoring interval during the episode. The baby's *recovery* from distress during each recovery episode was calculated as the proportion of the total number of scoring intervals before the baby's affective expressions returned to and remained at a nondistressed level. These summary measures were calculated independently for each of the pertinent Strange Situation epi-

---

[3] Although Episodes 3 and 6 present the baby with an identical social situation—interaction with an unfamiliar adult in the mother's absence—we consider them psychologically different to the baby because of the contexts within which they occur. Episode 3 is the mother's first departure from the room, and even though the stranger is supportive it is clearly a distressing separation to most infants. Episode 6 is also distressing, but, following a three-minute episode in which the baby has been alone (Episode 5), the stranger's return often has a more comforting effect upon the baby than it did earlier. For this reason, it seems more appropriate to regard Episode 6 as a recovery episode, even though mother has not yet returned to the room.

sodes, and for each expressive modality (i.e., facial and vocal) (see Thompson, 1983, for further details).

Ratings for the security of attachment, stranger sociability, and separation distress and recovery were performed by highly trained research assistants with whom a criterion level of interrater agreement during training was achieved. In addition, independent reratings of a subsample of the videotapes yielded exact agreement on discrete ratings (i.e., sociability assessment subscales, interval-by-interval affect ratings) exceeding 85% each time. Raters worked independently without knowledge of the infants' scores on the other measures.

Following the laboratory procedures at each age, mothers were asked to complete and return a questionnaire. Part of the questionnaire was designed to yield descriptive demographic data as well as information concerning family and caregiving circumstances and changes in these circumstances over the baby's lifetime. The latter included specific inquiries concerning changes in residence, paternal employment, major separations between mother and baby, maternal employment, the occurrence of regular nonmaternal care (e.g., day care, babysitters, paternal care, relatives, etc.), and similar variables. For purposes of analysis, we determined whether changes in any of these circumstances occurred either during the first 12 months of the baby's lifetime or during the subsequent 7 months (i.e., between our two observations).

The second part of the questionnaire consisted of an *infant temperament* measure—specifically, the Infant Behavior Questionnaire (IBQ) (Rothbart, 1981; Rothbart, Furby, Kelly, & Hamilton, 1977). The IBQ is a behavior-based measure in which the parent is asked to rate the baby's characteristic response to commonly occurring situations during the preceding week (or, in some instances, two weeks). Scoring the 92-item questionnaire entails computing mean values for each of the six temperament subscales: activity level, distress to limitations, fear, duration of orienting, smiling and laughter, and soothability.

RESULTS AND DISCUSSION

In general, the results confirmed our expectations regarding the importance of considering mediating influences in assessing socioemotional continuity. Both family circumstances and the interrelationships among variables contributed to a picture of consistency and change which could be meaningfully interpreted. We discuss these findings here in light of the two primary questions which guided our investigation.

Stability and Change in the Security of Attachment

The distribution of infants across attachment groups (A, B, C) was very similar at both assessments (see Table 1) and accorded with the normative findings reported by Ainsworth (Ainsworth et al., 1978) and others. However, only 23 out of 43 infants (53%) were classified in the same overall group (i.e., A, B, or C) at both ages (see Table 1). Temporal consistency in the classification subgroups (e.g., B_1, B_2, etc.) was only 26%.

The degree of stability we found (53%) is comparable to that observed for Vaughn et al.'s (1979) study of socioeconomically disadvantaged families described earlier (62%). Both studies indicate that the security of attachment, as assessed in the Strange Situation, reflects the *current* status of the mother–infant relationship, which may change over time. In contrast to the view that the first year of life constitutes a critical period during which the security or insecurity of mother–infant attachment is determined, therefore, these data indicate that the security of the mother–infant relationship may change substantially during the first two years of life.

But can we predict when these changes in attachment status will occur? Vaughn and his colleagues (1979) showed that changes from secure to insecure attachment relationships occurred when there were frequent family changes and stresses. To determine whether a similar relationship was evident in our middle-class sample we turned to the information obtained from the maternal questionnaires. Two (albeit related) variables were strongly associated with changes in attachment status from one group (A, B, or C) to another: maternal employment and

Table 1. Temporal Stability of Attachment Classifications

12½-month classification		A		B				C	
		A_1	A_2	B_1	B_2	B_3	B_4	C_1	C_2
A	A_1	1	0	0	0	0	1	1	0
	A_2	0	0	2	1	1	0	0	0
B	B_1	0	0	1	0	0	0	0	0
	B_2	1	0	2	3	3	2	2	0
	B_3	1	3	0	1	3	2	1	0
	B_4	0	0	0	0	2	1	2	0
C	C_1	0	0	2	1	0	0	1	0
	C_2	0	0	1	0	0	0	0	1

(column group heading: 19½-month classification)

the beginning of regular nonmaternal care for the baby (see Table 2). Between the two assessments, five of the mothers in the sample returned to work, and in four of these cases the security of attachment changed. More than half (12 of 20) of those whose attachment status changed had mothers who had returned to work by the time of the 19½-month assessment, while maternal employment was characteristic of less than 15% (3 of 23) of those whose classification remained the same ($\chi^2 = 10.28$, $df = 1$, $p < .005$). Interestingly, of those 12 infants whose attachment status changed, five shifted from secure to insecure, six moved from insecure to secure, and one baby shifted from one insecure group to another.

Similarly, all four infants who entered into a regular (i.e., ≥15 hours weekly) nonmaternal care arrangement between the two assessments (such as regular caregiving in a day-care center, or by a babysitter or relative or the father) changed in attachment status. By the 19½-month assessment, significantly more babies whose attachments changed experienced regular nonmaternal care (11 of 20) than did those whose attachment status was consistent (2 of 23) ($\chi^2 = 11.09$, $df = 1$, $p < .001$). Again, bidirectional changes were characteristic of these 11 babies: 6 moved from secure (Group B) to insecure (Groups A or C), 4 shifted in the reverse direction, and one baby moved from one insecure group to the other. By contrast, other events—such as a major (≥24 hours) separation from mother or moving to a new home—were not associated with changes in the security of attachment (see Thompson, Lamb, & Estes, 1982).

The family circumstances most strongly associated with changes in attachment status, therefore, were those which were most likely to affect directly the ongoing quality or quantity of mother–infant interaction. Our findings, like those of Vaughn and his associates, indicate that changes in the security of mother–infant attachment are consistently related to family events which are likely to affect the mother–infant relationship. Because it is a dynamic relationship, the attachment bond is appropriately responsive to changes in the life experiences of infant and adult.

One important difference between these findings and those of Vaughn and his colleagues concerns the direction of change in attachment status. In Vaughn's disadvantaged sample, family changes and stresses were associated primarily with changes from security to insecurity. By contrast, family changes in our middle-class sample were associated with *bidirectional changes* in the baby's attachment status. In other words, maternal employment and regular nonmaternal care appeared to produce shifts from insecurity to security, as well as the reverse. These events thus appeared to cause a renegotiation of the mother–infant

Table 2. Maternal Employment and Regular Nonmaternal Care: Temporal Relationships to Stability of Attachment Classifications

	Maternal employment		Regular nonmaternal care	
	Before 12½ months	Between 12½ and 19½ months	Before 12½ months	Between 12½ and 19½ months
Initially insecure—change (N = 10)	5 (50%)	2 (20%)	3 (30%)	2 (20%)
Initially secure—change (N = 10)	3 (30%)	2 (20%)	4 (40%)	2 (20%)
Initially insecure—stable (N = 3)	0	0	2 (67%)	0
Initially secure—stable (N = 20)	2 (10%)	1 (5%)	0	0

relationship without the bias toward insecurity which was evident in Vaughn's study.

One reason for this difference may be that the kinds of life experiences encountered by the families in each sample were very different. In Vaughn's disadvantaged sample, families experienced a variety of legal, financial, ecological, and health-related difficulties which were probably more severe and enduring than the stresses encountered by the middle-class families in our sample. As a result, it is likely not only that these families had fewer resources for coping adaptively with these stresses, but also that the events were amenable to fewer constructive solutions (Thompson & Lamb, 1983a). By contrast, events such as mother's return to work in a middle-class family may initiate a difficult transition for mother and baby but still permit a variety of constructive responses. For example, Pedersen, Cain, Zaslow, and Anderson (1982) report that some working mothers devote special time to enjoyable play with their babies following their return from work each evening, and this may have a positive effect on the mother–infant relationship.

As mentioned earlier, events occurring during the *first* year of the infants' lives (change in maternal work status, entry into regular non-maternal care) were associated with changes in the security of attachment between 12½ and 19½ months. This suggests that short-term changes in the quality of mother–infant interaction during the first year may have influenced the initial Strange Situation assessment, but not the later one. For example, infants who are placed in day care for the first time may experience anxiety which affects the security of infant–mother attachment. These initial reactions are likely to be short-lived (Blanchard & Main, 1979), however, and after a period of time babies and mothers may resume earlier, more familiar patterns of interaction. This would mean that changing family circumstances can have either short-term or long-term effects on the security of attachment. It would thus be wise, in future investigations, to obtain information concerning family events occurring at various points in the baby's lifetime in order to explore these influences.

In sum, the security of attachment changed between 12½ and 19½ months in nearly half the dyads in our middle-class sample; these were associated with family events which were likely to affect the ongoing quality or quantity of mother–infant interaction. Thus the stability of the caregiving environment affected the *robustness* of the mother–infant relationship, whether that relationship was secure or insecure. Clearly, to understand patterns of stability and change in the security of attachment, one must view mother and infant in the context of the stability or consistency of their social environment.

Stability of Attachment and Socioemotional Continuity during the Second Year

In this section we consider the relationships between security of attachment and other socioemotional variables in contemporaneous assessments, particularly where these relationships help to explain patterns of stability and change in socioemotional development.

Attachment and Stranger Sociability. There were consistent and meaningful relationships between the security of attachment and stranger sociability scores at 12½ and 19½ months. At each age, the infants who obtained the highest sociability scores were those who were securely attached and in subgroups B_1 or B_2. These infants were most likely to use distal behaviors when greeting their mothers upon reunion, and our results indicated that this interactive style facilitated positive distal interactions with unfamiliar adults as well. As a result B_1 and B_2 infants could interact comfortably with strangers without the wariness which often accompanies close contact with a stranger. By contrast, infants in secure classification subgroups B_3 and B_4 and babies in both the Group A and Group C classifications received lower sociability scores at each age, perhaps because these infants had developed interactive styles with the mother which either required close contact or were more negative in emotional tone. Generalized to interactions with a stranger , these styles would likely result in less sociable encounters (see Thompson & Lamb, 1983b).

Mean stranger sociability scores were very similar at each age (12½ months: 34.53; 19½ months: 34.95), and were significantly correlated over the two assessments ($r = .40$, $p < .01$). However, the temporal stability of each individual's scores varied depending on whether attachment status was stable or not. When attachment classification (A, B, or C) was stable ($N = 23$), sociability scores correlated very highly over time ($r = .74$, $p < .0001$). When the security of attachment changed ($N = 20$), however, sociability scores were not significantly correlated ($r = -.18$). Measures of the change in stranger sociability over the two assessments revealed a similar pattern of results. When attachment classifications were temporally consistent, the average absolute change in sociability was 4.22; when attachment status changed, the average absolute change in sociability scores was 7.45 ($t = 2.30$, df = 41, $p < .03$).

In other words, when the security of attachment remained consistent, so too did the infant's sociability scores. When attachment status changed, stranger sociability responses did also. Although it is possible that both sociability and attachment were independently influenced by family changes, our analyses indicated that family conditions were not strongly or consistently related to changes in sociability scores over time (see Thompson & Lamb, 1982a). More likely, as patterns of mother–infant interaction changed, this change was generalized to the baby's way of interacting with strangers. Thus individual differences in the security of mother–infant attachment and the quality of stranger responsiveness appear to be related developmental phenomena, as reflected in their relationships at 12½ and 19½ months and in the intercoordinated pattern of change in each variable over time. For a thorough understanding of stability and change in stranger responsiveness, in other words, one must take into account concurrent stability in the infant–mother relationship.

Attachment and Separation Reactions. Like the sociability measure, the measures of separation distress and recovery revealed a great deal of heterogeneity within the secure classification, so we again partitioned the B group into two, with the B_1 and B_2 subgroups distinguished from the B_3 and B_4 subgroups for purposes of analysis. When we compared the four derived groups (i.e., A, B_1B_2, B_3B_4, C) on the basis of their emotional reactions in the Strange Situation, we found that at each age infants who were classified in groups B_3B_4 or C displayed significantly more intense distress during the separation episodes and required a longer period to soothe during the recovery episode than did infants in groups B_1B_2 and A. By contrast, there were fewer overall differences between securely attached (Group B) and insecurely attached (Groups A and C) infants. These findings were consistent for both vocalization and facial expression measures (see Thompson, 1983).

These analyses confirmed our expectations concerning differences in emotional responding between avoidant and resistant infants based on the earlier work of Ainsworth et al. (1978). However, they also revealed significant variation within the secure classification. This variability indicates that secure reunions with the mother may be preceded by separation reactions of varying intensity. Some securely attached infants become highly distressed whereas others are much more placid during separations. Moreover, the fact that the patterns of emotional responsiveness which emerged from these analyses cut across the A, B, and C classification groups indicates that differences between securely and insecurely attached infants are not simply determined by the intensity of separation distress. Some babies who were highly upset during separation exhibited resistant behaviors upon reunion (i.e., the C infants), whereas others greeted their mothers positively and sought close contact with them (i.e., the B_3B_4 infants). Similarly, while some of the infants who showed little or no separation distress were avoidant (i.e., the A infants), some greeted their mothers with smiles, vocalizations, or approaches (i.e., the B_1B_2 infants). In sum, although there was a relationship between separation reactions and the baby's attachment classification, the important variations in separation distress cut across securely and insecurely attached groups. Rather, the intensity of separation distress appeared to determine the extent to which infants actively sought contact with mother or were satisfied with distal interactive modes (see Thompson, 1983).

For the sample as a whole, there were few significant correlations between the 12½- and 19½-month measures of separation distress and recovery (mean $r = .20$). However, Table 3 indicates that for those 23 infants whose attachment classifications (A, B, or C) were stable at both ages, cross-age correlations on the separation measures were generally higher (mean $r = .44$, $p < .05$), whereas only one significant positive correlation out of ten emerged in cross-age analyses for those infants who changed from one attachment group to another (mean $r = .07$). Change scores for measures of separation distress and recovery revealed a similar pattern, with greater change on the emotion measures evident when security of attachment changed. Thus, as with the sociability scores, the temporal consistency of measures of separation distress and recovery was significantly affected by whether the infant obtained the same attachment classification on both assessments. To understand patterns of stability and change in separation reactions, one must consider also current stability in the security of attachment.

Attachment and Temperament. In contrast to the relationships between security of attachment and stranger sociability or separation

Table 3. Consistency Across Age of Measures of Separation Reactions

	Overall	By attachment stability	
		Stable	Change
Vocalizations			
Episode 3 Distress	.21	.52[a]	−.13
Episode 5 Distress	.09	.46[a]	−.26
Episode 4 Recovery	.15	.42[a]	−.15
Episode 6 Recovery	.31[a]	.51[a]	.08
Episode 7 Recovery	−.08	.06	−.19
Facial expressions			
Episode 3 Distress	.17	.43[a]	−.14
Episode 5 Distress	.46[b]	.65[c]	.48[a]
Episode 4 Recovery	.06	.46[a]	−.41
Episode 6 Recovery	.51[c]	.79[c]	.10
Episode 7 Recovery	.09	.09	−.09

[a] $p < .05$.
[b] $p < .005$.
[c] $p < .001$.

reactions, there were very few significant relationships between the security of attachment and the six dimensions of temperament yielded by the IBQ; none were consistent across the two assessments. This was true whether overall attachment classifications (i.e., A, B, or C groups) or the derived subgroup clusters (i.e., A, B_1B_2, B_3B_4, C) were considered (Thompson & Lamb, 1982b).

In view of this, it was perhaps unsurprising to find that the consistency of temperament characteristics from 12½ to 19½ months was unaffected by stability in the security of attachment over the same period. Temperament dimensions were highly autocorrelated over time regardless of consistency or change in attachment status (see Table 4). In fact, mean correlation coefficients were the same for the sample as a whole (mean $r = .61$), for those infants who obtained the same classification both times (mean $r = .63$), and for those whose classification changed (mean $r = .61$).

The fact that maternal-report measures of temperament were weakly associated with the security of attachment is consistent with other findings which suggest that temperament has little direct association with the security of attachment. First, infants may have relationships of different quality with their mothers and fathers (Grossman, Grossman, Huber, & Wartner, 1981; Lamb, 1978; Lamb, Hwang, Frodi, & Frodi, 1982; Main & Weston, 1981). This indicates that the security of attachment is not an individual characteristic (like temperament) but is instead relationship-specific. Second, Waters, Vaughn, and Egeland (1980) found that scores on the Neonatal Behavioral Assessment Scale (Brazelton, 1973) at 7 and 10 days of age were weakly and inconsistently associated with the security of attachment assessed at one year of age. Thus attachment security does not appear attributable to early constitu-

Table 4. Consistency Across Age of IBQ Temperament Dimensions

| | Overall | By attachment stability | |
		Stable	Change
Activity level	.53[c]	.59[b]	.50[a]
Distress to limitations	.61[d]	.55[b]	.69[c]
Fear	.65[d]	.73[d]	.60[b]
Duration of orienting	.67[d]	.63[b]	.67[b]
Smiling and laughter	.76[d]	.75[d]	.81[d]
Soothability	.46[b]	.54[b]	.38

[a] $p < .05$.
[b] $p < .01$.
[c] $p < .001$.
[d] $p < .0001$.

tional differences. Third, the fact that the security of mother–infant attachment quite commonly changes during the second year also argues against viewing security of attachment as an epiphenomenon of temperament. Of course, genetically based characteristics like inhibition or activity level probably affect mother–infant interaction and thus the development of attachment relationships, but this influence is one of a network of factors influencing the mother–infant relationship (including maternal characteristics, transient and enduring family conditions, paternal influences, etc.) and is thus unlikely to evince a direct relationship to attachment status at one year of age.

Summary.   Taken together, these findings concerning the ratings of stranger sociability, separation distress and recovery, and temperament yield a coherent picture of socioemotional continuity during the second year. In each case, patterns of consistency and change in these variables could be meaningfully interpreted in light of their relationships to the security of attachment at each age. When variables were strongly related to contemporaneous assessments of attachment security, their temporal consistency was strongest when the security of attachment was stable over time and weakest when attachment status changed. Such was the case for measures of sociability and separation reactions. On the other hand, maternal ratings of infant temperament were weakly related to attachment security at each age, and thus the temporal consistency of temperament ratings was unaffected by stability or change in the security of attachment. Perceived temperament was highly stable over time in any case.

In other words, the conditions in which continuity was discerned differed across variables. For variables like separation reactions, significant temporal stability was found only when the security of attachment was stable. For variables like temperament, by contrast, the conditions fostering temporal consistency were somewhat different: stability in the security of attachment was not important. In each case, continuity in socioemotional development was evident; what varied were the conditions in which continuity could be discerned.

SUMMARY AND CONCLUSION

To summarize, we propose that it is inappropriate to render a verdict on the stability of individual differences on the basis of existing research, since developmentalists have tended to examine simple, straightforward relationships between early and later assessments of behavior. Rather, a variety of important mediating influences must be considered which may make more interpretable the observed discontinuities as well as continuities in behavior. From this perspective, understanding the origins of instability may be as useful as detecting consistency, since both help us to understand the conditions in which behavioral continuity may reasonably be expected and when it should not be anticipated.

To illustrate, we presented findings from a short-term longitudinal study of socioemotional development during the second year. Two kinds of mediating influences were found to be significant factors in our search for stable individual differences in behavior across the seven-month period. The first was consistency or change in the child's interpersonal environment, especially in family and caregiving circumstances. Although individual differences in the security of attachment changed in nearly half our sample, these changes were associated with changes in caregiving conditions which were likely to affect mother–infant interaction. Thus the stability of attachment relationships was tied to the stability of the caregiving environment. Second, we found that the interrelationships among socioemotional behaviors also influenced our search for behavioral continuity. Specifically, when behaviors were strongly linked to the security of attachment in contemporaneous assessments, they were most stable over time when attachment status was also stable and were least consistent temporally when attachment status changed. On the other hand, behaviors which were essentially independent of the security of attachment in contemporaneous assessments showed patterns of stability and change which were not associated with consistency in the baby's attachment status. In short, contemporaneous associations among socioemotional behaviors enabled us to detect interrelated patterns of stability and change in behaviors over time.

If we had studied developmental continuity in the typical fashion, we would have concluded that temperament alone was stable. The remaining variables would have yielded a confusing picture of substantial discontinuity over the brief assessment period. However, by taking into consideration two mediating influences—stability in the child's interpersonal environment and the interrelationships among socioemotional behaviors at each age—we were able to identify the conditions in which significant stability in each variable occurred, as well as the conditions in which change and discontinuity occurred.

The importance of these mediating influences is becoming increasingly appreciated in developmental theory and research. For example, the transactional perspective (Meisels & Anastasiow, 1982; Sameroff, 1975; Sameroff & Chandler, 1975) views development in terms of the ongoing interaction between the child and various aspects of the interpersonal surround. Originally applied to predicting long-term outcome in cases of developmental disability or trauma, this perspective is equally important to an understanding of stability in normative developmental processes. With respect to behavioral interrelationships, McCall (1979; McCall, Eishorn, & Hogarty, 1977) has drawn upon data from the Berkeley Growth Study to demonstrate how an understanding of the organization of behaviors at each assessment can yield an appreciation of both short-term consistencies in behavior as well as qualitative changes in behavioral interrelationships over time. Both factors, of course, are essential to interpreting stability data. Future studies of the consistency of individual differences in behavior must take both kinds of mediating influences into account.

Exploring these influences in the context of short-term longitudinal studies enables researchers to avoid many of the problems which typically hamper longer-term studies of continuity. In our investigation, for example, the morphology of the behaviors of interest did not change enough over the seven-month period to create interpretational difficulties. To use Kagan's (1971) terms, there was "complete continuity" in our response measures, in which both the behavior and the processes underlying it remained stable over the assessment period. In contrast, longer-term studies often encounter the phenomenon of "homotypic continuity," in which the response remains consistent but the underlying processes governing it have changed (e.g., the meaning of crying in a 12-month-old compared to a 3-year-old), or "heterotypic continuity," in which the process has remained stable but the behavioral manifestations have changed (e.g., the behaviors reflecting child–parent attachment in a 12-month-old compared to a 3-year-old) (see also Bell, Weller, & Waldrop, 1971). In short, we could use the same measures at each age and infer similar underlying processes in the infant.

In addition, the interrelationships among the socioemotional variables were generally similar at both ages, and this eased the task of identifying meaningful associations between stability and change in related variables. If, for example, stranger sociability and the security of attachment had been associated in different ways at 12½ and 19½ months, if would have been more difficult to determine how the two variables were related over time. Thus another advantage of using short-term longitudinal studies in the initial exploration of the conditions affecting developmental continuity is that the researcher is dealing with behavioral interrelationships which are more easily detected over time.

Even so, the patterns of consistency and change in socioemotional development that we identified in this study were not simple, and complexity is likely to be the rule for long-term developmental investigations as well. Mediating influences like family changes and behavioral interrelationships are only two of the factors which are likely to influence the stability of individual differences in development. Nevertheless, our general conclusion is likely to be confirmed in future studies: Developmental consistency probably does exist, but in varying and complex circumstances. Identification of these conditions is an important task for students of child development.

Acknowledgments

We are very grateful for the assistance provided by Lisa Colvin, Susan Dickstein, Melinda Feller, Leslie Fried, Michael Iskowitz, Margaret Madden, Susan McFarlin, Susan Piconke, Jane Stein, Jamie Steinberg, Susan Strzelecki, and Michele Tukel. David Estes and Catherine Malkin provided special help with testing and scoring. We are also appreciative of technical assistance provided by Lee Davis. Finally, we wish to express our thanks to the families who participated in this study and to acknowledge the permission of the Society for Research in Child Development to reprint some previously published material.

REFERENCES

Ainsworth, M. D. S. The development of infant–mother attachment. In B. Caldwell and H. Ricciuti (Eds.), *Review of child development research* (vol. 3). Chicago: University of Chicago Press, 1973.

Ainsworth, M. D. S., Blehar, M. C., Waters, E., & Wall, S. *Patterns of attachment.* Hillsdale, N. J.: Lawrence Erlbaum, 1978.

Bayley, N. Consistency of maternal and child behaviors in the Berkeley Growth Study. *Vita Humana*, 1964, *7*, 73–95.

Bayley, N. Development of mental abilities. In P. H. Mussen (Ed.), *Carmichael's manual of child psychology* (Vol. 1). New York: Wiley, 1970.

Beckwith, L. Prediction of emotional and social behavior. In J. Osofsky (Ed.), *Handbook of infant development.* New York: Wiley, 1979.

Bell, R. Q., Weller, G. M., & Waldrop, M. F. Newborn and preschooler: Organization of behavior and relations between periods. *Monographs of the Society for Research in Child Development*, 1971, *36*, 1–2 (Serial No. 142).

Blanchard, M., & Main, M. Avoidance of the attachment figure and social-emotional adjustment in day-care infants. *Developmental Psychology*, 1979, *15*, 445–446.

Bowlby, J. *Maternal care and mental health.* Geneva: World Health Organization, 1951.

Bowlby, J. *Attachment and loss* (vol. 1): *Attachment.* New York: Basic Books, 1969.

Brazelton, T. B. *Neonatal behavioral assessment scale.* Clinics in Developmental Medicine, No. 50. Philadelphia: Lippincott, 1973.

Clarke, A. M., & Clarke, A. D. B. *Early experience: Myth and evidence.* New York: Free Press, 1977.

Connell, D. B. *Individual differences in attachment: An investigation into stability, implications, and relationships to structure of early language development.* Unpublished doctoral dissertation, Syracuse University, 1976.

Escalona, S. K. *The roots of individuality.* Chicago: Aldine, 1968.

Escalona, S. K., & Heider, G. M. *Prediction and outcome.* New York: Basic Books, 1959.

Grossman, K. E., Grossman, K., Huber, F., & Wartner, U. German children's behavior towards their mothers at 12 months and their fathers at 18 months in Ainsworth's Strange Situation. *International Journal of Behavioral Development,* 1981, *4,* 157–181.

Hiatt, S. W., Campos, J. J., & Emde, R. N. Facial patterning and infant emotional expression: Happiness, surprise and fear. *Child Development,* 1979, *50,* 1020–1035.

Hollingshead, A. B. *Four factor index of social status.* Unpublished manuscript, Yale University, 1975.

Kagan, J. *Change and continuity in infancy.* New York: Wiley 1971.

Kagan, J., & Moss, H. A. *Birth to maturity: A study of psychological development.* New York: Wiley, 1962.

Kagan, J., Kearsley, R. B., & Zelazo, P. R. *Infancy: Its place in human development.* Cambridge: Harvard University Press, 1978.

Lamb, M. E. Qualitative aspects of mother– and father–infant attachments. *Infant Behavior and Development,* 1978, *1,* 265–275.

Lamb, M. E. The origins of individual differences in infant sociability and their implications for cognitive development. In H. W. Reese and L. P. Lipsitt (Eds.), *Advances in child development and behavior* (Vol. 16). New York: Academic Press, 1982.

Lamb, M. E., Hwang, C-P, Frodi, A. M., and Frodi, M. Security of mother– and father–infant attachment and its relation to sociability with strangers in traditional and nontraditional Swedish families. *Infant Behavior and Development,* 1982, *5,* 355–367.

Lamb, M. E., Thompson, R. A., Gardner, W., Charnov, E. L., and Estes, D. Security of attachment: Its origins, assessment, interpretation, and consequences. *Brain and Behavioral Sciences,* 1984, in press.

Lewis, M., & Starr, M. D. Developmental continuity. In J. D. Osofsky (Ed.), *Handbook of infant development.* New York: Wiley, 1979.

Main, M., & Weston, D. R. The quality of the toddler's relationship to mother and to father: Related to conflict behavior and the readiness to establish new relationships. *Child Development,* 1981, *52,* 932–940.

McCall, R. B. The development of intellectual functioning in infancy and the prediction of later I.Q. In J. D. Osofsky (Ed.), *Handbook of infant development.* New York: Wiley, 1979.

McCall, R. B., Eichorn, D. H., & Hogarty, P. S. Transitions in early mental development. *Monographs of the Society for Research in Child Development,* 1977, *42,* 3 (Serial No. 171).

Meisels, S. J., & Anastasiow, N. J. The risks of prediction: Relationships between etiology, handicapping conditions, and developmental outcomes. In S. G. Moore & C. R. Cooper (Eds.), *The young child: Reviews of research* (Vol. 3). Washington, D.C.: NAEYC, 1982.

Moss, H. A., & Kagan, J. Report on personality consistency and change from the Fels longitudinal study. *Vita Humana,* 1964, *7,* 127–138.

Murphy, L. B., & Moriarty, A. E. *Vulnerability, coping and growth: From infancy to adolescence.* New Haven: Yale University Press, 1976.

Pedersen, F. A., Cain, R. L., Zaslow, M. J., & Anderson, B. J. Variation in infant experience associated with alternative family roles. In L. Laosa & I. Sigel (Eds.), *Families—Research and practice* (Vol. 1): *Families as learning environments for children.* New York: Plenum Press, 1982.

Rothbart, M. K. Measurement of temperament in infancy. *Child Development,* 1981, *52,* 569–578.

Rothbart, M. K., Furby, L., Kelly, S. R., & Hamilton, J. S. *Development of a caretaker report temperament scale.* Paper presented to the biennial meeting of the Society for Research in Child Development, New Orleans, March, 1977.

Rutter, M. Psychological development—Predictions from infancy. *Journal of Child Psychology and Psychiatry,* 1970, *11,* 49–62.

Rutter, M. Maternal deprivation, 1972–1978: New findings, new concepts, new approaches. *Child Development*, 1979, *50*, 283–305.

Sameroff, A. J. Early influences on development: Fact or fancy? *Merrill-Palmer Quarterly*, 1975, *21*, 267–294.

Sameroff, A. J., & Chandler, M. Reproductive risk and the continuum of caretaking casualty. In F. D. Horowitz (Ed.), *Review of child development research* (Vol. 4). Chicago: University of Chicago Press, 1975.

Schaefer, E. S., & Bayley, N. Maternal behavior, child behavior, and their intercorrelations from infancy through adolescence. *Monographs of the Society for Research in Child Development*, 1963, *28* (Serial No. 87).

Stevenson, M. B., & Lamb, M. E. Effects of infant sociability and the caretaking environment on infant cognitive performance. *Child Development*, 1979, *50*, 340–369.

Thomas, A., & Chess, S. *Temperament and development*. New York: Brunner/Mazel, 1977.

Thompson, R. A. *Assessing qualitative dimensions of emotional responsiveness in infancy*. Paper presented to the biennial meeting of the Society for Research in Child Development, Detroit, April, 1983.

Thompson, R. A., & Lamb, M. E. Stranger sociability and its relationships to temperament and social experience during the second year. *Infant Behavior and Development*, 1982, *5*, 277–287. (a)

Thompson, R. A., & Lamb, M. E. *Temperamental influences on stranger sociability and the security of attachment*. Paper presented to the annual meeting of the American Psychological Association, Washington, D.C., August, 1982. (b)

Thompson, R. A., & Lamb, M. E. Infants, mothers, families, and strangers. In M. Lewis & L. Rosenblum (Eds.), *Beyond the dyad*. New York: Plenum Press, 1983. (a)

Thompson, R. A., & Lamb, M. E. Security of attachment and stranger sociability in infancy. *Developmental Psychology*, 1983, *19*, 184–191. (b)

Thompson, R. A., Lamb, M. E., & Estes, D. Stability of infant–mother attachment and its relationship to changing life circumstances in an unselected middle-class sample. *Child Development*, 1982, *53*, 144–148.

Vaughn, B., Egeland, B., Sroufe, L. A., & Waters, E. Individual differences in infant–mother attachment at twelve and eighteen months: Stability and change in families under stress. *Child Development*, 1979, *50*, 971–975.

Waters, E. The reliability and stability of individual differences in infant–mother attachment. *Child Development*, 1978, *49*, 483–494.

Waters, E., Vaughn, B., & Egeland, B. Individual differences in infant–mother attachment relationships at age one: Antecedents in neonatal behavior in an urban, economically disadvantaged sample. *Child Development*, 1980, *51*, 208–216.

CHAPTER 14

Continuities and Change in Early Emotional Life

MATERNAL PERCEPTIONS OF SURPRISE, FEAR, AND ANGER

Mary D. Klinnert, James F. Sorce, Robert N. Emde, Craig Stenberg, and Theodore Gaensbauer

INTRODUCTION

During the past few decades, developmentalists have focused increasing attention on the substantial transformations which occur in early infancy (Emde, Gaensbauer, & Harmon, 1976; Kagan, Kearsley, & Zalazo, 1979; McCall, Eichorn, & Hogarty, 1977; Spitz, 1959). This orientation has prompted research concerning the *emergence* of various emotions in the developing infant. Bridges (1932) provided the foundation for this tradition by postulating the differentiation of more complex emotions out of simpler ones. For example, she suggested that by 4 months of age anger became differentiated out of a global distress which existed early in infancy. Recent versions of the differentiation hypothesis specify the emergence of discrete emotions as a function of the attainment of cognitive prerequisites (e.g., Sroufe, 1979). Sroufe, for example, views early rage as a precursor to anger, which he believes emerges in the latter half of the first year; only when the infant is capable

Mary D. Klinnert, Robert N. Emde, and Theodore Gaensbauer • Department of Psychiatry, University of Colorado School of Medicine, Denver, Colorado 80262.   **James F. Sorce** • Bell Laboratories, Holmdel, New Jersey.   **Craig Stenberg** • Department of Psychology, University of Denver, Denver, Colorado 80208. Robert N. Emde is supported by Public Health Service Grant MH 22803 and National Institute of Mental Health Grant MH 35808.

of perceiving the *cause* of an interruption to his plan can the previously diffuse rage reaction become the focused emotion of anger.

Other notions of transformation in emotional development encompass still more areas of infant functioning. Emde, Gaensbauer, and Harmon (1976) pinpointed two periods within the first year when qualitative changes take place in aspects of infants' functioning, including emotional responsivity. They marshalled evidence that affective behaviors such as the social smile at 2 to 3 months and fear reactions at 7 to 9 months are indexes of neurophysiological reorganizations, with concomitant surges in cognitive functioning. In yet another transformational view, Izard (1978) proposed that the major subsystems in the human personality become organized at progressively more complex levels, with the domination of sensory-affective processes giving way to affective-perceptual ones, and these in turn giving way in development to the dominance of affective-cognitive processes.

Despite differences among these conceptualizations, in each case the transformations are structural in nature; that is, they involve changes in the manner in which the various systems are organized and in how they relate to one another. With each transformation a different mode of functioning emerges and, perhaps as a consequence, so also does a new emotion.

The discontinuity suggested by these theoretical formulations contrasts sharply with recent findings coming from two directions: maternal reports and research on neonatal facial expressive capacities. Our own interviews with mothers of young infants consistently revealed that they describe a wider range of infant emotions than we had presumed possible. Commonly occurring statements such as "Ever since he was born he's been quick to get mad" suggested not only an early onset of certain emotions, but also a sense on the part of the caretakers that there was continuity in their infants' emotional lives. Further, empirical evidence is mounting that patterned facial expressions for a variety of emotions may be present in infants from the early months onward. For example, Oster and Ekman (1978) have reported that all but one of the component facial movements for the various emotional expressions are present in the newborn, although not necessarily in the configurations typical of adult expressions. Although links to environmental circumstances have not been definitively established, other investigators have reported patterned expressions in 1-month-olds of interest, joy, distress, and disgust (Izard, 1978), and by 4 months anger, sadness, and possibly fear have been observed (Gaensbauer & Hiatt, 1982; Izard, Hembree, Dougherty, & Spizzirri, 1983; Stenberg, 1982).

How does the continuity implied by these reports coexist with the discontinuity suggested by the emergence model of emotional development? In this chapter we will attempt to resolve this apparent contradiction by suggesting that the most important source of continuity may result from the infant's emerging abilities to send discrete (and increasingly elaborated) emotion signals and the mother's capacities to receive them, even in primitive, perhaps ambiguous forms.

DATA FROM OUR LABORATORY

We began these initial investigations of the ontogeny of emotions by looking at mothers' perceptions of their own infants' affective reactions. Our logic was as follows. Mothers spend more time with their infants than investigators possibly can and have more opportunities for seeing rare but salient expressions. We therefore expected mothers to perceive the various emotions in their infants slightly earlier than laboratory researchers because of the probable low base rate of discrete, full-blown infant reactions and because of the mothers' advantageous position as observers. However, we did not expect mothers to report observing as many emotions as early as they did in our first study.

Study 1 (Johnson, Emde, Stenberg, Pannabecker, & Davis, 1982) was a cross-sectional study which involved 611 mothers of infants ranging in age from birth to 18 months (approximately 33 mothers per each infant month). Mothers were asked whether they perceived each of the following emotions in their infants: interest, enjoyment, surprise, fear, anger, shyness, contempt, guilt, disgust, sadness, and distress. Table 1 presents the proportions of mothers who reported the presence of six of the emotions in their infants from 3 to 18 months of age. (Data were collapsed into 3-month time periods since there were no differences between the individual months of each age quarter.)

One of the most striking findings was that the mothers reported perceiving several discrete emotions in their infants by the time they reached 3 months of age. It was not surprising that almost all of the mothers described interest, enjoyment, and distress in their infants at this early age. Most investigators (cf. Izard, 1978; Sroufe, 1979) have described the onset of distress shortly after birth, interest at about 1 month, and delight or pleasure around 2 or 3 months. However, a majority of the mothers in this sample also reported seeing surprise (69%), anger (86%), and fear (69%) in their infants by 3 months of age. The reports of the latter three emotions at this age contrasted with previous findings from laboratory investigations. For example, Bridges (1932) had reported the onset of anger at 4 months and fear at 6 months, and others (e.g., Sroufe, 1979) suggested that the cognitive require-

Table 1. Proportion of Mothers Who Reported Six Emotions as Present in Their Infants: Cross-sectional and Longitudinal Studies

		Infant age in months					
		3	6	9	12	15	18
Interest	Cross-sectional[a]	1.00	1.00	1.00	1.00	1.00	1.00
	Longitudinal[b]	1.00	1.00	1.00	1.00	1.00	1.00
Enjoyment	Cross-sectional	1.00	1.00	1.00	1.00	1.00	1.00
	Longitudinal	1.00	1.00	1.00	1.00	1.00	1.00
Surprise	Cross-sectional	.69	1.00	.94	.87	1.00	1.00
	Longitudinal	.91	1.00	1.00	1.00	.97	.91
Fear	Cross-sectional	.69	.77	.71	.76	.82	.95
	Longitudinal	.65	.75	.85	.91	.84	.97
Anger	Cross-sectional	.86	.96	.94	.92	.96	.98
	Longitudinal	.91	.97	.97	1.00	1.00	1.00
Distress	Cross-sectional	.71	.73	.77	.71	.67	.76
	Longitudinal	.94	.91	.91	.88	.75	.88

[a]*N* varied from 27 to 50
[b]*N* varied from 32 to 34

ments for fear, anger, and surprise dictated that these emotions could not appear in infants before about 7 months.

How could we explain the high frequency with which mothers perceived these three emotions in their very young infants? One possibility was that mothers had become finely attuned to the subtleties and individual features of their own infants' emotional expressions and thus were able to discriminate emotional states which had previously not been apparent to investigators. Other possibilities were that the mothers were interpreting undifferentiated affective behaviors in the light of contextual information and supplying emotional labels, or that they were "adultapamorphizing," or projecting their own emotions onto the infants. Of course, the findings might also have been due to sampling error.

We conducted a second study to delimit the possibilities. There were three primary goals: (1) to determine whether the previous results regarding early onset were replicable; (2) to learn about the developmental course of the emotions in a longitudinal as opposed to a cross-sectional sample; and (3) to understand what the mothers were responding to when they reported the presence of the emotions at any particular age. A longitudinal design permitted the tracing of developmental changes in mothers' descriptions of their infants' emotions. Targeted ages were 3, 6, 9, 12, 15, and 18 months, and 34 mother–infant pairs were studied. A questionnaire was developed which asked whether each of the 11 emotions was present at the three-month intervals. For each emotion reported as present, mothers were asked to report in their own words (1) what *caused* their babies to show the emotion, (2) what baby *behaviors* let the mothers know that the emotion was present, and (3) what *response* the babies' emotion elicited from them. The following is a preliminary report of findings with respect to the emotions of surprise, fear, and anger. A complete report of the methodology and data is in preparation (Emde, Sorce, Klinnert, Stenberg, & Gaensbauer).

REPLICABILITY OF EARLY ONSET AND AGE COURSE FOR DISCRETE EMOTIONS

The mothers' reports on the presence or absence of each of the emotions provided a clear replication of the onset patterns shown in the cross-sectional study. The longitudinal data in Table 1 shows that both interest and enjoyment were present in 100% of the infants from 3 months onward. Surprise and anger showed similar onset patterns, in that at 3 months 91% of the mothers reported both of the emotions present, and the proportions fluctuated between 91% and 100% through the 18-month data point. In the case of fear, 65% of the mothers reported that the emotion was present at 3 months, and the proportion increased steadily to 97% at 18 months. Distress was reported among a slightly higher proportion of mothers in the longitudinal as opposed to the cross-sectional study, but in both the proportions were high and remained constant.

The replication had two methodological implications which gave meaning to both data sets. First, it suggested that the longitudinal sample was similar to the cross-sectional one, and decreased the possibility that the previous results were due to sampling error. The high agreement between samples suggests that mothers reliably report perceiving emotions in their very young infants (although the validity of these reports requires further documentation). Second, the replication of results across age indicated that "practice effects" from repeated measures were not a major problem in our longitudinal sample. Since there was little difference between the two samples in reports of the presence of emotions, any sensitization that may have occurred in early phases of the longitudinal sample apparently did not influence the frequency with which the various emotions were reported at later points.

Most importantly, the results addressed the central question which had motivated the second study. The data replicated the earlier findings that mothers perceived a number of emotions in their 3-month-old infants including not only interest, enjoyment, and distress, but also surprise, fear, and anger. These last three have been relegated by previous investigators to the second half of the first year of life. Further, the information obtained in the longitudinal study allowed us to investigate an additional issue. What were the mothers responding to that led them to report the presence of these emotions in their very young infants?

THE BASIS FOR MATERNAL REPORTS

The qualitative information provided by mothers proved to be invaluable in clarifying the stimuli which prompted them to report the presence of emotions in their infants at the various intervals during the 18 months. The following generalizations can be made: By at least 3 months, the mother–infant signaling system was intact. That is, infants showed specific patterns of behavior which signaled particular states to their mothers, and mothers in turn responded with appropriate caretaking behaviors. Thus, for each emotion that a mother reported as present, she reported what could be interpreted as a unique patterning of interactive behavior that was based on the signaling system. Furthermore, the prominent features of each emotion signaling system (eliciting situations, infant behaviors, usual maternal response) remained constant across development. This was true in spite of developmental changes in several areas. Such changes included increasing complexity of those events which elicited the emotions, the disappearance of reflexive behavior, and an increased sophistication of instrumental behavior.

An interesting exception to this remarkable picture of continuity concerned anger. Although the category of events which elicited anger remained the same across time, at the beginning of the second year infant anger behavior underwent a qualitative change. Correspondingly, maternal responses also changed. Thus, from the point of view of maternal perceptions, this shift in the quality of interaction between mother and infant could be considered a major discontinuity in the development of infant anger between 12 and 18 months of age.

The following descriptions present overviews for the emotions of surprise, anger, and fear.

Surprise

When mothers said surprise was present in their 3-month-olds, the majority of them based their report on infant facial expressions. Further, when facial expressions were mentioned, mothers almost always spontaneously described one or more of the components of the classic surprise expression, such as wide open eyes or mouth opened in an "O"

(Ekman, Friesen, & Ellsworth, 1972; Izard, 1971). Facial expressions remained the most commonly reported behavioral indicator of surprise from 3 through 18 months of age. The second most common behavioral response indicating surprise was infants' motor reactions. At 3 months of age, motor responses were reported almost as frequently as facial expressions. Some mothers noted jumps, but most either stated "the Moro" or listed various elements of the Moro such as "arms thrown back" (the most common) or "arches back." The proportion of mothers reporting motoric responses as indicators of surprise dropped to about one-half when infants reached 6 months. At this time no mothers reported seeing Moro responses or their components, although jumps and startles remained important. Despite the decrease in the proportion of mothers who mentioned them, motoric responses remained prominent in mothers' descriptions of surprise behavior throughout the 18-month time period. Thus, facial expressions and motoric responses remained constant as display features on the infant side of the signaling system.

On the maternal response side, a pattern which appeared well adapted to the infants' needs was present at 3 months and remained constant throughout the time period studied. Mothers indicated that at the moment of surprise, when they were uncertain as to whether the babies' follow-up reaction would be positive or negative, they tried to influence the valence of the babies' subsequent response. They did this by providing positive emotional signals, most often by talking in a calm or pleasant voice in order to prevent a negative reaction. This timing of maternal responses at the moment of uncertainty appears consistent with speculation that emotional signals are especially likely to influence appraisal when infants are in a state of uncertainty (see Klinnert, Campos, Sorce, Emde, & Svejda, 1983).

These maternal interventions illustrate the mothers' sensitivity to the sequential nature of surprise (i.e., the "resetting" phase followed by the reaction elicited by the stimulus event, as described by Tomkins, 1962). That awareness of the two phases of surprise was also apparent in the mothers' reports that, following the initial stage, they observed their babies for evidence of subsequent positive or negative reactions before they decided on appropriate responses. Descriptions would typically read "If it's a happy surprise, I would . . . but if it's a negative surprise, I would. . . ."

The consistency through time in the surprise signaling system was noteworthy, but mothers also described developmental changes in the appearance of this emotion. We have already mentioned their notation of the disappearance of the complete Moro reflex and its replacement by less extreme startle responses. A developmental progression was also evident in the reports of causes for surprise. Across age, the most frequently mentioned cause was sudden or unexpected events, but at 3 months these events took the form of sudden sounds, whereas at later infant ages it appeared that increased cognitive capacities allowed for more complexity in unexpected events.

Previous investigators have focused on the infant's developing central nervous system as well as on the increasing cognitive capacity for appreciating discrepancies as criteria for determining when the infant is able to experience surprise. Our data suggest that although the infant's capacity for this level of surprise may await further organization and development, already at 3 months there exists an interpersonal signaling system which, when put into action, leads mothers to label their infants' responses as surprise. The pattern involves a general category of an environmental event which leads to an emotional display on the infant's part, and this appears to be responded to by the mother with a particular pattern of behavior. At this early age, whatever the developmental inadequacies of the infant, there is an emotional communication system between infant and mother which allows a mutual regulation of behavior. We infer that this emotional signaling system remains constant across development, providing continuity despite the infant's constantly changing capabilities.

Fear

Mothers' perceptions that their infants had shown fear at 3 months of age appeared to reflect a phenomenon similar to that described for surprise: namely, that a particular emotional communication system between mother and infant was already present at the early age. The behavior most frequently reported as signaling fear to the mother was the infant's cry. In fact, the cry was predominant in signaling fear at 3 months, and remained so for the subsequent 15 months. Mothers described the cry as "screaming," "piercing," or "frantic." Surprisingly, fearful facial expressions were not salient in mothers' judgments of their infants' fear.

In their descriptions of fear behavior, mothers' reports of crying were invariably accompanied by information from at least one other channel of expression. At 3 months of age the second most frequently reported class of behavior was motoric and, like surprise, many of these responses were aspects of the Moro reflex. Related to this, only the cry distinguished surprise from fear at 3 months. As noted above, mothers often described surprise in terms of temporal components, with a later component involving either a positive or a negative outcome. Negative surprises were identified by the presence of the cry, and in their narrative descriptions of surprise-turned-negative, mothers often referred to the emotional display as fear.

Motor responses were still the second most frequently reported category of behavior at 6 months of age, although instead of reflexive jumps or startle responses they were apt to take the form of increased tension. From 9 months to 18 months, inferred purposeful or instrumental behaviors replaced generalized motoric reactions as the second most salient indicator to mothers that their infants were experiencing fear. The vast majority of instrumental behaviors mentioned involved attempts on the infant's part to make physical contact with mother; however, avoidance of the feared object was also mentioned frequently. The advent of meaningful fear behavior at 9 months is an important aspect of one of the apparent discontinuities in infants' emotional development.

The eliciting situations which caused fear in the infants also changed at the 9-month point, reflecting the infants' increased cognitive capacities. For example, at 3 and 6 months loud sounds (as opposed to the unexpected sounds reported as causes for surprise) were reported as primary causes of fear in infants, but in subsequent months strangers became increasingly central as elicitors of fear. This developmental progression in the inferred causes for fear as well as in observed fear behaviors is consistent with the prevalent research view that the period of 7 to 9 months marks the emergence of this emotion. The emergence of these cognitive and behavioral indexes of fear during the infant's second half-year constitutes an important qualitative change, one which is prototypical of the type of transformations which occur in infancy.

Despite these qualitative changes in the infant, the mother–infant signaling system appeared to remain fairly constant. Recall that within the pattern of infant behaviors which signaled fear to the mothers, the infants' cries provided continuity across ages. In addition, the maternal responses to their infants continued much as they were when their infants were 3 months of age. Whether their infants were 3 or 18 months old, all mothers who perceived fear in their infants reported that their response was to comfort. For the 3-month-olds, virtually all of the mothers specified that this comfort involved physical touching. While all of the mothers continued to report soothing their fearful infants, somewhat fewer of them specified physical comfort as the means for comforting their infants as they got older (e.g., two-thirds reported that their response was physical soothing when their infants signaled fear at 12 months). It is of interest that the instrumental behavior reported for these older, mobile infants was primarily seeking contact with mother. It is tempting to speculate that the mothers were refraining from initiating such contact because their locomoting infants were coming to them. Whether or not this is true, there was clearly an infant–mother signaling system, intact at 3 months and continuing through 18 months, where the infant cried, the mother soothed, and the goal of the system appeared to be the achievement of physical proximity between the two of them.

Anger

The developmental sequence revealed by the mothers' reports on anger proved to have a twist that contrasted with the continuity evident in the surprise and fear systems. In the earliest months mothers identified a unique group of behaviors which they labeled as "anger." As was true with fear, crying was always one element in this group. Unlike the crying which occurred with fear, however, infant vocalizations were described as "hard," "loud," or "forceful" cries. Further, most mothers noted at least one other behavior besides crying. For about half of the 3-month-olds, mothers described facial expressions (mostly red-faced) as signaling anger. Equally prominent were motoric responses such as kicking and back-arching. The pattern of crying, facial expressions, and motoric responses continued over the next 15 months. However, as infants became increasingly capable of instrumental behavior, this became the most salient feature in mothers' perceptions of their infants' anger. By 9 months about a fourth of the mothers identified anger by the babies' intentional behaviors, such as pushing mom away. These mildly aggressive behaviors heralded a major qualitative change that was to take place at the beginning of the second year. Crying still occurred in every case, but behavior was now viewed as involving willful aggressiveness or tantrums. For example, at 15 months, every mother but one described her infant as showing anger either by aggression toward another person (such as kicking or hair-pulling) or by throwing him or herself onto the floor. Mothers also reported an increase in the sheer number of angry incidents which occurred.

When their infants reached 18 months of age, mothers were asked to look back on the development of each emotion and to decide whether they saw the emotion as evolving through a smooth progression, or whether they felt that there were times of major change. The reports for anger were different from those for surprise and fear. Whereas mothers perceived a continuity in the behaviors which specified surprise and fear, they reported a discontinuity for anger. Corresponding to the behaviors noted above, mothers described a surge of aggressiveness between 12 and 18 months.

Maternal responses paralleled infant behaviors; when the infants' expressions of anger changed, so did the mothers' behavior. Across all ages, the primary cause which mothers reported for anger was frustration of wants or needs. Therefore, when infants cried angrily at 3 months of age, mothers' most frequently reported response was a sense of urgency to alleviate the cause or, if it could not be alleviated, to comfort the infants. But by the time infants had reached 12 months of age the mothers' responses had shifted. Instead of hurrying to remove the frustration or meet the need, mothers reported that they most frequently distracted, ignored, or disciplined the infants. A few of the mothers reported feeling or responding angrily to the infants at these times, and we speculate that, given the social undesirability of this response, it was probably underreported. What caused this shift in the entire interactive system to occur? Let us review what had happened in terms of signaling.

From the earliest months, the infants' behavioral pattern of hard crying, red face, and sharp motor movements signaled to mothers that their needs were being frustrated and that they needed instant help. In turn, the mothers did their best to help. But as the infants developed, things changed. While the frustration of a need or want remained the primary cause of anger, there was a shift in the type of event which frustrated the infants. Situations such as being diapered and dressed or placed in a carseat suddenly led to resistance whereas previously the same procedures had been passively endured. Presumably, the infants began to see their mothers as the source of frustration, while the mothers began to experience the infants as obstructive. Also of great import was the infants' increased capacity for goal-directed behavior, which allowed them to show their anger in a manner that was previously not possible. Whereas in the fear system this development simply allowed infants to do their part in making contact with mother, in the anger system the babies' new instrumental skills put them into conflict with an environment that was heretofore primarily nurturant. By nature, the instrumental or intentional behaviors that characterize anger are aggressive acts, and as the aggression began to show itself mothers initiated a deliberate socialization process by attempting to decrease or eliminate such emerging intentional behaviors. Perhaps more importantly, the infant behavior was often directed at the mothers. This development is consistent with the cognitive view that anger becomes organized when a cause is recognized. Consider the impact that this cognitive-affective development within one individual has on the emotional interaction between two human beings. The nurtured partner suddenly turns the previously unfocused anger on the heretofore primarily nurturing partner, eliciting either patient restraint or outright anger! Thus it is not surprising that the development of anger is felt as discontinuous in a manner that is different from the other emotions. The individuating infant probably experiences a similar discontinuity.

Other Emotions with Early Onset: Interest, Enjoyment, and Distress

We have focused on surprise, fear, and anger because their early onset was unexpected. Interest, enjoyment, and distress also emerged during the earliest months, as would be expected from previous research (e.g., see reviews in Emde, Gaensbauer, & Harmon, 1976; and Campos, Goldsmith, Lamb, Caplovitz, & Stenberg, 1983). For these three emotions, the signaling system manifested a fairly continuous progression over the 15 months of the study. For interest, babies' visual attention to stimuli alerted mothers that the emotion was present, whereas vocalizations, smiling, and motoric behaviors such as wriggling were reported by mothers as signaling enjoyment. In general, the causes of interest were objects or noninteracting people, while interacting with people was the most common cause of enjoyment for infants. For both of these emotions, the stimuli which elicited them increased in complexity relative to the infants' cognitive development. These two emotions were also similar in that mothers responded by attempting to prolong the state; with interest this meant keeping the objects available, and with enjoyment it meant continuing the interactions. Distress, also present from early infancy, was perceived when infant cries were unaccompanied by further cues such as red faces or startle reactions. The quality of the cries tended to be whimpers or wails, and the majority of the mothers attributed these emotional reactions to physical illness or discomfort. Mothers tended to be less sure of what it was that distress referred to than they were for any of the other emotions.

DISCUSSION

Past research on infant emotional development has emphasized the apparent discontinuities which result from physiological integration and cognitive acquisitions. In this view, infancy is characterized by the emergence of new emotions which were not present before. Prominent among these emergent emotions were surprise, fear, and anger. This picture of discontinuity, however, contrasted with mothers' perceptions of emotional continuity in their infants as well as with reports of patterned facial expressions observed in very young infants. We have suggested in this chapter that, although certain aspects of emotional development are characterized by plateaus and spurts, there is an emotional signaling system between mothers and infants which provides a picture of continuity across the span of infancy.

We have found that mothers perceived consistent patterns of affective behaviors in their infants. These occurred in certain situations and not in others, were separable into a number of emotional categories, and led to particular sets of maternal responses. Thus, from the mothers' point of view, both a *clarity* of infant signaling patterns and a *regularity* of maternal response patterns occurred. The signaling function of the system was intact by at least 3 months and remained consistent throughout the subsequent 15 months. The fact that such patterning occurred in this longitudinal study adds to our belief that this is an adaptively important and biologically based signaling system.

Given developmentalists' current emphasis on facial expression patterning, the finding that mothers' perceptions of infant emotions were based on information from a number of communication channels provides a noteworthy caution. Consistent with previous research, facial expression patterning was important in signaling some emotions, such as surprise. However, for other emotions different communication channels were involved. Fear, for example, was signaled by the infant's cry and motoric response, with almost no mothers mentioning facial expressions as indicators of the emotion. (The latter finding is consistent with the difficulty in several studies—e.g., Hiatt, Campos, & Emde, 1979—with evoking or detecting coherent fear faces in young infants.)

In the past, evidence for patterned facial expressions in early infancy has been criticized on two points. First, many have felt that the expressions occurred too infrequently to represent a real phenomenon. These mothers' reports suggest that facial expressions comprise only a partial aspect of their infants' signaling repertoire. Other channels of communication are important. Second, the ecological validity of patterned facial expressions in the early months has been questioned. The present data suggest that such facial expressions may well have ecological validity as components or subunits of a larger emotion signaling system, even though isolated facial expressions may not carry functional messages without other cues. The ways in which infant facial and other expressive parameters contribute to judgments of the presence of discrete emotions deserves careful attention.

Prior discussions of emotional development have tended to neglect the communication aspect of the emotional system. There has been some acknowledgment of the importance of infant emotional expressions for signaling state and getting needs met. In addition, however, it appears likely that interpersonal interactions contribute a great deal to infant experience, not only in terms of contingencies and reciprocity, but also in the quality of emotional experience. Since there are relatively distinct signaling systems for the various emotions, the experiences of the infant from the earliest months onward may well be characterized by sensations akin to adults' experience of the various emotions—prototypical experiences, to be sure, but continuous over time. Early experiences could, then, be organized around an "affective core," such as that proposed by Emde (1982) as responsible for a sense of personal continuity extending from infancy through later life.

But what of the emotion of anger, which was perceived by mothers as changing somewhat abruptly around 12 months of age? Recall that, although the change caught our attention, there *was* evidence of *continuity* in anger. This was particularly evident in the mothers' reports of the causes for anger, where frustration of wants or needs remained constant, and in expressive aspects of anger behavior, such as crying and a red face. Furthermore, despite the changes that occurred, mothers persisted in labeling this behavioral constellation as *anger.* A sense of *discontinuity* was related to a seeming change in the interactive system. Mothers reported a shift from cooperation between themselves and their infants to more angry interactions, often with infant anger targeted at mother. This anger toward mother, as well as descriptions of infant resistance to being manipulated in previously routine caretaking situations, would appear to indicate an increased sense of separateness. Mothers' descriptions of infant angry behaviors as "willful" and "intentional" adds to the picture. Facilitating the development of autonomy through tolerating the expression of anger while at the same time helping to regulate the energy of the diffuse lashing out that often occurs is no easy task. Research is needed to examine the differential impacts of various caregiving approaches to socializing infant expressions of anger. Differing parental styles may be related to the development of individual differences in infants' experience of both anger and the sense of self.

The further study of the emotional signaling system is important, therefore, not only because it may itself be characterized by continuity, but because it may be a means by which affect contributes to the development of an experiential sense of continuity—the sense of self. If infant emotional signals elicit consistent caretaker responses, then there may well be at least a rudimentary level of continuity in the experience of the infant. But continuity is, of course, not the whole story. We have also attempted to illustrate, through the emotion of anger, another aspect of the emotional transactions that occur between infants and their caretakers—the fact that a change in one part of the system results in a shift within the entire system and thereby feeds back into individual experiences of the participants. This feedback loop, at least for infants at critical points in development, may result in emotional growth within the individual.

Acknowledgments

We wish to acknowledge the considerable help of Dr. Joseph J. Campos in many helpful criticisms of earlier drafts of this chapter, and of Dr. Michelle Lampl, who worked long hours in data analysis.

REFERENCES

Bridges, K. M. Emotional development in early infancy. *Child Development*, 1932, *3*, 324–341.

Campos, J. J., Goldsmith, H., Lamb, M., Caplovitz, K., & Stenberg, C. Socioemotional development. In M. Haith & J. Campos (Eds.), *Infancy and developmental psychobiology*, Vol. 2 of P. Mussen (Series Ed.), *Handbook of child psychology*. New York: Wiley, 1983.

Ekman, P., Friesen, W., & Ellsworth, P. *Emotion in the human face: Guidelines for research and an integration of findings*. New York: Pergamon, 1972.

Emde, R. N. The prerepresentational self and its affective core. *The psychoanalytic study of the child*, 1983, *38*, 165–192.

Emde, R. N., Gaensbauer, T. J., & Harmon, R. J. Emotional expression in infancy: A biobehavioral study. *Psychological Issues, A Monograph Series, Inc.* (Vol. 10, No. 37). New York: International Universities Press, 1976.

Emde, R. N., Sorce, J. F., Klinnert, M. D., Stenberg, C., & Gaensbauer, T. J. Patterned emotions in infancy: Mothers' perceptions during the first 18 months. In preparation.

Gaensbauer, T. G., & Hiatt, S. Facial communication of emotion in early infancy. In N. Fox & R. Davidson (Eds.), *Affective development: A psychobiological perspective*. New York: Lawrence Erlbaum, 1983.

Hiatt, S., Campos, J. J., & Emde, R. N. Facial patterning and infant emotional expression: Happiness, surprise, and fear. *Child Development*, 1979, *50*, 1020–1035.

Izard, C. *The face of emotion*. New York: Appleton-Century-Crofts, 1971.

Izard, C. On the ontogenesis of emotions and emotion–cognition relationships in infancy. In M. Lewis & L. A. Rosenblum (Eds.), *The development of affect*. New York: Plenum Press, 1978.

Izard, C. E., Hembree, E. M., Dougherty, L. M., & Spizzirri, C. C. Changes in facial expressions of 2- to 19-month-old infants following acute pain. *Developmental Psychology*, 1983, *19*, 418–426.

Johnson, W., Emde, R. N., Pannabecker, B. J., Stenberg, C., & Davis, M. Maternal perception of infant emotion from birth through 18 months. *Infant Behavior and Development*, 1982, *5*, 313–332.

Kagan, J., Kearsley, R., & Zalazo, P. *Infancy: Its place in human development*. Cambridge: Harvard University Press, 1978.

Klinnert, M. D., Campos, J. J., Sorce, J. F., Emde, R. N., & Svejda, M. Emotions as behavior regulators: Social referencing in infancy. In R. Plutchik & H. Kellerman (Eds.), *Emotion: Theory, research, and development* (Vol. 2: *Emotions in early development*). New York: Academic Press, 1983.

McCall, J. B., Eichorn, D. H., & Hogarty, P. S. Transitions in early mental development. *Monographs of the Society for Research in Child Development*, 42, Serial No. 171, 1977.

Oster, H., & Ekman, P. Facial behavior in child development. In A. Collins (Ed.), *Minnesota symposium on child psychology* (Vol. 11). Hillsdale, N.J.: Lawrence Erlbaum, 1978.

Spitz, R. *A genetic field theory of ego formation*. New York: International Universities Press, 1959.

Sroufe, A. L. Socioemotional development. In J. D. Osofsky (Ed.), *Handbook of infant development*. New York: Wiley, 1979.

Stenberg, C. *The development of anger in infancy*. Unpublished doctoral dissertation. Denver: University of Denver, 1982.

Tomkins, S. S. *Affect, imagery, consciousness*, Vol. 1: *The positive affects*. New York: Springer, 1962.

CHAPTER 15

Early Experience Effects

EVIDENCE FOR CONTINUITY?

Linda S. Crnic

The concepts of continuity and discontinuity apply to a variety of issues in developmental psychology today. Concern with these concepts reflects a continuing interest in the genetics–environment issue—the question of the extent to which behavior is a product of inheritance or of environmental influences. As with the nature–nurture issue, it is important to see that hypotheses about continuities and discontinuities in development are not necessarily in conflict. However, this conclusion does not weaken the importance of understanding the role of *both* processes in any particular developmental task.

CONTINUITY AND DISCONTINUITY: ISSUES IN DEVELOPMENTAL PSYCHOBIOLOGY

The continuity–discontinuity question can be applied to at least two fields of current interest in developmental psychology. First is the issue of whether temperamental or personality traits show continuity over time. This issue has two facets, addressed by two contributors to this volume. Kagan has explored the extent to which temperamental traits and their physiological correlates persist during development and has concluded that inhibition alone, of all traits measured, appears to have

---

Linda S. Crnic • Departments of Pediatrics and Psychiatry, University of Colorado School of Medicine, Denver, Colorado 80262. This work was supported by Grant HD08315 from the National Institutes of Child Health and Human Development.

some continuity over time (Chapter 2). Rutter has explored the extent to which the effect of early environment persists and the question of the extent to which the persistence is mediated by environmental invariance (Chapter 3). Thus the first issue involves the continuity of both inherited and acquired behavioral traits and the physiological or environmental mediators of their continuity. The second issue involves the question of whether emerging competencies are dependent upon prior experiences or are a product of brain maturation and relatively free of specific experiential input. It is easy to see that it is impossible to take an extreme position on the continuity–discontinuity issue; for example, even achievements which are dependent upon brain maturation for their origins may be susceptible to improvement through experience within a given stage of brain development. On the other hand, the obvious educational implications of continuity–discontinuity issues illustrate the importance of determining their roles in any developmental task. Answers to questions of the roles of continuity–discontinuity in the development of particular functions must ultimately depend upon knowledge of the genetics of behavior and the development, structure, and function of the brain. Simple appeals to genetic or maturational determination of development beg the question: maturational events can be cited as evidence for continuity or discontinuity in behavior.

EARLY EXPERIENCE RESEARCH

Definitions

Early experience research in animals has investigated the continuity of the effects of traits acquired through experience. The lessons learned from this research may be useful in the design of future human research.

Early experience effects refers to a diverse group of phenomena in which an event during development has a lasting influence on later (usually adult) behavior. By definition, research in this field contains a strong assumption of continuity of influence of experience over a lifetime. It offers an arena in which ideas about continuity from human research can be investigated with some hope of gaining information of a causal rather than correlational nature. In addition, mechanisms mediating effects of early experience can be explored.

The early experience studies are of two types: those that explore stimuli which, according to Gottlieb's (1976) schema, maintain, facilitate, or induce normal development and those which explore unusual stimulation such as handling or shock. These two types of study have different roots and different goals and, it has been argued (Henderson,

1980), different rates of success. The work on stimulation necessary for development has, because of the well defined nature of the stimuli, outcome measures, and hypotheses, led to substantial progress in defining not only what input the nervous system needs to arrive at its optimal structure and function, but how that is achieved. An outstanding example of research in this field is the studies of the stimuli necessary for the development of the visual system reviewed by Greenough and Schwark in this volume (Chapter 4). The second type of early experience study has asked more difficult questions, used complex stimuli, and measured subtle changes in behavior. Progress in this field has been disappointingly slow, but it is this latter approach to early experience effects which I will discuss here because this approach to early experience research is more relevant to the question of whether and how human traits can be acquired through experience.

The study of unusual experiences arose, according to Denenberg (1977), from a desire to test the analytic concept that early events are potent determiners of adult personality. The ethological tradition of careful analysis of behavior provided the tools for this endeavor, and Hebb's (1949) notion that experience might serve as the organizer of the development of behavior and neurophysiological processes provided further impetus to the field.

This research has largely involved rodents presented with experiences in the prenatal, neonatal, or immediate postweaning periods. Effects of interest have been behavioral and physical sequelae measured in adulthood. With few exceptions, the immediate or intermediate effects of experience have not been the focus of interest. Among the types of stimulation which have been explored are stress of the mother during pregnancy, effects of environmental manipulations during infancy, including handling and shocking the infants, and manipulation of maternal characteristics.

As noted above, the study of unusual early experiences has not led to much progress relative to the amount of effort expended. Exploring the reason for this will perhaps aid in future studies.

Reasons for Slow Progress

Complex Stimuli, Weak Conceptual Bases. According to Henderson (1980), part of the reason why some early experience studies have had so little success is the complexity of the stimuli presented and the lack of an empirical or theoretical basis for selecting stimuli. The same criticism applies to the outcome measures used. For example, in the early handling studies, the experimenter removed the infant mice from the nest and put them into individual containers for three minutes for several days early in life. This involved deprivation of the presence of the dam, exposure to cold and novel tactile stimuli, handling by the experimenter, and different treatment by the dam upon return to the nest: a very complex set of stimuli. Outcome measures have commonly been tests of learning or activity in the open field, with only vague justifications as to why alterations in these behaviors should result from early handling. To some extent this complexity and vagueness is necessitated by the nature of the questions being asked and to some extent results from the fact that it is impossible to manipulate neonatal rodents without having substantial effects upon the behavior of the dams.

Failure to Determine Whether Effect Is Early. A second problem with this research has been that seldom have the necessary controls been run to determine whether the effects are indeed *early* or whether they may not occur throughout life: critical periods for experience have been assumed without evidence. The necessary control is one in which an equivalent experience is presented in adulthood. If the effect is an early one, the adult experience should not have so potent an effect as the same experience during the hypothesized critical period. The importance of this control is illustrated by the case of a manipulation which was at one time considered to be an early experience effect: complex environments. In this research, rodents are kept in cages provided with a wider range of stimulus objects than the normal cage. Experiments with adult animals have shown that rodents are susceptible to the effects of complex environments throughout life, (e.g., Rosenzweig, Bennett, & Diamond, 1972), a finding which has had profound effects upon our ideas about neuronal plasticity.

Statistical Problems. One source of much confusion in this area of research has been that statistically significant effects are reported which are spuriously significant for any of a number of reasons. First, in some studies, significant mortality of infants under study resulted in the selection of an unrepresentative subpopulation of animals (e.g., Denenberg, Hudgens, & Zarrow, 1966). Second, it has only recently become standard to consider littermates nonindependent subjects. As littermates share a common genetic, intrauterine, and neonatal environment, they are likely to vary less among themselves than they vary from other litters. This is most obviously the case for variables such as body weight but may apply to behavioral parameters as well. The use of nonindependent littermates as independent subjects in an experiment exaggerates the significance of a finding because it reduces the variance and exaggerates the degrees of freedom for the test. The most appropriate treatment of results in which littermates are used is a nested analysis of variance in which litter is the nested factor. If the litter effect is significant, analyses must be computed with the litter as the unit of analysis

(Abbey & Howard, 1973). Nested analysis of variance requires equal numbers of subjects in each group and so is often not usable. Other options are to average the littermates together to arrive at an average for the litter which is used as the datum or to design experiments so that only one or two littermates are used for any one measure. Finally, as the interactive nature of early malnutrition with other experiences and with genotype have become apparent, experiments have become very complex so that many terms are tested in an analysis of variance. In addition, the use of multiple outcome measures also leads to the performance of many statistical tests and thus an increased probability that chance variation will be detected as significant. The use of experiment-wise rather than test-wise Type I error levels would help to alleviate this problem.

Contribution to Continuity–Discontinuity Issues

Despite these problems, what can this area of research offer to students of human early experience effects? Lasting effects of early experience have been documented; however, it is difficult to use these to address questions about how potent similar effects may be in humans. This is because of the restricted life experiences of laboratory rodents. Humans have many experiences which may interact with and moderate the effect of any target experience, whereas laboratory animals have few experiences to modulate the effects of an experimental manipulation.

One of the principal contributions of early experience research has been to show that the effects of early experiences depend upon what the organism brings to the experience through both past experiences and genetic determinants. This research has also shown that it must be proven rather than assumed that an experience is effective only early in life. Finally, these experiments have provided some evidence about how the effects of early experience may be mediated by physiological changes. Although few investigators have explored the intermediate effects of early experience, an outstanding example of such work is that of Grota on the hormonal mediators of early experience with handling and shock. He has shown that the reduced adrenocortical activity which sometimes results from early handling or shock does not depend upon secretion of corticosterone by the adrenals of the pups or the dam (Grota, 1975). The adult effect is not due to an alteration in the ability to metabolize corticosterone in adulthood nor the responsiveness of the adrenal cortex to ACTH (Grota, 1976). He has some evidence which is suggestive that serotonin circuits may mediate the effects of this early experience on later adrenocortical function (1979). Another line of experimentation has shown that the mechanism underlying continuity does not necessarily reside in the brain. It is this research that I will review in detail.

AN EXAMPLE OF EARLY EXPERIENCE RESEARCH: MALNUTRITION EFFECTS ON FOOD MOTIVATION

I will analyze in depth one of the early experience effects which finds its roots in analytic theory: food deprivation. The idea that denial of gratification in feeding leads to preoccupation with satisfying oral needs later in life led to some early studies on the effects of partial starvation on a variety of food-related behaviors. The study of early malnutrition has provided another body of evidence relevant to this issue.

Many lines of evidence indicate that rats who have experienced inadequate nutrition early in development behave differently toward food in adulthood. They appear to be more motivated to obtain and ingest food many months after they have been rehabilitated with normal diets.

Altered behavior toward food in rats who have been malnourished early in development (prenatally and/or during suckling) has been observed under two circumstances: during food deprivation and during satiation. To avoid confusion in terminology, I shall refer to rats who have experienced malnutrition early in development and then rehabilitated onto normal diets as *previously malnourished*. When animals are deprived of food in adulthood for short times in order to motivate them to perform a task, they will be called *food-deprived*.

Prior Findings

Most of the evidence for altered food-related behavior in previously malnourished rats has been obtained while they are food-deprived in adulthood. Many behaviors have been examined, including operant performance, competition for food, and food intake.

Several operant schedules are useful for assessing motivation. On variable interval schedules, the rate of bar pressing is normally low and steady. Increasing the rate of pressing has little effect on the rate of reward, and thus the response rate is thought to be a sensitive index of the level of motivation (Ferster & Skinner, 1957). On this schedule, previously malnourished rats press more rapidly than controls for both food (Smart, Dobbing, Adlard, Lynch, & Sands, 1973) and water (Smart & Dobbing, 1977) when maintained at 90% of their normal body weight. Higher rates of pressing on variable interval schedules have been established for previously malnourished rats for a variety of levels of food deprivation (Levitsky, 1975). The point at which rats cease to respond on a progressive ratio schedule (in which each reward requires progressively more responses) has also been used to assess motivation. Previously malnourished rats have higher "breaking points" than normal rats on such schedules (Halas, Burger, & Sandstead, 1980). Previously malnourished rats also press more rapidly on fixed-ratio schedules on which rapid pressing does increase the rate of reinforcement (Mandler, 1958). There is some evidence that unpredictability of reinforcement is the critical variable in determining higher rates of response in previously malnourished rats (Roberts, Smart, & Weardon, 1982).

When trained to run for food in an alley, previously malnourished rats run more quickly than normal rats (Amsel & Penick, 1962; Smart *et al.*, 1973). These animals also compete more successfully than normal rats for access to food (Seitz, 1954) and water (Whatson, Smart, & Dobbing, 1975). When given access to food for limited periods of time each day they eat more and spill more food (Barnes, Neely, Kwong, Labadan, & Frankova, 1968) and eat more rapidly (Mandler, 1958; Marx, 1952; McKelvey & Marx, 1951). These behavioral adjustments lead them to be better able to maintain their body weight when access to food is restricted to a short time each day (Barnes *et al.*, 1968).

In all of the studies cited so far, behavior was measured during food restriction. On the other hand, two lines of evidence were obtained from free-feeding rats. Free-feeding rats who were underfed as infants hoard more food pellets when housed singly but not when housed in groups during development (Seitz, 1954). In two of three studies, malnutrition initiated after weaning also increased hoarding of food pellets (Hunt, 1941; Hunt, Schlosberg, Soloman, & Stellar, 1947) while the third study found no evidence for this (McKelvey & Marx, 1951). The second line of research on free-feeding rats comes from my laboratory. I have found that rats experiencing severe malnutrition in infancy consume more of the most palatable of a variety of concentrations of sucrose than normal rats (Crnic, 1979). In this study, the procedure of Richter and Campbell (1940) for testing response to a range of sucrose solutions (called a palatability function) was used. Rats were exposed to a range of sucrose solutions, each for 24 hours at a time in their home cages with chow freely available.

Thus, when either food-deprived or normally fed, previously malnourished rats appear to be more eager than normal to obtain and ingest food. It is important to note that this altered behavior does not appear to be due to any alteration in the ability of these animals to regulate their body weight. My colleagues and I (Crnic, Bell, Mangold, Gruenthal, Eiler, & Finger, 1981) and others have shown that previously malnourished rats are smaller than normal throughout their lives, maintaining low but normally shaped growth curves. When their ability to regulate body weight is challenged by offering them an obesity-producing "supermarket" diet (Stephens, 1980) or high fat diet (Borer, 1979), their response is not different from that of normal rats.

Explanations for Altered Behavior

What accounts for the altered behavior of previously malnourished rats toward food? I (Crnic, 1976) and others (Smart & Dobbing, 1977) have argued that when previously malnourished rats are deprived of food by any of the commonly used methods, there is good reason to believe that their level of deprivation is more extreme than that of normal rats to which they are compared. Smaller animals have higher nutrient requirements because of their increased metabolic rate, which is due to the larger ratio of surface area to volume. In addition, they have a lower percentage of fat in their bodies (Stephens, 1980), thus a larger percentage of metabolically active tissue, and consequently higher nutrient needs than a normal rat of the same weight. Finally, although the evidence is mixed (reviewed by Crnic, 1976), the efficiency with which previously malnourished rats digest, absorb, and utilize nutrients may be different from normal. The fact that they are more motivated to ingest water in addition to food when deprived also argues for metabolic differences in previously malnourished rats (Smart & Dobbing, 1977).

Food deprivation to motivate task performance is achieved by either removing food for a set time every day, feeding an inadequate ration, or reducing the rat to a percentage of its free-feeding weight. Because of their higher metabolic rates, any period of food deprivation puts previously malnourished rats at a larger deficit than normals. Reduced rations, which are allocated per gram of body weight, similarly do not take into account the smaller, leaner animals' greater metabolic needs. When reduced to a percentage of free-feeding body weight, the leaner, previously malnourished rats lose a larger proportion of their body fat stores than do normal rats; when deprivation is severe, they may also lose lean tissue.

Thus, the explanation for the increased motivation of previously malnourished rats to obtain and ingest food when they are performing under food deprivation may be simply their small body size and low fat content rather than, or in addition to, any alteration involving the central nervous system. This points up the importance of considering other than brain characteristics underlying behavior and continuity in development. It should be noted that this effect has not been shown to require that malnutrition occur early in life. However, weight and body fat deficits produced by adult malnutrition do not result from loss of cell numbers (Widdowson & McCance, 1963) and thus are not permanent losses as are those which occur after malnutrition in infancy.

Palatability function testing was useful in establishing alterations in food intake in previously malnourished rats which are independent of food deprivation and the emotional disturbance of testing in a strange environment. However, it is a complex and dynamic testing situation and as such provides us with few clues about the mechanism behind the altered sucrose intake. Many possibilities exist: taste or alteration in the evaluation of taste; immediate postingestive factors in stomach, duodenum, or hormonal systems; long-term postingestive factors such as absorption and utilization of nutrients leading to altered diet preferences.

Collier and co-workers (Castonguay, Hirsch, & Collier, 1981; Hill, Castonguay, & Collier, 1980) have argued that the palatability function is a curve of response to dilution rather than palatability. They postulate that rat chow provides a protein-to-carbohydrate ratio higher than that preferred by the rat. In seeking to ingest the optimum ratio, the rat consumes sugars, with the amount of solution needed to do this varying with the concentration. The fact that rats will consume different amounts of 16% and 32% solutions of glucose, resulting in the consumption of 60% of their calories as sugar (Castonguay et al., 1981) and the fact that a nonpalatable source of carbohydrate, dextrinized starch, is consumed in the same proportion as glucose (Hill et al., 1980) are evidence for this theory. It is possible that previously malnourished rats prefer a different ratio of protein to carbohydrate than normal rats, because of their altered body composition or altered efficiency of nutrient uptake and utilization. Were this the case, however, the increased sucrose intake of previously malnourished rats would be expected to extend across all sucrose concentrations, and it does not. However, because of the evidence that food intake and utilization is altered in previously malnourished rats (reviewed in Crnic, 1976) and the possibility that this could lead to altered nutrient requirements, it is important to explore this possibility and to see whether the ability to choose intake levels increases the efficiency of food utilization and rate of growth.

New Findings

In order to study the general question of whether the behavior of previously malnourished rats toward food may be based on their different ability to handle nutrients, it seemed important to explore two variables, malnutrition and weaning age. It is common practice to wean rats at 21 days of age; however, because malnourished pups are retarded in their development, weaning at the normal age constitutes an early weaning manipulation. Malnourished rats will continue to suckle until about 35 days of age when allowed to stay with their dams (Crnic, unpublished observation). In this experiment, Sprague Dawley rats were malnourished by providing their dams with a diet which contained 8% casein with .05% methionine added; the control diet contained 25% casein. We have shown that low protein diets cause a drop in the nitrogen content of the dams' milk and an elevated lipid content in addition to lowering the amount of milk produced (Crnic & Chase, 1978). From 21 days, all dams and pups were allowed access to laboratory chow and half of each litter was weaned while the other half remained with the dam until 35 days of age, at which time none of the dams showed evidence of suckling their pups. At 21 days of age, all pups and dams were allowed access to normal laboratory chow.

One or two male rats from each litter were studied at 7 or 15 months of age, allowing ample time to recover from their early malnutrition. They were allowed access to low and high protein diets in separate food cups, and by taking in various ratios of the two foods they could adjust their intake to any level between 8% and 25% protein. In the low protein diet, starch replaced protein. The pattern of results was the same at the two ages, although fewer effects were significant at the later age. At 7 months, both of the early weaned groups ate more of the high protein diet ($p = .001$) while the interaction between weaning age and postnatal diet approached significance ($p = .069$), with the difference between weaning ages less for the rats who had had the high protein diet in infancy. In the rats tested at 15 months, the diet effect was significant ($p = .01$) with those whose dams were fed a low protein diet eating more of the high protein diet. Total food intake showed the same pattern at both ages, but only in the 15-month group was the interaction effect significant, with the early-weaned, low protein group eating more than the others. The same pattern occurred for the grams of protein consumed per gram of body weight where at both ages the early weaned, low protein diet animals ate more. This same group of rats gained more weight when allowed to choose their diet, although the difference reached significance only at the younger age. This heightened weight gain was due not solely to the increased food intake but also to a more efficient use of food, as the body weight increase per gram of food was higher in these rats too.

These results indicate that previously malnourished rats need more food per gram of body weight, probably because of their higher proportions of metabolically active tissue and the fact that smaller animals have a larger surface-to-volume ratio. They prefer higher levels of protein in their diets, particularly if they have also been weaned early, and gain weight more efficiently when allowed to eat their preferred diet. This experiment also indicates the importance of considering weaning age when studying the effects of malnutrition. These results cannot explain the effects of malnutrition upon the palatability function; the opposite results would have been expected if the previously malnourished animals had been drinking more sucrose solutions in an attempt to increase the ratio of carbohydrate to protein in their diets.

The fact that the previously malnourished rats grew more efficiently when they were allowed to choose a higher protein diet is congruent with some prior findings that these rats are not able to utilize proteins as efficiently as normal rats (reviewed in Crnic, 1976). We do not know how the need for a higher than normal level of protein in the diets affects any behavior other than food preference; however, it probably contributes to relatively more severe deprivation in previously malnourished rats placed on deprivation schedules to motivate test performance.

How does the research just described meet the requirements and criticisms of early experience research described earlier? First, the variable of interest, malnutrition, suffers of necessity from being a complex variable. It is not possible to malnourish neonatal rats without altering their social environment. Different methods of producing malnutrition can be used to circumvent some of these problems. The method used, feeding the dam a low protein diet, avoids problems such as competition between pups for access to the dam which occurs when malnutrition is produced by fostering large litters to a normally fed dam. The hypothesis that behavioral changes might occur because of alterations in body size, composition, and ability to utilize nutrients is explicit and grounded in sound past research. The outcome variables, food intake under various conditions, are clearly related to the concept of interest, easy to measure unambiguously, and usually free of excess variability. Although the necessary studies have not been run to prove that the effects of malnutrition on food intake depend upon the malnutrition occurring early in life, prior studies have indicated that the permanent effects upon body composition and size occur only with perinatal malnutrition (Widdowson & McCance, 1963). On any test, only one or two littermates were used from any litter, thus avoiding the problem of nonindependence of littermates for statistical analysis.

CONCLUSIONS

What can this work contribute to the continuity–discontinuity issue? Does the level of nutrition early in the lives of humans influence later behavior toward food? Little evidence is available on behavior; however, considerable research has addressed the question of whether early nutrition, and thus body weight, is predictive of later adiposity. The conclusions have generally been that within usual limits there appears to be little relationship between infant nutrition and later body weight (e.g., Dine, Gartside, Glueck, Rheines, Greene, & Khoury, 1979; Vuille & Mellbin, 1979). The hypothesis that early nutrition determines adipocite (fat cell) numbers and thus later potential for obesity has similarly not been supported (Roche, 1981). However, as in the studies reported here, human nutrition which is much more extreme than that normally encountered can have permanent effects upon body weight. Men who were *in utero* during the 1944–1945 Dutch famine were studied at 19 years of age. Those whose mothers were undernourished during the first two trimesters of pregnancy had a higher incidence of obesity than normal, and those whose mothers were malnourished during the last trimester had a significantly lower obesity rate than normal (Ravelli, Stein, & Susser, 1976). The former finding has been duplicated in the rat (Jones & Friedman, 1982).

Although there is nothing that can be generalized directly from the preliminary research reported here to humans, this research does raise an important possibility: that alterations in the structure and function of the body (as opposed to the nervous system) may be responsible for continuity of some acquired behavioral traits. Some genetic sources of continuity may have a similar substrate. It is true that any bodily condition can ultimately influence behavior only by influencing the central nervous system, but we have in the past overemphasized the central nervous system as the only route through which behavior may be influenced.

REFERENCES

Abbey, H., & Howard, E. Statistical procedure in developmental studies on species with multiple offspring. *Developmental Psychobiology*, 1973, 6, 329–335.

Amsel, A., & Penick, E. C. The influence of early experience on the frustration effect. *Journal of Experimental Psychology*, 1962, 62, 167–176.

Barnes, R. H., Neely, C. S., Kwong, E., Labadan, B. A., & Frankova, S. Postnatal nutritional deprivations as determinants of adult rat behavior toward food, its consumption and utilization. *Journal of Nutrition*, 1968, 96, 467–476.

Borer, K. T. Weight regulation in hamsters and rats: Differences in the effects of exercise and neonatal nutrition. In D. Novin, W. Wyrwicka, & G. A. Bray (Eds.), *Hunger: Basic mechanisms and clinical implications.* New York: Raven Press, 1979.

Castonguay, T. W., Hirsch, E., & Collier, G. Palatability of sugar solutions and dietary selection. *Physiology and Behavior*, 1981, 27, 7–12.

Crnic, L. S. The effects of infantile undernutrition on adult learning in rats: Methodological and design problems. *Psychological Bulletin*, 1976, 83, 715–728.

Crnic, L. S. Effect of infantile undernutrition on adult sucrose solution consumption in the rat. *Physiology and Behavior,* 1979, *22,* 1025–1028.

Crnic, L. S., & Chase, H. P. Models of infantile undernutrition in rats: Effects on milk. *Journal of Nutrition,* 1978, *108,* 1755–1760.

Crnic, L. S., Bell, J. M., Mangold, R. Gruenthal, M., Eiler, J., & Finger, S. Separation-induced early malnutrition: Maternal, physiological and behavioral effects. *Physiology and Behavior,* 1981, *26,* 695–707.

Denenberg, V. H. Assessing the effects of early experience. In R. D. Meyers (Ed.), *Methods in Psychobiology, Vol. 3: Advanced laboratory techniques in neuropsychology and neurobiology.* New York: Academic Press, 1977.

Denenberg, V. H., Hudgens, G. A., & Zarrow, M. X. Mice reared with rats: Effects of mother on adult behavior patterns. *Psychological Reports,* 1966, *18,* 451–456.

Dine, M. S., Gartside, P. S., Glueck, C. J., Rheines, L., Greene, G., & Khoury, P. Where do the heaviest children come from? A prospective study of white children from birth to 5 years of age. *Pediatrics,* 1979, *63,* 1–7.

Ferster, C. B., & Skinner, B. F. *Schedules of reinforcement.* New York: Appleton-Century-Crofts, 1957.

Gottlieb, G. The roles of experience in the development of behavior and the nervous system. In G. Gottlieb (Ed.), *Development of neural and behavioral specificity.* New York: Academic Press, 1976.

Grota, L. J. Effects of early experience and dexamethasone on adrenocortical reactivity. *Developmental Psychobiology,* 1975, *8,* 251–259.

Grota, L. J. Effects of early experience on the metabolism and production of corticosterone in rats. *Developmental Psychobiology,* 1976, *9,* 211–215.

Grota, L. J. Plasma corticosterone response to serotonin altering drugs in the 3-day-old rat. *Developmental Psychobiology,* 1979, *12,* 399–405.

Halas, E. S., Burger, P. A., & Sandstead, H. H. Food motivation of rehabilitated malnourished rats: Implications for learning studies. *Animal Learning and Behavior,* 1980, *8,* 152–158.

Hebb, D. O. *The organization of behavior.* New York: Wiley, 1949.

Henderson, N. D. Effects of early experience upon the behavior of animals: The second twenty-five years of research. In E. C. Simmel (Ed.), *Early experience and early behavior: Implications for social development.* New York: Academic Press, 1980.

Hill, W., Castonguay, T. W., & Collier, G. H. Taste or diet balancing? *Physiology and Behavior,* 1980, *24,* 765–767.

Hunt, J. McV. The effects of infant feeding frustration upon hoarding in the albino rat. *Journal of Abnormal and Social Psychology,* 1941, *36,* 338–360.

Hunt, J. McV., Schlosberg, H., Solomon, R., & Stellar, E. Studies of the effects of infantile experience on adult behavior in rats. I. Effects of infantile feeding frustration on adult hoarding. *Journal of Comparative and Physiological Psychology,* 1947, *40,* 291–304.

Jones, A. P., & Friedman, M. I. Obesity and adipocyte abnormalities in offspring of rats undernourished during pregnancy. *Science,* 1982, *215,* 1518–1519.

Levitsky, D. A. Malnutrition and animal models of cognitive development. In G. Serban (Ed.), *Nutrition and mental functions.* New York: Plenum Press, 1975.

Mandler, J. M. Effect of early food deprivation on adult behavior in the rat. *Journal of Comparative and Physiological Psychology,* 1958, *51,* 513–517.

Marx, M. H. Infantile deprivation and adult behavior in the rat. Retention of increased rate of eating. *Journal of Comparative and Physiological Psychology,* 1952, *45,* 43–49.

McKelvey, R. K., & Marx, M. H. Effects of infantile food and water deprivation on adult hoarding in the rat. *Journal of Comparative and Physiological Psychology,* 1951, *44,* 423–430.

Ravelli, G.-P., Stein, Z. A., & Susser, M. W. Obesity in young men after famine exposure in utero and early infancy. *The New England Journal of Medicine,* 1976, *295,* 349–353.

Richter, C. P., & Campbell, C. R. Taste thresholds and taste preferences of rats for five common sugars. *Journal of Nutrition*, 1940, *20*, 31–46.

Roberts, H. J., Smart, J. L., & Wearden, J. H. Early life undernutrition and operant responding in the rat: The effect of reinforcement schedule employed. *Physiology and Behavior*, 1982, *28*, 777–785.

Roche, A. F. The adipocyte-number hypothesis. *Child Development*, 1981, *52*, 31–43.

Rosenzweig, M. R., Bennett, E. L., & Diamond, M. C. Chemical and anatomical plasticity of brain: Replications and extensions. In J. Gaito (Ed.), *Macromolecules and behavior* (2nd ed.). New York: Appleton-Century-Crofts, 1972.

Seitz, P. F. D. The effects of infantile experiences upon adult behavior in animal subjects. I. Effects of litter size during infancy upon adult behavior in the rat. *American Journal of Psychiatry*, 1954, *110*, 916–927.

Smart, J. L., & Dobbing, J. Increased thirst and hunger in adult rats undernourished as infants: An alternative explanation. *British Journal of Nutrition*, 1977, *37*, 421–430.

Smart, J. L., Dobbing, J., Adlard, B. P., Lynch, A., & Sands, J. Vulnerability of developing brain: Relative effects of growth restriction during fetal and suckling periods on behavior and brain composition of adult rats. *Journal of Nutrition*, 1973, *103*, 1327–1338.

Stephens, D. N. Growth and the development of dietary obesity in adulthood in rats which have been undernourished during development. *British Journal of Nutrition*, 1980, *44*, 215–227.

Vuille, J.-C. & Mellbin, T. Obesity in 10-year-olds: An epidemiologic study. *Pediatrics*, 1979, *64*, 564–572.

Whatson, T. S., Smart, J. L., & Dobbing, J. Dominance relationships among previously undernourished and well fed male rats. *Physiology and Behavior*, 1975, *14*, 425–429.

Widdowson, E. M., & McCance, R. A. The effect of finite periods of undernutrition at different ages on the composition and subsequent development of the rat. *Proceedings of the Royal Society of Medicine B*, 1963, *158*, 329–342.

PART IV

CROSS-DISCIPLINARY STRATEGIES IN SEARCHING FOR CONTINUITIES

CHAPTER 16

Mathematical-Statistical Model Building in Analysis of Developmental Data

Robert Short, John Horn, and Jack McArdle

INTRODUCTION

A number of subtle and complex concepts of development are discussed in other chapters of this volume. As other authors note, study designs and data analyses must be carefully executed if concepts are to be properly represented. Kagan, for example (Chapter 2), has pointed to the need to distinguish between connectivity on the one hand and absence of connectivity and emergent phenomena on the other hand. Rutter (Chapter 3) has demonstrated the desirability of distinguishing within-environment determinism from within-person continuity. Fischer *et al.* (Chapter 5) have raised issues about quantum, or stage, development in contrast to near-continuous development. In almost all chapters there is discussion of how properly to identify stability when, as is inevitably the case, it is found in the company of instability.

The purpose of this chapter is to describe what we refer to as *MAthematical-Statistical MOdel Building (MASMOB)* procedures. MASMOB procedures are potentially useful in developmental research, both in the design of experiments and the analysis of data. MASMOB procedures are not a panacea for treating problems of design and data analysis, but if used properly they can help. Moreover, they can help in ways that are not available with other methods. The aim of this chapter is to show

Robert Short, John Horn, and Jack McArdle • Department of Psychology, University of Denver, University Park, Denver, Colorado 80210.

why and how MASMOB procedures augment and improve upon other methods of design and analyses.

There is not enough space here to provide a comprehensive discussion of MASMOB techniques. Only a few of the major ideas will be considered. But examples will be presented that should suggest wider implications and applications. One example will be aimed at indicating how lawful change along connected pathways of a simplex might be identified and distinguished from the emergence of entirely new pathways and the discontinuation of old pathways (or the start and end of Escher staircases, to use Kagan's metaphor). Another example will indicate formal ways of thinking about, and analyzing, questions relating to the locus of change: whether it is the organism that changes, or the measurement device, or both. Other examples will indicate that traditional correlations and tests of means simply do not tell a comprehensive story about data, particularly the interactions among variables of change.

MASMOB represents a very general way of thinking about a large number of design and data analysis methods—including analysis of variance, analysis of covariance, factor analysis, canonical correlation, multiple regression analysis, and path analysis. In this sense, most developmental researchers are already using MASMOB methods. Within MASMOB, however, the various methods can be seen to be part of, and variations of, a single method. Such integration of different methods not only allows an investigator to see how different methods are related, it also allows him to use the different methods in ways that reveal features of data that otherwise would not be revealed.

Current interest in MASMOB methodology owes a great deal to the work of Jöreskog (1966, 1969, 1973), Jöreskog and Sörbom (1978, 1982), and McDonald (1978, 1982), particularly for the development of computer programs (e.g., LISREL-V and COSAN-II) that efficiently perform the complex and time-consuming calculations of structural equation modeling. The methods also derive in important ways from the ideas of path analysis, as put forth by Duncan (1975), and partial least squares (PLS) procedures, developed by Wold (1975) and Lohmöller (1981). The reader is referred to McArdle (in press) and James, Mulaik, and Brett (1982) for reviews of recent developments in this area.

Our contributions in this area are represented by (1) demonstrations that the methods are more general, and more clearly related to other techniques, than previously was believed (McArdle, 1980; McArdle & McDonald, 1983); (2) simulation indications of important problems in using the methods (Horn & McArdle, 1980); (3) studies of particular applications of the methods for different kinds of problems (McArdle, 1978; McArdle & Horn, 1981b; McArdle, Goldsmith, & Horn 1981); and

(4) techniques for showing that it is relatively easy to translate the complex ideas of a substantive theory into a diagramatic representation that is isomorphic to a set of mathematical and statistical calculations (McArdle & Horn, 1981b; McArdle, McDonald, & Horn, 1984).

The latter contribution, the translation from theory to diagram to mathematical-statistical representation, is especially important. It assists an investigator in more nearly doing in data analysis what a substantive theory indicates should be done to express the theory. Achieving a close match between reality and representation—matching observations with a model of observations—may be the principal task of science (Medawar, 1981). The major value of MASMOB methods may be found in their usefulness in helping scientists to achieve this match.

STEPS FOR DOING A MASMOB STUDY

Before discussing model building *per se*, some ideas about the data set should be noted, for a scientific model must ultimately refer to data. Often, the basic data will consist of a number of different subjects (although MASMOB procedures are not restricted to this configuration of data). In Kagan's discussion of inhibition, for example, the subjects are children and the variables are the various behaviors Kagan describes as being relevant to inhibition. In Table 1 some computer-generated data are presented to illustrate such a data set. The data matrix contains scores for active memory (A), self-awareness (W), and inhibition (H) created for 102 subjects (Tom, May, Kim, etc.). Ordinarily, in modeling analysis, these raw data are first converted to an association matrix, usually a correlation matrix, as shown in Table 1, or a covariance matrix, but possibly a matrix of sums of squares and cross products. Modeling procedures involve finding a simplified set of equations that reproduce this association matrix as closely as possible. That is, the associations, represented here in the form of a correlation matrix, are described mathematically in terms a set of relations, relations specified in the D and U matrices in Table 1, which constitute a model. At a general level, then, mathematical-statistical model building represents a series of actions directed at finding a set of relations in the D and U matrices that will reliably, validly, and meaningfully account for the association matrix.

The ISEER Steps

Five major steps can be considered in attempting to find a set of relations which adequately reproduce the association matrix, namely: Initialization, Specification, Estimation, Evaluation, and Readjustment.

Table 1. Relationships Among the Three-variable Data Matrix, the Computed Correlation Matrix, and the Model Matrix Equation[a]

Subject number	Subject name	Active memory (A)	Self-awareness (W)	Inhibition (H)
1	Tom	7	8	8
2	May	6	7	7
3	Kim	7	5	4
4	Lil	6	7	10
.	.	.	.	.
.	.	.	.	.
.	.	.	.	.
.	.	.	.	.
102	Don	6	7	5

Scores are used to calculate the corelation coefficients:

$$\begin{array}{c} \\ A \\ W \\ H \end{array} \begin{array}{ccc} A & W & H \\ \left[\begin{array}{ccc} 1.00 & & \\ .40 & 1.00 & \\ .31 & .60 & 1.00 \end{array}\right] \end{array} = R$$

which can be expressed by the model equation[b]

$$R \cong \begin{bmatrix} 1 & 0 & 0 \\ 0 & 1 & 0 \\ -d_{HA} & -d_{HW} & 1 \end{bmatrix}^{-1} \begin{bmatrix} 1 & 0 & 0 \\ 0 & 1 & 0 \\ 0 & 0 & 1 \end{bmatrix} \begin{bmatrix} 1 & 0 & 0 \\ 0 & 1 & 0 \\ -d_{HA} & -d_{HW} & 1 \end{bmatrix}^{-t} = (\text{I-D})^{-1}\, U\, (\text{I-D})^{-t}$$

[a]The entries in the D and U matrices specify the model such that the correlation matrix is reproduced as closely as possible.
[b]For simplicity, scaling matrices have been omitted from this model equation. -1 is the inverse of the matrix; $-t$ is the inverse of the transposed matrix. For the complete model equation, see Figure 2.

These ISEER steps can be thought of as 5 distinct phases of design and analysis.

To *inititalize* a modeling study is to use a rationale (theory) to guide the gathering and arrangement of data. To *specify* a model is to designate unambiguously all of the variables (manifest and latent) and all of the relations (directed and undirected) among the variables of the model. To *estimate* a model is to obtain (by definition and calculation) numerical values for all the relations that are specified. To *evaluate* a model is to invoke mathematical and statistical theory to provide a basis for determining how well the specified equations reproduce the association matrix. To *readjust* a model is to use the results from the estimation and evaluation of one model to specify a modified model.

Ideally, the *initialization*, or rationale, should be explicit and unambiguous, but in practice often it is not; it might be little more than the investigator's intuitions. Even when theory is not explicit, decisions are made when data are gathered. These decisions determine what is observed, how the observations are made, how measurements are obtained, and so forth. Such decisions constitute the *rationale*. In exploratory studies this might best be referred to as an implicit or intuitive rationale.

Specification is a conversion of the rationale into a design for analysis. Specification includes expressing each statement (hypothesis) of the rationale as either an element to be fixed or an element to be estimated in the model. It is often convenient and clarifying to begin specification

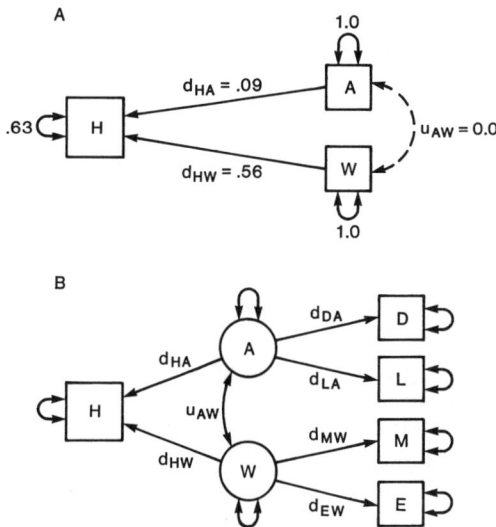

Figure 1. Specification of manifest variable and latent variable models that might represent Kagan's theory of inhibition. In both FIGMODs, elements such as d_{HA}, pictured as single-headed arrows, represent *directed* relations as when one variable (H) is estimated as a weighted combination of other variables (A and W). Elements such as u_{AW}, pictured as double-headed arrows, represent *undirected* relations as when one variable (A) is correlated with another (W). **A.** The manifest variable version specifies the hypothesis (from theory) that inhibition (H) is dependent on active memory (A) and self-awareness (W). This can also be represented in an equation as $H_i = d_{HA} A_i + d_{HW} W_i$. The $i = 1, \ldots, 102$ observations are those presented in Table 1. **B.** The latent variable FIGMOD represents a similar statement of the hypothesis, but the latent variable active memory (A) is indicated by the manifest variables recognition-after-delay memory (D) and retention–search memory (L = likelihood of finding), and the latent variable self-awareness (W) is indicated by manifest variables apprehension about unfamiliar child (M) and "seeing" own reflection in mirror (E). Additionally, A and W are correlated beyond their relation with H.

with a graphical representation—a *FIG*ural *MOD*el or *FIGMOD*—corresponding to the modeling equation. Examples of two FIGMODs are illustrated in Figure 1. These examples express a few possible interpretations of Kagan's theoretical ideas about the development of inhibition in young children.

Because measured variables are said to be *manifest* variables, the model in Figure 1A is a manifest variable representation of Kagan's ideas. The 102 scores for H (inhibition), A (active memory) and W (self-awareness) from Table 1 were used in computing the illustrated coefficients. The scores were first converted to a 3-by-3 correlation matrix and a model was then specified in an attempt to reproduce these correlations. In the FIGMOD, directed relations were specified from A to H and from W to H to indicate the parameters that were to be estimated. The single-headed arrows were drawn to specify an hypothesis that inhibition (H) is determined by a combination of active memory (A) and self-awareness (W). These directed relations, d_{HA} and d_{HW}, thus specified to be nonzero, were entered in the appropriate section of the D matrix (Table 1) and then estimated as linear regression coefficients.

An *undirected* relation, u_{AW}, is suggested in Figure 1A by a dashed two-headed arrow between A and W. The dashed-line symbolism is intended to indicate that the undirected u_{AW} *could* have been specified to be nonzero, and thus *could* have been entered in the U matrix illustrated in Table 1 and subsequently estimated, but in this particular expression of the model this undirected relation is specified to be zero. This makes the model equivalent to multiple regression analysis in which H is made up from independent portions of A and W. The main point is that multiple regression analysis represents a particular model for data, perhaps the model that one is thinking about in one's theory, but perhaps not. By using MASMOB one can be rather clear in specifying the model for analysis that represents the theory about which one is thinking.

In Figure 1B a model is specified that involves essentially the same hypotheses as were expressed in the manifest variable model, but the predictor variables are themselves specified in terms of other variables and are therefore said to be *latent* variables. The manifest variable inhibition in the face of the unexpected (H) is based on the latent variables active memory (A) and self-awareness (W). These two latent variables are to be estimated *within* the model. Hypotheses about some of the relations of these latent variables to the measured, or manifest, variables are also indicated in the figure. The latent variable of active memory (A) is specified to be responsible, in part, for performances indicating the amount of time over which a child can delay recognition memory (the manifest variable D) and the likelihood that a child will retain awareness in search of a hidden object (L). Similarly, the latent variable of self-awareness (W) is, according to this specification of Kagan's theory, responsible for manifest variable measures of apprehension in meeting an unfamiliar child (M) and displays of recognition of one's own image in a mirror (E).

Specification of the d_{DA}, d_{LA}, d_{MW}, and d_{EW} relations between manifest variables and latent variables can be referred to as the specification of the measurement model (Jöreskog & Sörbom, 1979). These relations can also be referred to as the specification of an outer *structure* of manifest variables in terms of latent variables (Wold, 1981). In contrast, relations among the latent variables (as A and W to H) are often referred to as the *inner* part of the model. Other distinctions, such as those between the exogenous and endogenous variables in a model (Duncan, 1975), may also be useful for describing relations among manifest and/or latent variables.

It is important to recognize that arrows *not* drawn in a FIGMOD represent specification hypotheses no less than arrows that *are* drawn. For example, a directed arrow might have been drawn from W to L in the latent-variable model of Figure 1B. That this arrow is not drawn signifies a hypothesis that the model relation between W and L is zero. Similarly, possible directed and undirected relations between D and E and between L and E are in each case specified to be zero. Indeed, in the latent variable model of Figure 1B, quite a few of the parameters that might plausibly be nonzero have been specified to be precisely zero.

To specify that the strength of a directed or undirected relation be a particular value is to *fix* that model parameter. Several parameters are fixed at zero in Figure 1, but fixed relations need not be zero. An undirected relation between D and E could be fixed to .30 or −.65, for example. When specifying a model, the determinations of which parameters are to be fixed, and what values they are to be fixed to, are as important as the values that are to be estimated.

It is a relatively simple matter of application of rules to convert a FIGMOD specification into a set of mathematical equations (Horn & McArdle, 1980; McArdle, 1980; McArdle et al., 1984). Each to-be-estimated and fixed relation of the FIGMOD is simply entered into particular sections of the model equation (the D and U matrices of Table 1). At this point the investigator is ready to make the third major step of a MASMOB study.

The third step is *estimation*. To understand MASMOB computer algorithms, which estimate the parameters of a model, is a major undertaking. We will not try to convey such understanding in this chapter (the interested reader may consult McArdle, in press). Under most conditions, the algorithms in such programs as LISREL V (Jöreskog & Sörbom, 1982), PLS (Lohmöller, 1981), COSAN (McDonald, 1978), and

RAM (McArdle & Horn, 1981a) are dependable and the investigator can assume that most of the results, most of the time, are reliable. The following, however, are two important things that an investigator should remember about the workings of MASMOB algorithms:

1. The procedures optimize estimates in a particular sample of data. This means that the estimation can capitalize on chance (Horn & Knapp, 1973; Wackwitz & Horn, 1971). Concretely, capitalization on chance means that a parameter estimate obtained in one analysis will not, with regularity, replicate in another sample of observations. The extent of this lack of replication is not always well indicated by the size of the calculated standard error associated with that parameter, particularly in small samples.

2. Parameters are estimated jointly, and thus the value estimated for a particular parameter depends in part on other parameters that are or are not estimated in an analysis. This means that the value estimated for a particular parameter may change drastically when a new variable is entered or a relevent variable is removed from the analysis. These are the problems of suppressor and enhancer influences. They are discussed more fully in the Horn and McArdle (1980) study and in the references therein (e.g., McFatter, 1979).

The fourth major step in MASMOB is *evaluation*. Evaluation is the calculation and interpretation of indicators of how well a model fits. Several indicators can be calculated to indicate goodness of fit. An indicator frequently used is the chi-square statistic corresponding to a likelihood ratio. The likelihood ratio is a measure of the proportion of the variation in the data that can be reproduced by the model.

To use MASMOB procedures intelligently it is not essential to have a detailed understanding of the rather technical matters of how likelihood ratios are formed. Assumptions made in forming these ratios are not different, basically, from the assumptions made in using most other widely used parametric statistics. Two aspects of evaluation should be mentioned, however: (1) A chi-square distribution is only an approximation of the likelihood-ratio distribution, and the approximation becomes close only as the sample size exceeds 30, and (2) the likelihood-ratio distribution is derived under the assumption that the residual variates are normally distributed.

In most MASMOB studies a fifth major step of analysis is taken. This is the step of *readjustment*. As noted elsewhere (Horn & McArdle, 1980), this step can be referred to as tinkering with the model. At the evaluation stage of analysis the standard errors for parameter estimates are obtained. If a standard error is large relative to the estimated value of a parameter, it suggests that that parameter is not significantly different from zero. This information might be used in readjustment by respecifying this parameter to be fixed at zero and then reestimating the modified model. Such changes can, of course, lead to a better model fit, but at a cost of losing the statistical basis for the indicators of goodness of fit. In making readjustments one can be capitalizing on chance variations in a particular set of data. Statistical theory, from which goodness-of-fit indicators are derived, does not contain provisions for the influences of such nonchance events; thus the indicators do not allow for exploratory tinkering. Most important, results obtained from notably readjusted models cannot be expected to replicate in new sets of data (Horn & Knapp, 1973; Horn & McArdle, 1980).

Before moving on to consider specific examples of modeling analyses, let us summarize the major points discussed thus far.

Summary of Rationale

Underlying any data analysis there is a model which a researcher adopts about how data (reality) can be represented. But researchers sometimes use statistical models without being clear about the nature of the model they are assuming. Yet models should reflect what researchers intend to say—the rationale for their studies, their hypotheses. A major value that can be realized with MASMOB methods is becoming explicit about the rationale for data analyses.

MASMOB represents procedures for expressing relations among variables in terms of mathematical equations. The equations are the model. Relations among variables are represented in an association matrix—usually a correlation or covariance matrix. The equations of a model are intended to reproduce the association matrix and in that sense account for it. These equations can represent assumptions, hypotheses, and beliefs about the relations that lie under (are latent in) the obtained association matrix.

MASMOB analysis is a simplification of data. There are fewer parameters in a model than measurements in the data, as such, or associations in the association matrix. The simplification not only can enable one to deal with complexity that otherwise is incomprehensible, it can also represent a more fundamental reality than the data itself: it can represent scientific laws that apply to data other than that of a particular analysis—data yet to be seen, even data that may never be seen.

The ISEER steps are summarized briefly below (expanded from Horn & McArdle, 1980):

I. *Initialization.* Initialize a rationale (theory) to guide the assembling of data.

S: *Specification.* Among latent and manifest variables, specify the directed and undirected, fixed and to-be-estimated relations that define the structural and inner parts of the model.

E: *Estimation.* By definition or calculation, obtain numerical values for all the specified parameters.

E: *Evaluation.* Evaluate the adequacy of the model using goodness-of-fit indicators.

R: *Readjustment.* Readjust the model in order to obtain a better representation of the data in hand.

It is to be hoped that a model represents an adequate theory about the reality which is represented by the data. This can be true only if investigators are careful not to abuse the procedure. It is particularly important to avoid capitalization on chance and to guard against misinterpretation of suppressor–enhancer influences. Actually, since no model of reality can really *fit* reality—that is why a model is called a model—it is advisable, whenever possible, to contrast plausible models, rather than look for *the* model. This approach is encouraged by studies of the logic of scientific research (Brody, 1970) as well as by studies of the practical value of the goodness-of-fit indicators that are used in model evaluation (Horn & McArdle, 1980; Jöreskog, 1969).

Summary of Symbolism

In Figure 2 we provide a summary of the symbols used in FIGMODs and a complete form of the model equation. The symbols used for drawing FIGMODs correspond in a one-to-one manner to elements of the model equation. Thus one can directly translate a particular FIGMOD into a particular instance of the model equation (Horn & McArdle, 1980; McArdle, 1980; McArdle *et al.*, 1984).

Manifest variables are represented in FIGMODs by rectangles, latent variables by circles. Each variable represents a set of N measures or scores. The measures corresponding to latent variables can be thought of as implicit. Usually latent variable scores are not estimated in modeling analysis; only summary statistics, such as means, variances, and covariances, are estimated.

An arrow in a FIGMOD represents a nonzero relation; absence of an arrow denotes a zero relation. A relation may be *directed,* in which case the arrow is one-headed, or *undirected,* in which case the arrow is two-headed. The magnitude and sign of the coefficients indicate the strength and charge (positive or negative) or a relation. A directed relation is a specification that the variable at the arrow point is to be estimated (in part) using the variable from which the arrow comes. An undirected relation is a specification that the variables at the two end points of the

Mnemonic Label	Graphic Symbol	Conceptual Representation	Algebraic Reference
Manifest Variables	M Square or rectangle	Data–derived variables; measured on N cases	Operationally defined observations, summarized by means, variances and correlations
Latent Variables	L Circle or ellipse	Theory–derived variables; not directly measured	Common residual variables, error variables, scaling variables
Undirected Parameters	u_{jk} Two–headed arrow	Two-way symmetric association among two variables; value u denotes strength of the association	Correlations, variances, covariances, residual variances
Directed Parameters	d_{jk} One–headed arrow	One-way asymmetric relationship from one input variable to one output variable; value d denotes the strength of the relation	Regression coefficients, transitional probabilities

Figure 2. Basic structural equation modeling notation (from RAM; Horn & McArdle, 1980; McArdle, 1980) used in specifying a model. Elements of FIG-MODs are inserted into the following equation:

$$C = X^t X = F(I - D)^{-1} U(I - D)^{-t} F^t$$

in which C is the obtained m-by-m moment (covariance) matrix; X is the data matrix of m manifest variables by N subjects; F is an m-by-r matrix in which most elements are zero but an element equal to 1.0 designates a manifest variable; r is the total number of variables (manifest plus latent); t represents "transpose" (a matrix), $-t$ represents the inverse of a transposed matrix; I is a r-by-r identity matrix; D is a r-by-r matrix in which a d_{jk} element specifies a *directed* relation from variable k to variable j; and U is a r-by-r matrix in which a u_{jk} element specifies an *undirected* relation between variables j and k.

arrow are correlated (if the variables are in standard-score form) or co-vary (if the variables have not been scaled to be standard scores). If a two-headed arrow for a variable turns back on itself (both arrow-points are on one variable), the relation represented by the arrow is for the variable with itself and thus is a variance. Often this variance is a residual variance representing what is left over after the variable is used in other relations.

A SUBSTANTIVE EXAMPLE

Using the symbols of Figure 2, any model can be specified for MAS-MOB analysis. To illustrate this point with a reasonably complex theory from developmental psychology, Kagan's theory of inhibition has been interpreted in some detail and expressed as a FIGMOD in Figure 3. Here it is assumed that the principal dependent variable that Kagan wishes to explain is inhibition in response to the unexpected (H). The principal explanatory variables are assumed to be self-awareness (W), active memory (A), and fear of strangers (F), each of which is thought (in this specification of the theory) to be indicated by several manifest variables. Developmental features of the theory are presented in terms of the relations to age, which have been hypothesized as having both linear and quadratic components (represented by age and the square of age respectively). Given data, one can evaluate the model of Figure 3. If the model is found to fit, it provides a powerful, objective basis for the kind of discussion Kagan has provided.

To indicate more clearly the meaning of the coefficients and symbols, a one-predictor manifest variable model is presented in Figure 4. In terms of bivariate correlation analysis (when B and S are set to standard score form) d_{BS} is the regression coefficient and in this case is also the correlation coefficient, r. The two-headed arrow with the value u_{BB} simply means that after B is estimated from S along pathway d_{BS} there will

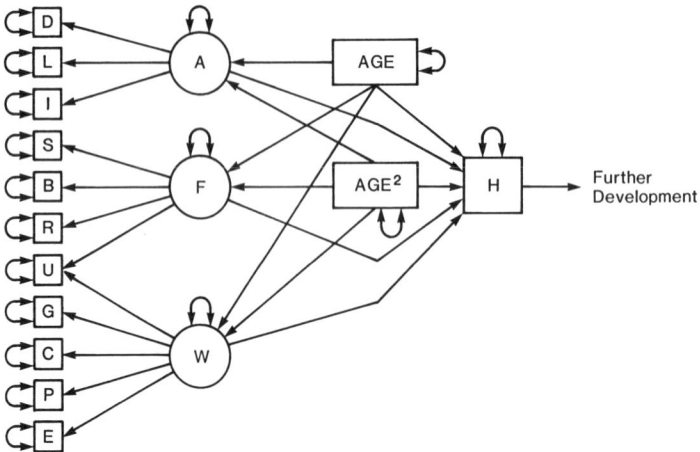

Figure 3. A simple (but possible) model representation of Kagan's theory of inhibition. Note that the coefficients have not been specified as they are, given data, to be estimated. *Manifest variables:* D = Delay to recognize; L = Likelihood of finding in search; I = Integration in speech; S = Stranger anxiety; B = Heart rate stability; R = Heart rate; U = Apprehension about unfamiliar child; G = Guilt when "bad"; C = Control (avoid aggression towards other child); P = Pride/anxiety in implicit imitation; E = Seeing self-reflection; H = Initial inhibition in response to unexpected. *Latent variables:* A = Active memory; F = Fear of strangers; W = Self-awareness.

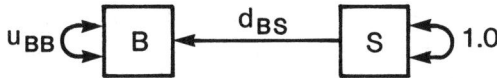

Figure 4. A one-predictor manifest variable model for bivariate regression or a t test. Both S and B have been put in standard score form (mean = 0.0 and standard deviation = 1.0). The FIGMOD shows the residual variance of B (u_{BB}), the regression coefficient (which in this case is equal to the correlation coefficient), and the variance of S. If S happened to be a dichotomous variable, the FIGMOD would be isomorphic to Student's t test.

be some variance left over in B. u_{BB} is the residual variance of B and in this case is equal to $1 - r^2$. Note that the total variance for both B and S has been scaled to equal 1.0 (for B the total variance equals $u_{BB} + d_{BS}{}^2 = (1.0 - r^2) + r^2 = 1.0$).

In the behavioral sciences a correlation or regression coefficient is usually evaluated using a fixed-variate ANOVA theory rather than a likelihood ratio theory. The chi-square for likelihood of fit has a reverse interpretation relative to the ANOVA F test and many other such statistics researchers are accustomed to using. In using the ANOVA theory, when the probability for an F ratio is small, the null hypothesis can be rejected and one's favored hypothesis can be regarded as tenable. In using a chi-square statistic for a maximum likelihood ratio, however, when the probability for the chi-square is small, it means that the data are not well-represented by one's favored hypothesis and the model is then rejected. A *large* probability, associated with a small chi-square indicates that one's favored model hypothesis is tenable; the model is a good representation of the data.

When both an ANOVA test and a chi-square test can be used to evaluate a particular model for data, the two tests support the same conclusion. They are just flip sides of the same coin. In the simple model presented in Figure 4 these theories can be directly translated one into the other. But in general this is not true. For many models that can be rather easily considered using MASMOB procedures it is not at all easy to see how, if at all, the problem could be formulated in terms of ANOVA. The reverse is not true: It is almost always easy to see how an ANOVA study can be specified for modeling analyses.

It is interesting to note that if variable S in Figure 4 happened to be a dichotomous variable, an evaluation statistic for the model is Student's t test. The t test is equivalent to a chi-square test in the same sense that the ANOVA test we previously considered is equivalent to the chi-square test. Dichotomous variables, as well as multichotomous or continuous variables can be used in modeling analyses. Of course, one must consider the implications of statistical assumptions for dichotomous variables, but given such thought there is no difference in principle between using dichotomous variables and using multichotomous variables in modeling analyses. Indeed, as we will show shortly, even constants can be usefully considered in modeling analyses.

Although the illustrations of Figure 4 do indicate several important features of modeling analyses, they do not indicate very much about how developmental issues might be studied with MASMOB methods. Let us now consider, therefore, a few ways of looking at development from a data analysis point of view.

SIMPLEX EXAMPLES

A simplex, in the manner that Guttman (1954) developed the concept, is a pattern among correlation coefficients. The concept can be of particular interest to those studying development because the simplex pattern can be found when, in repeated measurements of a set of organisms, there is both some stability and some instability in the data. These are precisely the conditions that prevail for many of the variables that developmental researchers study. Hence ideas about simplex patterns can be regarded almost as prototypes for development, as such. MASMOB methods are particularly useful for studying these patterns.

Consider an attribute measured repeatedly in a smaple of, say, 100 subjects. Suppose the weight of children is measured ten times between birth and 10 years of age. Denote these measures as Y_{ti}, where the i subscript represents individuals ($i = 1, 2, \ldots, 100$) and the t subscript stands for occasion of measurement (in this case $t = 1, 2, \ldots, 10$).

If weight is a stable characteristic in the sense that heavy infants become heavy children and relatively light infants become relatively light children, then the correlations between measurements obtained at two different ages will be (a) large and (b) much the same regardless of which ages are considered. These two outcomes will be referred to as the magnitude and the homogeneity conditions of association. Concretely, if the correlation between weight at birth and weight at age 10 is $r = .80$, the magnitude is large relative to a correlation of $r = .30$; if the correlations are, say, .82, .85, and .86 between the measurements of 10-year-olds and 1-, 3-, and 7-year-olds, respectively, this set of coefficients is homogeneous relative to a set in which the values are .32, .91, and .16. Both homogeneity among test–retest coefficients and large magnitudes of these coefficients can be indicators of trait stability in development.

Of the two indicators of stability, homogeneity might be the more important. This can be true because several conditions besides instability can be responsible for small test–retest coefficients. For example, if the measuring instrument is not reliable but the trait is stable the test–retest coefficients will be small. Homogeneity among the coefficients could be quite substantial under these conditions if the extent of unreliability was about the same on each occasion of measurement. But, of course, both the magnitude and the homogeneity of retest coefficients are relevant for the study of stability-instability and for the study of patterns of development more broadly concerned.

Magnitude and homogeneity in retest coefficients can reflect two rather different conditions: (1) an absence of change in a trait: for example, eye color remains much the same from the first to the 60th birthday; and (2) uniform change in a trait: the increase in weight from year to year might be about the same for all children and thus the rank-order of weight would remain much the same as children grow older. In both conditions, test–retest coefficients will be both large and homogeneous. Usually in developmental research evidence of stability is thought to indicate uniform change.

In Table 2 a matrix of retest correlations is presented that exemplifies a simplex pattern. These correlations might be thought of as the retest coefficients for children's weight measures obtained each year beginning on the first and ending on the tenth birthday. There are several things to notice about this set of coefficients. First, there is evidence for stability. The one-year-to-the-next coefficients are both large and homogeneous (all are .90). Even if two- and three-year lags are considered, the correlations are large and homogeneous (the range is from .80 to .90). There is also evidence of instability. The correlation from age 1 to age 10 years is only .50 even as the reliabilities of both

Table 2. A Simplex Pattern of Retest Correlations

		Year of measurement									
		1	2	3	4	5	6	7	8	9	10
	1	$r_1{}^a$	.90	.85	.80	.75	.70	.65	.60	.55	.50
	2		r_2	.90	.85	.80	.75	.70	.65	.60	.55
	3			r_3	.90	.85	.80	.75	.70	.65	.60
	4				r_4	.90	.85	.80	.75	.70	.65
Year of mea-	5					r_5	.90	.85	.80	.75	.70
surement	6						r_6	.90	.85	.80	.75
	7							r_7	.90	.85	.80
	8								r_8	.90	.85
	9									r_9	.90
	10										r_{10}

[a] r_j = reliability in the year j; assume all are .99.

scores are .99. Looking at the matrix as a whole, one sees a lack of homogeneity in the coefficients. But the most remarkable characteristic of this matrix is the regularity of the decrease in correlation with the increase in the number of years between measurement occasions. The coefficients get smaller uniformly as the number of years between measurements increases. There is clearly a suggestion that if the occasions for measurement were extended on either side of the matrix—for example, by including measures obtained in the years between 11th to 20th birthdays—the correlations over the longer periods of time would be even smaller than the smallest coefficient in this matrix. The evidence then would be more indicative of instability than of stability. But in fact this is not really the case; a simplex pattern indicates both stability and instability—both must be present for the pattern to appear. Therefore it is not accurate to say that the pattern indicates only instability or only stability. We can see this by considering how, at the level of measurement as such, a simplex pattern of retest coefficients can come about.

Suppose there are individual differences in yearly increments to weight. That is, suppose in the year between the first and second birthdays the weight increment is I_{1i} ($i = 1, \ldots, 100$ subjects as before). Some children gain quite a bit, some even more, and some less. Similarly, suppose there are individual differences in weight increments from the second to the third birthday, the third to the fourth, and so forth. This set of conditions can be symbolized like this:

$$Y_{1i} = Y_{1i} \qquad\qquad\qquad \text{(the first-year measurement)}$$
$$Y_{2i} = Y_{1i} + I_{1i} \qquad\qquad \text{(the second-year measurement)}$$
$$Y_{3i} = Y_{2i} + I_{2i} \qquad\qquad \text{(the third-year measurement)}$$
$$\quad\; = Y_{1i} + I_{1i} + I_{2i} \qquad \text{(replacing } Y_{2i} \text{ with its equivalent)}$$
$$Y_{4i} = Y_{3i} + I_{3i}$$
$$\quad\; = Y_{1i} + I_{1i} + I_{2i} + I_{3i}$$
$$Y_{5i} = Y_{1i} + I_{1i} + I_{2i} + I_{3i} + I_{4i}$$
$$\cdots\cdots$$
$$Y_{10i} = Y_{1i} + I_{1i} + I_{2i} + I_{3i} + I_{4i} + I_{5i} + I_{6i} + I_{7i} + I_{8i} + I_{9i}$$

Suppose, finally, the variances and reliabilities for the I_{ji} increments are uniform, and the correlations between every I_{ji} and I_{ki} pair of increments and between these increments and the first year measures, Y_{ij}, are all zero. Under these conditions a pattern of correlations like the pattern shown in Table 2 can arise.

Notice that the correlations between weights obtained in adjacent years can be large—for example, if reliabilities are large and the amounts of variance contributed by the I_{ji} variables are small relative to the variance of Y_{ji} variables. Notice, too, that measures obtained over separations of several years have relatively few I_{ji} components in common and so can have low correlations. Indeed, correlations between measurements obtained at widely separated times may become zero or even negative.

In Figure 5 we present FIGMODs for simplex models like the example we have just considered. The first model, Figure 5A, may be called an equal interval simplex. The increments, that is to say the variances and the reliabilities of the increments, are equal. Each variance, shown as a two-headed arrow pointing around and back to a simple variable, is for a residual—that is, the variance that is left over after all that can be predicted by a previous measurement has been removed (in this case each residual has a magnitude of $1 - a^2$). Notice that in this equal-interval model only the magnitude of the relations between observations (the a's) are assumed to be equal. Time lags between measurements could vary or be varied in an effort to make the coefficients equal. The second model, Figure 5B, represents a first-order simplex. Here there is a difference in the extent to which the increments can influence the variable into which they enter. This complicates things relative to the equal-interval case, but it is still a straightforward matter to write out the equations indicating what each covariance in a matrix would be if this simplex model were a good representation of reality. Such a covariance

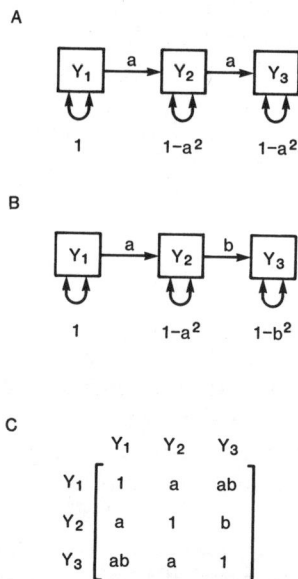

A

$$1 \qquad 1-a^2 \qquad 1-a^2$$

B

$$1 \qquad 1-a^2 \qquad 1-b^2$$

C

$$\begin{array}{c} \\ Y_1 \\ Y_2 \\ Y_3 \end{array} \begin{array}{ccc} Y_1 & Y_2 & Y_3 \\ \left[\begin{array}{ccc} 1 & a & ab \\ a & 1 & b \\ ab & a & 1 \end{array}\right] \end{array}$$

Figure 5. Some manifest variable simplex models. **A.** An equal interval simplex FIGMOD. **B.** A first-order simplex FIGMOD. **C.** The predicted correlation matrix from the first-order simplex in B.

breakdown for three measurement occasions is shown in Figure 5C. It is a straightforward matter to specify a model in which the increments are allowed to be unequal and are to be estimated.

The simplex models we have considered thus far are sometimes called sequential mediation models because each increment I_{ji} operates on only the one occasion, j, where it is said to mediate the sequence of measurements. As abstract ideas are formulated about what might be seen if extraneous influences were well controlled, these very restricted models can be of major value in representing theories of development.

But of course models based on simplex ideas need not be quite so restrictive as those just considered. For example, a first-order simplex model might be modified to represent an assumption that some of the increments that operated in an early period of development no longer have influence in measurement obtained at later times. In our example the I_{1i} and I_{2i} increments to weight might be dropped out of the Y_{10i} representation for weights obtained at 10 years of age—baby fat increments need not be represented in the weights of 10-year-olds. Even the initial measurement, Y_{1i}, might be lost in later measurements. These possibilities represent models for phenomena such as Kagan discusses in Chapter 2—for example, when he considered situations in which a prominent character in the first act of a developmental drama did not appear at all in the second act. A main point of this discussion is that such ideas can be considered along with the idea that some of the characters in the first act do indeed make an appearance in the last act; all such ideas can be studied in models one can specify.

Many ideas about human development are not well represented with a single manifest variable, such as weight. No one really believes that something as varied and complex as human intelligence, for example, is well represented by a single kind of test. Intelligence has several important and different features, each of which must be indicated by a different kind of measurement device (cf. Horn, 1982). To address the question of how intelligence develops, therefore, multiple indicators of the quality must be considered. Such considerations move us to study multiple variable simplex models, thence to the study of latent variable models.

MULTIVARIABLE SIMPLEX MODELS

In Figure 6 a few multivariable simplex models are illustrated. The first model, Figure 6A, is for a two-variable extension of the first-order simplex considered in Figure 5B. The coefficients between different occasions of measurement are not equal for either of the two variables

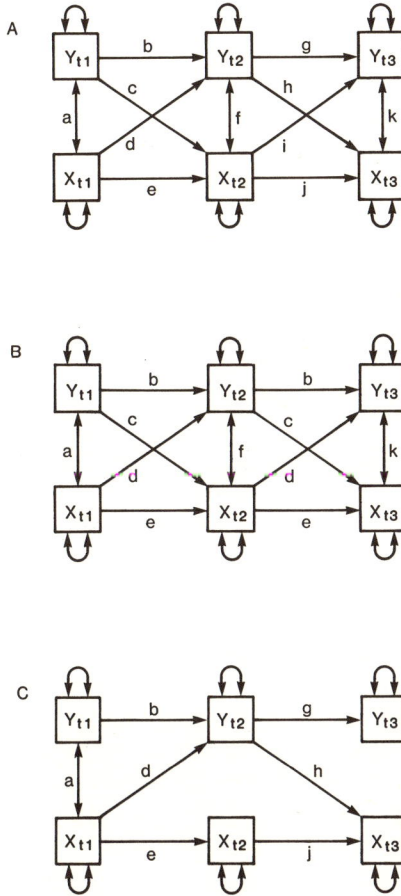

Figure 6. Dynamic process models: crossed and lagged analysis. **A.** A first-order crossed and lagged FIGMOD in which the influences from one occasion to the next vary. **B.** An equal-process model in which the influences from one occasion to the next occation remain constant. **C.** A simplified-process model in which some relations are hypothesized to be zero (i.e., some arrows removed).

(*b,c,d,e* versus *g,h,i,j*), indicating little homogeneity in the relations among the variables from occasion to occasion.

Figure 6B is an equal-process simplification of the 6A model. The coefficients between occasions 1 and 2 and between 2 and 3 are the same (*b,c,d,e*). Notice in Figure 6B that the undirected relations between the X and Y variables (coefficients *a*, *f*, and *k*) on the different occasions of measurement are not specified to be equal: Heart-rate stability and stranger anxiety, for example, might become less and less correlated as children become older even as both variables, and the changes in both variables, were uniform indicators of initial inhibition in response to the unexpected, as discussed in Kagan's chapter.

The model of Figure 6C is a further simplification (a more highly restricted version) of the first-order simplex of 6A. Notice particularly the absence of arrows in prominent places in this model—relations that have been specified to be zero. There is no directed relation from Y_{t1} to X_{t2}, for example, even as a directed relation is specified between X_{t1} and Y_{t2} (coefficient d). This latter relation is sometimes said to represent a crossed influence. In a study by Schmid and Crano (1974), for example, fluid intelligence at a young age was hypothesized to influence the development of crystallized intelligence at an older age, but crystallized intelligence at the younger age was not expected to influence the development appearing in later-age measures of fluid intelligence.

A condition of lagged relations is also illustrated in FIGMOD 6C: Whereas there is no Y_1 to X_2 directed relation, there is a Y_2 to X_3 relation (coefficient h). In substantive theory this might be interpreted to mean that Y_1 indirectly (not directly) influences X_3 through its influence on Y_2. The precocious sexual behavior of some of the subjects of Rutter's studies, reviewed in Chapter 3, could result, in part, from influences which in the early years of schooling led to disenchantment with school, the disenchantment then leading to early drop-out from school in adolescence, which then set the stage for pregnancy at a relatively young age. Perhaps the most important influence in such a chain of events is the early one, even if it is only a lagged, indirect influence.

Each of the models we have just considered can be expanded to involve any number of variables obtained on any number of occasions. MASMOB methods, particularly those of Lohmöller's (1981) PLS program and those embodied in McArdle and Horn (1981a), can handle large numbers of such variables and relations. In the study of Noonan and Wold (1982), for example, models with hundreds of variables were analyzed.

As noted before, it is important to realize that the inclusion or exclusion of particular variables and relations in a model can notably affect results; the values estimated for relations should be thought of in terms of the company they keep (cf. Cook & Campbell, 1979). Crossed and lagged analyses have been criticized, for example, because, in the view of the critic, relations that should have been estimated were not estimated (Duncan, 1975; Ragosa, 1979).

Incidentally, disagreements between investigators about inclusion of variables or relations are sometimes most constructively handled by framing the disagreement as a difference between two models and then examining the evidence of goodness-of-fit under the two conditions. As we mentioned earlier, a design to compare plausible models has a better epistemological basis than a design to see if a particular model fits. However, in order for comparisons between different models to be un-ambiguous it is necessary that they differ in respect to only one rela-tion—or in some cases a few well chosen relations—and it is usually very difficult in the behavioral sciences to narrow debates to a point at which they differ in respect to only one or a few relations. It is often good science to think in these terms, however, even when there is no chance for a completely unambiguous test. It helps one to focus on crucial, potentially testable hypotheses (cf. Horn & Donaldson, 1980; Horn, McArdle, & Mason, 1983; McArdle & Horn, 1981b).

The several variables considered in the simplex models of Figure 6 might all be thought to be separate indicators of a multiple-process entity that emerges over time. In this sense the models represent a concept of a latent variable. However, the idea of a latent variable is also, and more frequently, discussed as a set of interrelated processes that operate in one segment of time. Such an idea is at the core of many important scientific concepts. Let us therefore consider some examples of how such latent variables can be studied with modeling techniques.

STRUCTURAL MODELS SEEN THROUGH DEVELOPMENT

Reliability as a Latent Structure Model

In our examples of simplex models we saw a Y_{ji} variable being incremented by an I_{ji} variable. In a simplex such an increment is as-sumed to persist, at least for awhile; it stays around in the next measure-ment and the next until it is decided to modify the model to deal with disjunction issues, such as Kagan has discussed. An increment that can be either positive or negative (in a random sense) and that does not stay around in subsequent measures of a trait can be thought of as test–retest error of measurement. Error of measurement is a concept derived from theory. Error is an unobserved, latent variable, the effects of which can only be estimated.

In theory of reliability of measurement, the basic idea is that in order to measure something properly—distinguish between error and the true score—and to have a basis for inferring reliability, it is desirable to observe that thing several times, preferably from several different vantage points (e.g., Cattell, 1957, 1966; Cronbach, 1970; Horn, 1971). For example, in order properly to determine a child's own motivation for schoolwork, it is desirable not only to consider the child's assessments of his motivation, but also the assessments of the child's teachers, parents, and peers and to consider each kind of assessment at several points in time. Each of these several ways of assessing the attribute poses a threat to reliability in the sense that if measurements obtained in one way do not agree with measurements obtained in other ways, error of measurement is indicated. On the other hand, reliability is indicated by the agreement of assessment among the variables. The reliable component of a set of scores can be obtained by calculating a linear combination of the several relevant variables. Thus the reliable component of a set of measurements (the true score) is a latent variable defined in terms of several different manifest variables.

Overlapping Process Models

The idea of an overlapping process is similar to the concept of reliability. A number of manifest variables are again specified to be related to a latent variable, but this time each component represents something different. Figure 7A is a model of measurement derived from notions (e.g., Horn, 1982) that a latent variable called *crystallized intelligence, Gc,* involves multiple indicators, each of which is a somewhat separate process. The indicators are breadth of knowledge, quality of judgment, and skills of retrieval of acquired information. *Gc* is thus specified as involving several distinct processes working together and in this sense overlapping. One process cannot work entirely independently of the others, but the processes are separate too. Some things affect one process but not the other. These are the kinds of ideas a scientist often has about phenomena, and thus it is important to consider models for such theories. There are, however, several matters that should be carefully considered in designing a study to reveal that variables are organized in accordance with this model (Carroll & Horn, 1981; Horn, 1979; Horn & Goldsmith, 1981).

In Figure 7B a model for a first-order simplex for a single latent variable is presented. This model is quite similar to the manifest variable model presented in Figure 5B. This similarity between a latent variable model and a manifest variable model can alert the reader to realize that almost all of what was said about simplex models for manifest variables can be applied to considerations of latent-variable models. Thus, for example, just as the single variable simplex models were expanded to multiple variable models for manifest variables, so might the single latent variable simplex model of Figure 7B be expanded to include several latent variables identified on several different occasions. Similarly, the ideas of crossed and lagged influences apply as well to latent variable models as to the manifest variable models discussed earlier.

But a number of new issues and opportunities come to the fore

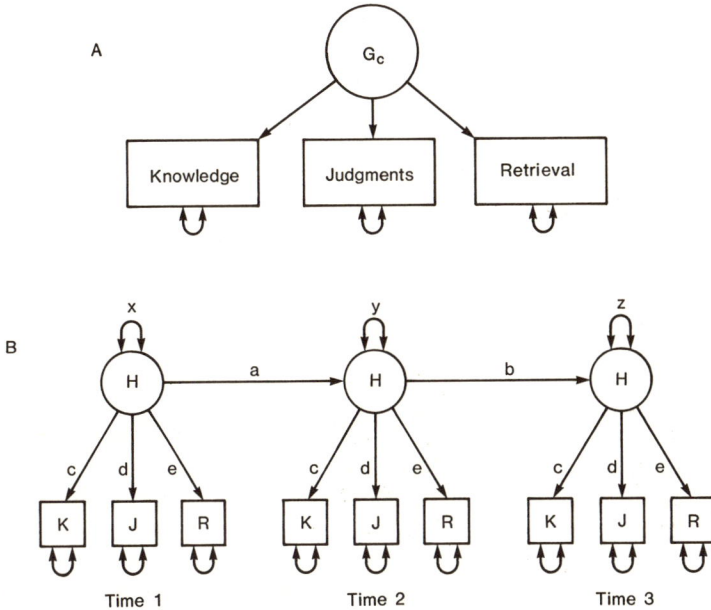

Figure 7. Multiple indicator models. **A.** A single-factor, or multiple-indicator, model in which overlapping processes are indicated. **B.** A single-factor outer structure with invariant loadings across occasions coupled with an inner simplex model.

when multiple-variable, multiple-occasion latent variable models are introduced. Only a few of these matters can be considered in the space available here.

Changing Functions and Changing Subjects

In the earlier discussion of the ideas of a simplex model an important issue was ignored. If a simplex model is reasonable, what has changed: that which is measured or those who are measured? When the ideas of a simplex were introduced, the I_{ji} increments were treated as if they represented things happening to the subjects. It was implied that the *thing* measured on each occasion was the same. But it could just as well have been implied that each increment represented change in the measurement operations. For example, if the thing measured was fluid intelligence (*Gf*), then at preschool ages such tests as "Draw-a-Child" or "Conservation with Clay" might be in the battery of measures, but by adolescence these tests would no longer be appropriate and Matrices and Letter Series would be used to measure *Gf*. If a simplex is found for repeated measures of *Gf* under conditions of gradual shift from one to another set of subtests, then the simplex can be due entirely to this change in the measurement definition of the attribute and not at all reflect changes in the subjects. Subtle (and sometimes not so subtle) shifts with age occur in the measurement operations of most variables studied by developmental researchers. Thus the discovery that a simplex fits data, although surely an exciting and important event, does not in itself indicate which of these two kinds of interpretations should be preferred (Humphreys, 1962).

One can conceive of situations in which this problem can be effectively addressed with latent variable modeling studies. Such a situation is suggested in Figure 7B. This figure depicts a case in which the *same* indicators of a latent variable are present on each of several occasions of measurement. The model stipulates that the directed relation of the latent variable to these indicators is invariant across occasions. Perhaps heart function could be expected to represent this kind of invariance: that is, the relations of overall function to right auricle, left auricle, left ventricle, and right ventricle processes might remain the same as children developed from age 5 to age 10 to age 15 years. If so, a case could be made that such invariance would not occur if the function itself had changed or if the operations of measurement for indicating the function had changed, and therefore if a simplex result is found, it can be unambiguously interpreted as indicating a change in the children.

This example is terribly hypothetical, of course, and perhaps not a realistic example of many of the attributes that are of interest to developmental researchers; but it should be useful in indicating a major problem about any developmental research and in suggesting one kind of benchmark with respect to which models can be compared.

Several other complexities of developmental research are adumbrated in Figure 7B. This FIGMOD suggests that the simplex among invariant indicators of the latent attribute is not an equal-process simplex. Similarly, the variances of *H* on the different occasions of measurement have not been specified to be equal. When examining the *c, d,* and *e* parameters it can be seen that a hypothesis of invariance of the process indicators of the latent variable has been made. These specifications raise questions about where it is reasonable and where it is not reasonable to expect invariance to occur with development. For example, suppose the intervals of time between occasions of measurement are equal: Would this make it most reasonable to expect that $a = b$ in Figure 7B and that an equal-increment simplex should obtain? If so, then very possibly a model cannot fit when both this restriction and the process invariance restrictions (on the *c, d,* and *e* parameters) are imposed. Similarly, if it seems most reasonable to require that the variances of *H* be equal ($x = y$

= z) on each occasion of measurement, then imposition of this reason-able restriction may make it impossible to find an acceptable fit when the equal-increment simplex and/or the invariant process restrictions are also stipulated.

Obviously, there is a sense in which if all of these sets of restrictions obtain simultaneously when a model fits, the results can be impressive indeed. But such a result would be impressive largely because it is unbelievable in research on genuine phenomena in the behavioral sciences (perhaps any of the sciences). We simply do not know enough about anything worthwhile to be able to specify it this precisely. Yet again, as a benchmark the example can be useful.

Incidentally, the problems indicated in this discussion are not problems of modeling, as such; they are problems of developmental research. Thinking in terms of MASMOB analyses can help one to see the problems more clearly, but the problems existed before ideas about modeling came on the scene (see Cattell, 1950, for example).

The means for variables often change as development proceeds, and thus far possibilities of how means can be incorporated into MASMOB analyses have not been mentioned. Largely due to the work of McArdle (1978, 1980; McArdle & Horn, 1981b), techniques are now available for modeling relations for means while simultaneously modeling the relations considered earlier. Let us now look at some examples of this feature of MASMOB.

SIMULTANEOUS ANALYSIS OF MEANS AND STRUCTURE

In Figure 8A, the percentage correct on a manifest variable called picture completion (PC) has been plotted for three different age groups (see McArdle & Horn, 1983, for a complete description of these data). In Figure 8B the same information is shown in a model. The long thin rectangle across the top of the FIGMOD is a representation of a base point with respect to which means can be compared. The single arrow from the thin rectangle to the rectangle for occasion 1 indicates that the mean for the first occasion has the indicated (.82) relation to the base point. This same relation is also represented by an arrow from the thin rectangle to the other PC-squares for each occasion. Additional arrows indicate increments to the basepoint. The arrow marked .09 that ends on the PC26 indicates that from age 16 to age 26 years there was increase (by .09) to .91 over the basepoint mean; the arrow marked −.24 from the thin rectangle to PC36 indicates that from age 26 to age 36 years there was decrease (by −.24) to .67 from the second occasion mean (a .09 + −.24 = −.13 increment from the basepoint). Figure 8B thus represents

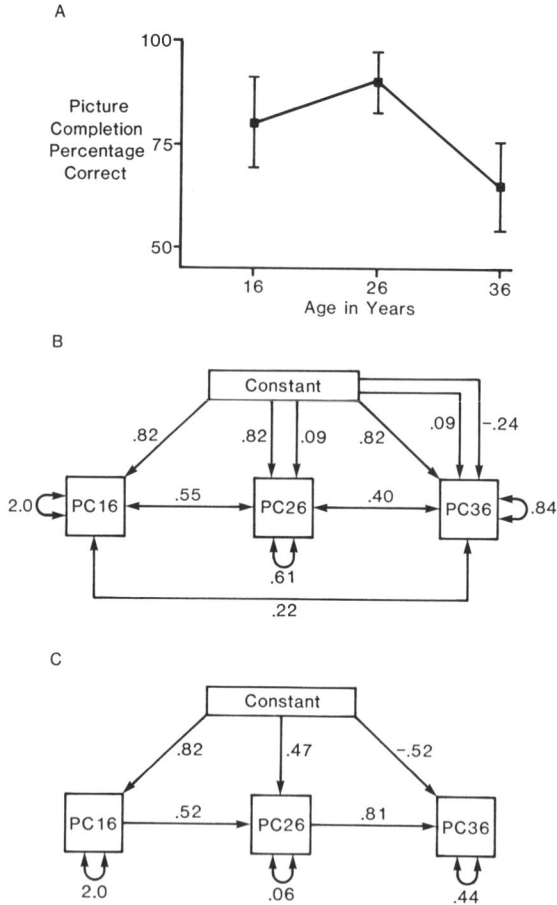

Figure 8. Modeling means in group shift, or growth curve, analysis. **A.** Means and variance of data obtained on 3 occasions. **B.** Growth curve analysis as typically done. **C.** Structural equation modeling of growth curve.

the kind of mean change that is commonly considered to be indicative of growth and decline of an attribute. The undirected relational arrows in Figure 8B represent the obtained correlations among the *PC* measurements obtained on the different occasions.

In Figure 8C, results from fitting a particular structured model to the interrelations shown in Figure 8B are presented. The aim has been to see whether the information of the three double-headed arrows of Figure 8B can be represented by the more parsimonious model of two directed relations shown in Figure 8C. The differences in means over the basepoint are again indicated by the arrows from the thin rectangle to the squares. Here it can be seen that the mean differences for the modeled variable are notably different from the mean differences for the manifest variable, as such. One advantage of including the analysis of means in the model is that all parameters can be assessed statistically in a simultaneous fashion.

A SUMMARIZING EXAMPLE

In Figure 9 several of the ideas of our previous discussions are brought together in one model. McArdle and Horn (1983) have developed this particular model on the basis of the adult WAIS data from the Berkeley Growth Study (Eichorn, Clausen, Haan, Honzik, & Mussen, 1981). The model includes the following kinds of basic hypotheses:

1. The WAIS variables PC, IN, and MS are structural manifestations (to different degrees) of a single latent variable G.
2. The structured relations for G (to the manifest variables) are invariant over the three occasions of measurement.

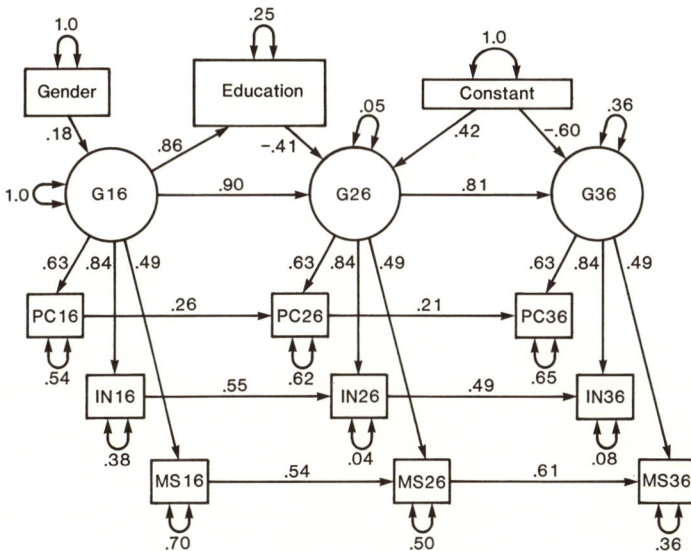

Figure 9. An example of a comprehensive structural equation model (taken from McArdle & Horn, 1983) illustrating many of the features described in earlier sections.

3. There are indications of a first-order simplex for the latent variable.
4. There are college education and gender differences (here controlled) in the latent variable.
5. There are latent variable differences in the means for the three occasions.
6. There are occasion-to-occasion relations among the manifest variables that are not fully represented in the simplex and structural relations for *G*.

This list of hypotheses is not exhaustive. Other hypotheses about what is *not* related to what and hypotheses about variances have not been listed. And there are other explicit hypotheses that might be specified or specified in a different manner. However, the list is perhaps sufficient to illustrate that many intricately interwoven features of a theory can be examined using modeling analyses. The interested reader is referred to McArdle and Horn (1983) for a more complete account of this model.

SUMMARY AND CONCLUSIONS

Space limitations do not allow discussion of several other important issues, including (1) which parameters should theoretically remain invariant during development (Horn *et al.*, 1983); (2) design differences, as between cross-sectional, longitudinal, and single subject-repeated measures (McArdle & Horn, 1981); and (3) how MASMOB procedures handle problems encountered in multivariable time series. The MASMOB procedures and the examples that were included in this chapter may, however, help the investigator approach these and other issues of developmental research, including questions related to continuity and discontinuity in development. The close tie between specification, utilizing figural representations of models (FIGMODs), and analyses can be of significant value in explicating hypotheses to be statistically tested and in seeing where adjustments might be made in models in order more nearly to represent reality. Graphic representation is also beneficial in conceptualizing multiple processes and interrelations among these processes. Indeed, many questions in developmental research are more easily defined by the adroit use of specification modeling. Many of the ambiguities concerning definitions of developmental phenomena can be reduced in number and kind by enabling an investigator to test empirically whether the inclusion of a relation does or does not improve the fit of a model. Precise theoretical statements can be specified and subsequently tested using MASMOB procedures.

REFERENCES

Brody, B. A. *Readings in the philosophy of science.* Englewood Cliffs, New Jersey: Prentice-Hall, 1970.

Carroll, J. B., & Horn, J. L. On the scientific basis of ability testing. *American Psychologist, Special Issue,* 1981, *36,* 1012–1020.

Cattell, R. B. *Personality.* New York: McGraw-Hill, 1950.

Cattell, R. B. *Personality and motivation structure and measurement.* New York: World Book, 1957.

Cattell, R. B. (Ed.) *Handbook of multivariate experimental psychology.* Chicago: Rand McNally, 1966, 553–561.

Cook, T. D., & Campbell, D. T. *Quasi-experimentation: Design and analysis issues for field settings.* Boston: Houghton Mifflin, 1979.

Cronbach, L. J. *Essentials of psychological testing* (3rd ed.). New York: Harper & Row, 1970.

Duncan, O. D. *Introduction to structural equation models.* New York: Academic Press, 1975.

Eichorn, D. H., Clausen, J. A., Haan, N., Honzik, M. P., & Mussen, P. H. *Present and past in middle life.* New York: Academic Press, 1981.

Guttman, L. A new approach to factor analysis: The Radex. In P. F. Lazarsfeld (Ed.), *Mathematical thinking in the social sciences.* Glencoe, Ill.: Free Press, 1954, pp. 216–348.

Horn, J. L. Integration of concepts of reliability and standard error of measurement. *Educational and Psychological Measurement,* 1971, *31,* 57–74.

Horn, J. L. Some correctable defects in research on intelligence. *Intelligence,* 1979, *3,* 307–322.

Horn, J. L. The theory of fluid and crystallized intelligence in relation to concepts of cognitive psychology and aging in adulthood. In F. I. M. Craik & S. Trehum (Eds.), *Aging and cognitive processes.* New York: Plenum Press, 1982.

Horn, J. L., & Donaldson, G. Cognitive development in adulthood. In O. G. Brim & J. Kagan (Eds.), *Constancy and change in human development.* Cambridge: Harvard University Press, 1980, pp. 445–529.

Horn, J. L., & Engstrom, R. Cattell's scree test in relation to Bartlett's chi-square test and other observations on the number of factors problem. *Multivariate Behavioral Research,* 1979, *14,* 283–300.

Horn, J. L., & Goldsmith, H. Reader be cautious: A review of *Bias in mental testing. American Journal of Education,* 1981, *89,* 305–329.

Horn, J. L., & Knapp, J. R. On the subjective character of the empirical base of Guilford's structure-of-intellect model. *Psychological Bulletin,* 1973, *80,* 33–43.

Horn, J. L., & McArdle, J. J. Perspectives on mathematical/statistical model building (MASMOB) in research on aging. In L. W. Poon (Ed.), *Aging in the 1980's: Psychological issues.* Washington, D.C.: American Psychological Association, 1980.

Horn, J. L., McArdle, J. J., & Mason, R. When is invariance not invariant: A practical scientist's look at the ethereal concept of factor invariance. *The Southern Psychologist,* 1983, in press.

Humphreys, L. G. The organization of human abilities. *American Psychologist,* 1962, *17,* 475–483.

James, L. R., Mulaik, S. A., & Brett, J. M. *Causal analysis: Assumptions, models, and data.* Beverly Hills, Calif.: Sage, 1982.

Jöreskog, K. G. Testing a simple structure hypothesis in factor analysis. *Psychometrika,* 1966, *31,* 165–178.

Jöreskog, K. G. A general approach to confirmatory maximum likelihood factor analysis. *Psychometrika,* 1969, *34,* 183–202.

Jöreskog, K. G. A general method for estimating a linear structural equation system. In A. S. Goldberger & O. D. Duncan (Eds.), *Structural equation models in the social sciences.* New York: Seminar Press, 1973, 85–112.

Jöreskog, K. G., & Sörbom, D. *LISEREL—IV: Analysis of linear structural relationships by the method of maximum likelihood.* Chicago: National Educational Resources, 1978.

Jöreskog, K. G., & Sörbom, D. *Advances in factor analysis and structural equation models* (J. Magidson, ed.). Cambridge, Mass.: Abt Books, 1979.

Jöreskog, K. G., & Sörbom, D. *LISEREL—V: Analysis of linear structural relationships by maximum likelihood and least squares method.* Chicago: National Educational Resources, 1982.

Lohmöller, J. B. *LVPLS 1.6 program manual: Latent variables path analysis with partial least squares estimation.* Forschungsbericht 81.04, Fachbereich Pädogogik, Hochschule der Bundeswehr, München, 1981.

McArdle, J. J. A structural view of longitudinal repeated measures. *Proceedings of the Social Sciences Section of 1978 American Statistical Association Annual Meeting.* Washington, D.C., 1978, 155–160.

McArdle, J. J. Causal modeling applied to psychonomic systems simulation. *Behavior Research Methods and Instrumentation,* 1980, *12,* 193–209.

McArdle, J. J. A review of Kenny's "Correlation and Causality." *Applied Psychological Measurement,* 1981, *5*(2), 275–280.

McArdle, J. J. Multivariate software systems. In J. R. Nesselroade & R. B. Cattell (Eds.), *The handbook of multivariate experimental psychology* (2nd ed.). New York: Plenum Press, in press.

McArdle, J. J., & Horn, J. L. *MULTITAB 81: Computer programs for multivariate data analysis.* Denver, Colorado: Psychology Department, University of Denver, 1981. (a)

McArdle, J. J., & Horn, J. L. Structural equation modeling of adulthood aging of GF-Gc intelligence. APA presentation, Los Angeles, August, 1981. (b)

McArdle, J. J., & Horn, J. L. *Modeling longitudinal systems of intellectual development: Structural equation models of the Berkeley Growth WAIS data.* Unpublished manuscript, Department of Psychology, University of Denver, 1983.

McArdle, J. J., & McDonald, R. P. The RAM logic for structural equations. *British Journal of Mathematical and Statistical Psychology,* 1983, in press.

McArdle, J. J., Goldsmith, H. H., & Horn, J. L. Genetic structural equation models of fluid and crystallized intelligence. *Behavioral Genetics,* 1981, *11*(6): 607.

McArdle, J. J., McDonald, R. P., & Horn, J. L. A simple graphic and algebraic representation for structural equation models. *Multivariate Behavior Research,* 1984, in press.

McDonald, R. P. A simple comprehensive model for the analysis of covariance structures. *British Journal of Mathematical and Statistical Psychology,* 1978, *31,* 59–72.

McDonald, R. P. A note on the investigation of local and global identifiability. *Psychometrika,* 1982, (1), 101–104.

McFatter, R. The use of structural equation models in interpreting regression equations including suppressor and enhancer variables. *Applied Psychological Measurement.* 1979, *3,* 123–135.

Medawar, P. B. *Advice to a young scientist.* New York: Harper & Row, 1981.

Noonan, R., & Wold, H. *Evaluating school systems using partial least squares.* Uppsala, Sweden: University of Uppsala, 1982.

Ragosa, D. Causal models in longitudinal research: Rationale, formulation, and interpretation. In J. R. Nesselroade & P. B. Baltes (Eds.), *Longitudinal research in the study of behavior and development*. New York: Academic Press, 1979.

Schmid, F. L., & Crano, W. D. A test of the theory of fluid and crystallized intelligence in middle and low SES children. *Journal of Educational Psychology*, 1974, *66*, 255–261.

Wackwitz, J. H., & Horn, J. L. On obtaining the best estimates of factor scores. *Multivariate Behavioral Research*, 1971, *6*, 389–408.

Wold, H. *Modeling in complex situations with soft information*. Toronto, Canada: Third World Congress of Econometrics, 1975.

Wold, H. Model construction and evaluation when theoretical knowledge is scarce. In J. Kmenta & J. Ramsey (Eds.), *Model evaluation in econometrics*. New York: Academic Press, 1981.

CHAPTER 17

Continuity of Personality

A GENETIC PERSPECTIVE

H. H. Goldsmith

There exists a largely unrecognized correspondence between two alternative views of the nature of longitudinal continuity of behavior and the dual nature of the discipline of genetics. The significance of this correspondence lies in its heuristic value for conceptualizing the genetic basis of behavioral continuity and change. The two alternative perspectives on longitudinal continuity are probably apparent in contributions to this volume as well as earlier conceptualizations (e.g., Emde, 1978; Kagan, 1971, 1980; Lewis & Starr, 1979; McCall, 1977; Overton & Reese, 1981; Riegel, 1976; Wohlwill, 1973). At the risk of oversimplifying these contributions, two major themes of continuity can be discerned: (1) *linear stability*, in the sense of maintenance of rank order of interindividual or intraindividual differences in the same, related, or derivative behavioral dimensions; and (2) functional continuity at the level of behavioral systems, including continuity of goals of behavior across periods of reorganization. The latter view of continuity, rapidly gaining adherents because it offers explanations for change and emergent behavior as well as constancy, will be referred to as the *organizational continuity* perspective.

The subfields of quantitative genetics and developmental genetics are the disciplines that show a correspondence with the two major themes of continuity. Quantitative genetics, along with the related fields of population genetics and biometrics, arose from practical concerns of agriculture and animal breeding and focuses on explaining observed phenotypic variation in traits. Developmental genetics arose primarily

H. H. Goldsmith • Department of Psychology, University of Texas, Austin, Texas 78712.

from work in molecular biology and focuses on issues of intraindividual genetic processes and their regulation.

Table 1 summarizes some of formal characteristics of the linear stability and organizational continuity perspectives. The entries in Table 1 attempt to capture the dominant characteristics of the typical applications of each model. There is no claim that Table 1 reflects the perspectives of each individual variant of the two broad approaches. The characteristics tabulated in Table 1 are chosen to highlight distinguishing features of the two models; a more complete account would reveal more common ground.

Elaborate exposition of the distinctions in Table 1 is unnecessary, having been accomplished in many of the chapters of this volume. What Table 1 does *not* do is specify the sort of evidence necessary to convince one to accept either of the models. Without clarity on this issue, mistaken inferences have occurred. The key point is this: Negative evidence concerning the linear stability of behavior cannot be taken as positive

Table 1. Characteristics of Two Types of Models of Continuity of Behavior

Characteristic	Linear stability	Organizational continuity
Target of inference	Populations	Individuals and networks of individuals
Phenomena studied	Behavioral indicators of traits	Behavioral systems
Aspect of development studied	Stability of rank order of interindividual or intraindividual differences	Patterns of change
Explanatory structures	Traits and their genetic and environmental underpinnings	Developmental stages; functional goals of development
Nature of research methodology	Correlational; structural modeling approaches	(Varies)
Does model admit biological influences?	Yes	Yes
Is model conducive to mathematical specification?	Yes	Not readily, given the present state of knowledge
Is model additive?	Yes, in typical applications	No, it may involve substitution, transformation, or mediation (Flavell, 1972)
Is model easily applicable to study of human development?	Yes, particularly within stages or under stable environmental conditions	Yes, particularly across stages or under changing environmental conditions

evidence for the organizational continuity perspective. Negative evidence for the linear stability model may result from difficult-to-resolve methodological factors, such as inadequate sampling of behavior, low reliability of measures, and selective factors in gathering and maintaining a population of persons for longitudinal studies. Using either equations from psychometric theory (Lord & Novick, 1968) or empirical results from parametric studies (Eaton, 1983; Epstein, 1979, 1980), one can easily construct cases in which observed correlations for longitudinal stability of, say, .30, would rise to the level of .60 if measurement problems were resolved.

There are, of course, various lines of investigation that can be pursued under either the linear stability or organizational continuity approaches. Both perspectives have a place for biological influences, and a substantial amount of investigation has been directed toward elucidation of the genetic subset of these biological influences (see Goldsmith, 1983, for review).

When we turn to genetic theory for guidance in investigating continuity, we find not one but two sets of principles. The characteristics of the two genetic approaches are outlined in Table 2.

Even without drawing out the comparisons, the reader will note striking parallels with views of behavioral continuity presented in Table 1. The linear stability model is compatible with the quantitative genetic approach whereas the organizational continuity model appears to be better served by the developmental genetic perspective. The two genetic approaches require further exposition in order to determine whether they contain object lessons for the study of behavioral continuity.

THE QUANTITATIVE GENETIC APPROACH

While the traditional application of quantitative genetic methods is to estimate genetic and environmental components of phenotypic variance at a given point in time, these methods can be focused on the issue of continuity. Investigators are attempting to move beyond the genetic analysis of trait *levels* to issues such as the genetics of individual differences in the stability of traits over time and individual differences in the consistency of behavior across situations. Since practically all such attempts using human subjects have employed some version of the twin method, we shall treat it exclusively here. Basically, two types of twin designs have been employed to investigate the genetics of longitudinal stability. The first is a cross-sectional design utilizing multiple groups of twins in different age groups (e.g., Eaves & Eysenck, 1976b; Floderus-Myrhed, Pedersen, & Rasmuson, 1980; Young, Eaves, & Eysenck, 1980).

Table 2. Characteristics of Two Types of Genetic Models That Can Be Used to Study the Developmental Continuity Issue

Characteristics	Quantitative genetic models	Developmental genetic models
Target of inference	Populations	Individuals
Phenomena studied	Traits and their structure	Biological systems and their components
Aspect of development studied	Individual differences	Normative development
Explanatory structures or mechanisms	Genetic and environmental variance components	Regulatory processes controlling gene expression (e.g., transcription, translation)
Nature of research methodology	Statistical analysis of individual differences measures	Laboratory experiments involving genetic manipulation
Does model focus on actual biological processes?	No	Yes
Is model conducive to mathematical specification?	Yes	Not readily
Is additivity of development assumed?	Yes, in typical applications	No
Is model easily applicable to empirical study of human development?	Yes	No, although results can be *interpreted* within this framework

Since there are well-known ambiguities (sampling variation, secular effects, etc.) in interpreting cross-sectional differences as being due to individual developmental changes, we will consider only the second type of twin design, longitudinal studies. Within the realm of personality, examples of such studies include those of Dworkin, Burke, Maher, and Gottesman (1976, 1977), Pogue-Geile and Rose (1981; see also Pogue-Geile, Carey, & Rose 1982), Eaves & Eysenck (1976a), Matheny (1980; and references therein for earlier reports from the Louisville Twin Study), Torgersen (1981; see also Torgerson & Kringlen, 1978), Lytton and Watts (1981), and Goldsmith and Gottesman (1981).

An intuitive way to conceive of a longitudinal genetic analysis using the twin design would be as follows: If we measure one member of a pair of identical cotwins at age 1 and the second member at age 2, repeat the process for a large number of pairs, and compute the across-twin, across-time correlations, the magnitude of the resulting correlation would reflect the effect of 100% gene overlap and the portion of the environment that is jointly (or equivalently) experienced by cotwins.

The same procedure done with fraternal twins would result in longitudinal correlations reflecting 50% gene overlap and, we assume, the same degree of common environment. Thus, doubling the difference between the identical and fraternal correlations would yield an estimate of the magnitude of genetic effects on individual differences in longitudinal stability.

The procedure just described is far from optimal for providing the best estimates of genetic influence. Current quantitative genetic approaches to this issue employ one or another version of structural equation modeling. In order to understand more fully the potential of quantitative genetic approaches for the continuity issue, we must examine one of these models. (See Chapter 16 in this volume for a more general

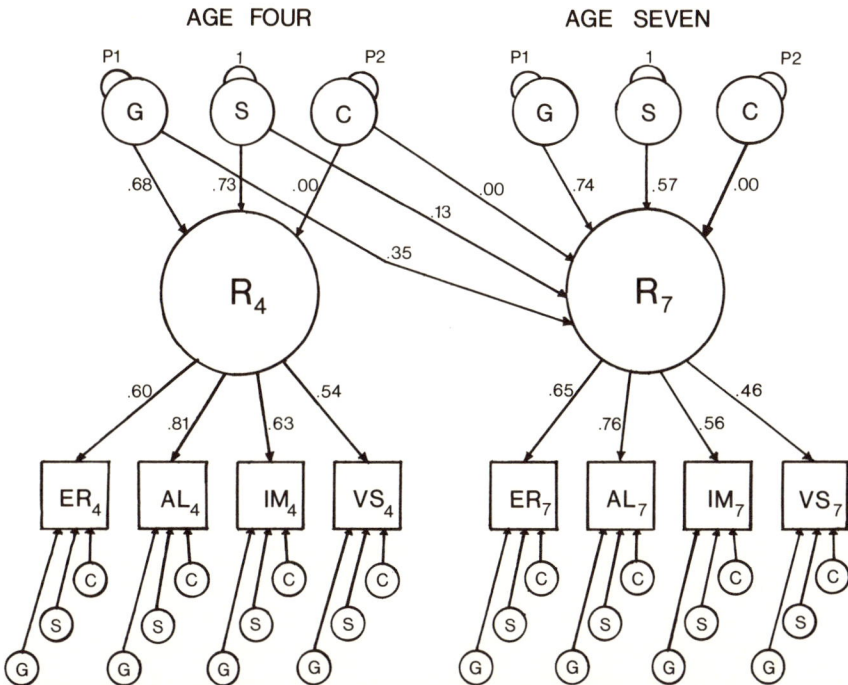

Figure 1. Longitudinal biometric model for the latent trait "reactivity." The figure depicts a simplified structural equation model with some parameters omitted for simplicity (see McArdle et al., 1980, for the full model). All parameters shown are greater than twice their standard errors with the exception of the .13 path from S4 to R7. The model shown, with no common environmental effects (C), was superior to an alternative incorporating C effects. In the solution shown, no effort was made to attain optimal fit. The observed goodness-of-fit chi-square was 262 with 119 degrees of freedom.

treatment of structural equation modeling.) Figure 1 (after McArdle, Connell, & Goldsmith, 1980) illustrates the potential of the method.

For present purposes, we are interested only in the conceptual information that this type of quantitative genetic model can provide; thus Figure 1 contains only a subset of estimates actually obtained.

The actual data for the analysis depicted in Figure 1 are psychologists' ratings of emotional reactivity (EM), activity level (AL), impulsivity (IM), and verbal spontaneity (VS), each measured longitudinally at ages 4 and 7 years on a large sample of identical and fraternal twins (see Goldsmith & Gottesman, 1981). The covariance among these observed, or manifest, variables (denoted by squares in Figure 1) is represented by a latent variable (denoted by a circle). In these data, this latent variable is interpreted as a *reactivity* factor, symbolized by R. The arrows from R to ER, AL, IM, and VS represent standardized maximum likelihood estimates of factor loadings. To this point, we have depicted a one-factor longitudinal model that could represent the observed data. However, the elements in the upper portion of Figure 1 add a quantitative genetic model to the longitudinal factor model. Consider first the three circles, G-4 (the genetic effect), S-4 (the effect of the specific, within-family environment that is *not* shared by cotwins growing up together), and C-4 (the common, between-family, or shared, environmental effect). The standardized partial regression coefficient of .68 indexes the strength of the genetic influence in accounting for the variability in 4-year reactivity. Thus, 46% $(.68)^2$ of the variance in reactivity at 4 years is accounted for by genetic factors. Similarly, the .73 value estimates the influence of unshared (by cotwins) sources of environmental variance (see Rowe & Plomin, 1981, for a treatment of the nature of unshared environmental effects). In these data, there is no effect of shared environmental variance. Note that, in accordance with path analytic rules (e.g., Li, 1975), $(.68)(.68) + (.73)(.73) = 1.00$, and thus all of the variance in R-4 is explained (S-4 includes the error variance).

Thus far the genetic analysis is static. However, we next ask whether the genetic and environmental factors at age 4 also influence variability in an identically composed reactivity factor at age 7. (Note: In the unstandardized solution from which the standardized values in Figure 1 were derived, the loadings of ER, AL, IM, and VS were constrained to be the same on R-4 and R-7. Thus, R-4 and R-7 are identically composed.) The genetic factor at age 4 does account for a significant portion of the variability in R-7 (the path coefficient is .35), but the longitudinal contribution of S-4 is insignificant (path = .13).

The residual variance in R-7, that not predictable from age 4, can also be analyzed into its genetic and environmental components, represented as G-7, S-7, and C-7. Note again that the common environmental effect does not enter into the model, and that $(.35)^2 + (.13)^2 + (.74)^2 + (.57)^2 = 1.00$.

The lower portion of Figure 1 illustrates similar biometrical analyses of the residual variances in each of the eight manifest measures after the reactivity factor has been extracted. Although 14 of the possible 24 path coefficients were greater than twice their standard errors, the actual estimates are not shown for the sake of simplicity of the diagram (see McArdle, Connell, & Goldsmith, 1980). The discussion of Figure 1 has glossed over a number of important statistical issues involving the estimation of the model and its fit to the observed data (see Horn & McArdle, 1980, and references therein, for rectification). The reader should realize, however, that (1) the P1s, 1s, and P2s affixed to the genetic and environmental latent variables are proportionality constants (which vary for data derived from identical versus fraternal twins) derived from quantitative genetic theory and are subject to certain assumptions (see, e.g., Goldsmith, 1983); and (2) this model was specified *a priori* and all estimates were obtained by the method of maximum likelihood using Jöreskog and Sörbom's (1978) LISREL-IV program.

Stated simply, the conclusion is that, although reactivity evinces moderate heritability at both ages, much of the genetic variance at age 7 is *not* common to that at age 4. Thus the evidence suggests newly arisen genetic variance at age 7. A conclusion just as pertinent to development is that the bulk of the observed stability in reactivity is accounted for by stable genetic, rather than environmental, factors. At the very least, such results caution that uncritical equation of the notions of "genetic influence" and "behavioral stability" is unwarranted.

Although other interpretations are possible, these results are consistent with the idea that the nature of genetic influence changes over time. Temporal change in genotypic expression is an idea consonant with the organizational continuity perspective, and it leads us to a consideration of the developmental genetic approach.

THE DEVELOPMENTAL GENETIC APPROACH

The implications of the rapidly growing field of developmental genetics for psychological development have seldom been drawn (cf. Gottesman, 1974; Wilson, 1978), perhaps because few scientists are conversant with both molecular biology and psychology. Although there are examples of the importance of gene regulation for behavior in the realm of human biochemical genetics (see Harris, 1976), the complexity of the gene-to-behavior pathway has not been penetrated sufficiently to yield clear *biological* evidence of the effects of gene regulation on human behavior, with the exception of some aspects of medical genetics. As work in developmental genetics progresses, we may expect to learn more of the biology of discontinuities in development. We might speculate that early advances will elucidate genetic influences on the timing of puberty, and perhaps, the variable age of onset of various psychoses.

We cannot state a comprehensive model of the regulation of gene expression in humans. We do know that the mechanisms of genetic regulation are considerably more complex in higher animals that the early research in viri and bacteria indicated. The prototype of the early research was the elucidation of the functioning of the *lac* operon, which codes for genetic information for the synthesis of lactose in *E. coli*, a species of bacteria found in the intestinal tract. This research demonstrated that the control of gene expression occurred at the level of transcription of DNA to messenger RNA through a feedback mechanism involving the production of a repressor substance that inactivated the series of genes responsible for lactose synthesis.

During the past 15 years there has been a tremendous amount of genetic research aimed at understanding the regulation of gene action in higher organisms. The questions are such as these: Of the 50,000 to 100,000 structural genes estimated to exist in, say, humans, what mechanisms determine which 1%–15% are operative at a given time? And how do genes become activated and deactivated in an orderly manner during ontogeny?

Various models have been proposed to answer these questions (e.g., Davidson & Britten, 1979), but as yet no model seems to have gained widespread acceptance. Perhaps the "flavor" of this research can best be conveyed by listing, in simplified language, some of the principles and phenomena currently accepted as valid.

1. In higher animals, much of the DNA codes not for protein-specifying structural genes but for genes that regulate the expression of structural genes. In fact, genetic differences between humans and the great apes are largely due to differences in regulatory genes.
2. Gene expression is subject to regulation at the levels of transcription, translation, and enzyme action (Leighton & Loomis, 1980).
3. Genes switch on and off in a programmed fashion during normal development and differentiation. The nature and regulation of *developmental switches* is under investigation.
5. There exist *movable genetic elements* (Cold Spring Harbor Symposia on Quantitative Biology, 1980) that control the expression of genes located nearby.
6. A given gene does not always necessarily have single start and

stop points for transcription. Nor is all the DNA that codes for the amino acid sequence of protein continuous in the gene; that is, there are intervening sequences that provide boundaries between gene subregions. Posttranscription assembly of messenger RNA allows intraindividual flexibility for such phenomena as antibody production in that variants of a given genetic product can be produced during development.

Thus basic research on such varied processes as transitions from one to another mating type in yeast (Nasmyth, 1983) and the initial commitment of mouse genes to the determination of coat color (Mintz, 1978) promises to provide developmental biologists with a way to study the relationship between specific genetic events and developmental change.

Unfortunately, we cannot yet determine whether the organizational continuity perspective on human behavior will find biological underpinnings in the mechanisms of regulation of gene expression. As emphasized by Hinde (1982), "we have to be wary about ascribing things that turn up at a particular age to timed gene action" (p. 292). Systematic environmental effects as well as the unfolding of a genetic program can result in predictable patterns of behavioral reorganization. It is likely that the two sets of influences have become intercoordinated, both during the evolution of the species and during individual development in social contexts.

REFERENCES

Cold Spring Harbor Symposia on Quantitative Biology (Vol. 45, parts 1 and 2). *Movable Genetic Elements*. Cold Spring Harbor, New York: Cold Spring Harbor Laboratory, 1980.

Davidson, E. H., & Britten, R. J. Regulation of gene expression: Possible role of repetitive sequences. *Science*, 1974, *204*, 1052–1059.

Dworkin, R. H., Burke, B. W., Maher, B. A., & Gottesman, I. I. A longitudinal study of the genetics of personality. *Journal of Personality and Social Psychology*, 1976, *34*, 510–518.

Dworkin, R. H., Burke, B. W., Maher, B. A., & Gottesman, I. I. Genetic influences on the organization and development of personality. *Developmental Psychology*, 1977, *13*, 512–521.

Eaton, W. O. Measuring activity level with actometers: Reliability, validity, and arm length. *Child Development*, 1983, *54*, 720–726.

Eaves, L. J., & Eysenck, H. J. Genetic and environmental components of inconsistency and unrepeatability in twins' responses to a neuroticism questionnaire. *Behavior Genetics*, 1976, *6*, 145–160. (a)

Eaves, L. J., & Eysenck, H. J. Genotype × age interaction for neuroticism. *Behavior Genetics*, 1976, *6*, 359–362. (b)

Emde, R. N. *Searching for perspectives: Systems sensitivity and opportunities in studying the infancy of the organizing child of the universe.* Invited address at the International Conference on Infancy Studies, New Haven, Connecticut, March, 1978.

Epstein, S. The stability of behavior: I. On predicting most of the people much of the time. *Journal of Personality and Social Psychology,* 1979, *37,* 1097–1126.

Epstein, S. The stability of behavior: II. Implications for psychological research. *American Psychologist,* 1980, *35,* 790–806.

Flavell, J. H. An analysis of cognitive-developmental sequences. *Genetic Psychology Monographs,* 1972, *86,* 279–350.

Floderus-Myrhed, B., Pedersen, N., & Rasmuson, I. Assessment of heritability for personality, based on a short-form of the Eysenck Personality Inventory: A study of 12,898 twin pairs. *Behavior Genetics,* 1980, *10,* 153–162.

Goldsmith, H. H. Genetic influences on personality from infancy to adulthood. *Child Development,* 1983, *54,* 331–355.

Goldsmith, H. H., & Gottesman, I. I. Origins of variation in behavioral style: A longitudinal study of temperament in young twins. *Child Development,* 1981, *52,* 91–103.

Gottesman, I. I. Developmental genetics and ontogenetic psychology: Overdue detente and propositions from a match maker. In A. Pick (Ed.), *Minnesota Symposia on Child Psychology.* Minneapolis: University of Minnesota Press, 1974.

Harris, H. *The principles of human biochemical genetics.* Amsterdam: North Holland, 1975.

Hinde, R. Comment in published discussion. In R. Collins & G. M. Collins (Eds.), *Ciba Foundation symposium 89: Temperamental differences in infants and young children.* London: Pitman, 1982.

Horn, J. L. & McArdle, J. J. Perspectives on mathematical statistical model building (MASMOB) in research on aging. In L. W. Poon (Ed.), *Aging in the 1980's: Selected contemporary issues in the psychology of aging.* Washington, D.C.: American Psychological Association, 1980.

Jöreskog, K. H. & Sörbom, D. *LISREL: Analysis of linear structural relationships by the method of maximum likelihood.* Chicago: National Educational Resources, 1978.

Kagan, J. *Change and continuity in infancy.* New York: Wiley, 1971.

Kagan, J. Perspectives on continuity. In O. G. Brim & J. Kagan (Eds.), *Constancy and change in human development.* Cambridge: Harvard University Press, 1980.

Leighton, T., & Loomis, F. (Eds.). *The molecular genetics of development* (Molecular Biology Series). New York: Academic Press, 1980.

Lewis, M., & Starr, M. D. Developmental continuity. In J. D. Osofsky (Ed.), *Handbook of infant development.* New York: Wiley, 1979.

Li, C. C. *Path analysis.* Pacific Grove, Calif.: Boxwood, 1975.

Lord, F. M., & Novick, M. R. *Statistical theories of mental test scores.* Reading, Massachusetts: Addison-Wesley, 1968.

Lytton, H., & Watts, D. The social development of twins in longitudinal perspective: How stable is genetic determination over age 2 to 9? In L. Gedda, P. Parisi, & W. Nance (Eds.), *Twin research 3: Intelligence, personality, and development.* New York: Liss, 1981.

Matheny, A. P. Bayley's Infant Behavior Record: Behavioral components and twin analyses. *Child Development,* 1980, *51,* 466–475.

McArdle, J. J., Connell, J. P., Goldsmith, H. H. Latent variable approaches to measurement structure, longitudinal stability, and genetic influences: Preliminary results from the study of behavioral style. *Behavioral Genetics,* 1980, *10,* 487 (abstract).

McCall, R. B. Challenges to a science of developmental psychology. *Child Development,* 1977, *48,* 333–344.

Mintz, B. Gene expression in neoplasia and differentiation. *Harvey Lectures,* 1978, *71,* 193–246.

Nasmyth, K. A. Molecular genetics of yeast mating type. *Annual Review of Genetics*, 1982, *16*, 439.

Overton, W. F. & Reese, H. W. Conceptual prerequisites for an understanding of stability–change and continuity–discontinuity. *International Journal of Behavioral Development*, 1981, *4*, 99–123.

Pogue-Geile, M. F., & Rose, R. J. A longitudinal twin study of the MMPI: Genetic analyses of personality stability and change. *Behavior Genetics*, 1981, *11*, 608 (abstract).

Pogue-Geile, M. F., Carey, G., & Rose, R. J. *The stability of genetic and environmental influences on personality: A longitudinal twin study of the MMPI.* Paper presented at the Behavior Genetics Association meetings, Ft. Collins, Colorado, June, 1982.

Riegel, K. F. The dialectics of human development. *American Psychologist*, 1976, *31*, 689–700.

Rowe, D. C., & Plomin, R. The importance of nonshared (E1) environmental influences in behavioral development. *Developmental Psychology*, 1981, *17*, 517–531.

Torgersen, A. M. Genetic aspects of temperamental development: A follow-up study of twins from infancy to six years of age. In L. Gedda, P. Parisi, & W. Nance (Eds.), *Twin research 3: Intelligence, personality, and development.* New York: Liss, 1981.

Torgersen, A. M., & Kringlen, E. Genetic aspects of temperamental differences in infants. *Journal of the American Academy of Child Psychiatry*, 1978, *17*, 433–444.

Wilson, R. S. Synchronies in mental development: An epigenetic perspective. *Science*, 1978, *202*, 939–947.

Wohlwill, J. F. *The study of behavioral development.* New York: Academic Press, 1973.

Young, P. A., Eaves, L. J., & Eysenck, H. J. Intergenerational stability and change in the causes of variation in adult and juvenile personality. *Journal of Personality and Individual Differences*, 1980, *1*, 35–55.

Index